DATA STRUCTURE AND SOFTWARE ENGINEERING

Challenges and Improvements

DATA STRUCTURE AND SOFTWARE ENGINEERING

Challenges and Improvements

James L. Antonakos

Distinguished Professor of Computer Science,
Broome Community College, State University of New York,
Binghamton; Online Instructor and Faculty Advisor, Excelsior College,
Albany, New York and Sullivan University, Kentucky, U.S.A.

Apple Academic Press

Data Structure and Software Engineering: Challenges and Improvements

First Published in the Canada, 2011
Apple Academic Press Inc.
3333 Mistwell Crescent
Oakville, ON L6L 0A2
Tel. : (888) 241-2035
Fax: (866) 222-9549
E-mail: info@appleacademicpress.com
www.appleacademicpress.com

The full-color tables, figures, diagrams, and images in this book may be viewed at www.appleacademicpress.com

First issued in paperback 2021

ISBN 13: 978-1-77463-255-0 (pbk)
ISBN 13: 978-1-926692-97-5 (hbk)

James L. Antonakos

Cover Design: Psqua

Library and Archives Canada Cataloguing in Publication Data
CIP Data on file with the Library and Archives Canada

CONTENTS

INTRODUCTION

What I find most remarkable about the field of computer science is its vast scope. Practically any topic you might imagine falls in some way under the umbrella of computer science. Many topics may seem to naturally belong there, such as research into advanced computer architectures, distributed and cloud computing (and their associated high-speed networking components), computer forensics, operating systems, and the details of many different programming languages. But these areas are just a few of a wider array of topics and activities found in computer science. Computer scientists spend a great deal of time and energy studying compression algorithms for images, video, and data; encryption techniques; efficient hardware computation pipelines; computer gaming and its associated artificial intelligence; networking protocols that enable secure and reliable transmission of information; image processing; database technologies; and new ways of sharing information over the Internet.

Of course, many areas of computer science require a good foundation in mathematics. Here it is remarkable to know that we use mathematics to prove that some things are possible and that other things are impossible. Some concepts or problems have not yet been proved either way, even with a great many researchers looking into them. When and if these open problems are eventually solved, the solutions will usher in a new age in computer science and also new challenges. For example, if we gain a deeper insight and understanding of random numbers, what will be the effect on the security algorithms we use every day to encrypt our private communication over the Internet?

Let us also take note of the astounding visual reality now available, of computer graphics algorithms so complex they render stunning visual effects in video games in real time and produce "Is it real or computer generated?" effects in motion pictures. Again, here the computer scientist must have programming skills, knowledge of mathematics, physics, optics, and the hardware details of the processor or processors rendering the image. We can see for ourselves the fruits of many researchers' labors over the years.

The time spent by computer scientists examining arcane topics that appear to have little practical application is very misleading. Advances in medical imaging, understanding biological processes, recognizing human speech, mining data and distinguishing patterns, and exploring the nature of memory via neural networks have all been made possible by computer scientists toiling away in their labs.

Today the line between software and hardware is becoming blurred. A computer scientist crafting a new optimizing compiler must have detailed knowledge of the internal hardware workings of a processor in order to efficiently schedule instructions and generate code that utilizes the processor pipeline, registers, and cache memory to provide maximum performance. Even something as simple as extending the life of the battery in a laptop computer is a combined effort between the hardware designers and the software writers.

Perhaps the most important quality a computer scientist can possess is curiosity, a constant desire to understand how things work. When this curiosity is coupled with determination, the end result is often useful in ways that were not originally intended. Keep this curiosity in mind as you read the papers on *Data Structure and Software Engineering* contained within this book.

—James L. Antonakos

An Architecture to Support Learning, Awareness, and Transparency in Social Software Engineering

Wolfgang Reinhardt and Sascha Rinne

ABSTRACT

Classical tools for supporting software engineering teams (collaborative development environment, CDE) are designed to support one team during the development of a product. Often the required data sources or experts reside outside of the internal project team and thus not provided by these CDEs. This paper describes an approach for a community-embedded CDE (CCDE), which is capable of handling multiple projects of several organizations, providing inter-project knowledge sharing and developer awareness. The presented approach uses the mashup pattern to integrate multiple data sources in order to provide software teams with an exactingly development environment.

Keywords: Learning Systems, Knowledge Management, Cooperative Development Environments, Learning Communities

Introduction

Traditional clichés about software developers lose their validity more and more. Times, when programmers sat in dark cellars and tried to solve all problems on their own are over once and for all. In the meantime software engineering has become a very knowledge-intensive [5] and communicative process (not only but also triggered by agile methods for software development) where the actors heavily exchange data (see Google-Code), connect with like-minded (see Google Summer of Code), blog about experiences in their own weblogs, provide code snippets free of charge (see Django-Snippets) or help novices with words and deeds in large mailing lists. This social software engineering—the creation of software and related artifacts within a social network—gained a lot of attention in recent software engineering research [1,17]. Besides the improvements of integrated development environments (IDE, e.g. Eclipse) or procedure models (e.g. eXtreme Programming [3]) research is addressing improvements of the daily working and learning environments of the developers. The main function of collaborative development environments (CDE) [2] is to support the whole development process of a team of software developers from start to finish. This includes version control of code artifacts as well as process documentation, coordination of tasks or support for division of labour.

CDEs usually are set up for one specific project; the possibilities for inter-project-collaboration within an organization with multiple software projects are very limited because the single CDEs are not able to exchange data.

Furthermore many developers are using data pools (bulletin boards, developer communities, mailing lists and a lot more) outside the organization in order to solve a specific problem. Furthermore existing CDEs lack in providing a transparent view on the progress of a project, awareness of developers' competencies and support for individual informal learning processes.

This paper describes an approach for a community-embedded CDE (CCDE), which is capable of handling multiple projects of several organizations, providing inter-project knowledge sharing and developer awareness. The presented approach uses the mashup pattern to integrate multiple data sources in order to provide software teams with an exactingly development environment. Furthermore we present requirements for a community of developers and sketch a first prototypical architecture for such a CCDE.

Related Work

The goal of this section is to behold the main aspects enlisted in the conception and implementation of a CCDE in order to derivate functional and technical requirements. Furthermore this section serves for definition and dissociation of the used terms.

Knowledge Management and Learning in Software Engineering

The different facets of the concept of knowledge have been discussed for over 2000 years now. Based on a fuzzy understanding of knowledge several theories for knowledge management came up and raised the idea of simply exchanging knowledge between individuals or organizations (among others [8]). It is probably the most important assessment to be made in this context that "you cannot store knowledge"[7] as in interpersonal communication only data is exchanged. Information emerges by interpreting this data with own prior knowledge in the personal context. Information then is the foundation for personal actions and decisions. So knowledge is first of all a rational capacity and not a transferable item. POLYANI distinguishes between tacit and explicit knowledge, whereas explicit knowledge is stored in textbooks, software products and documents, while tacit knowledge is in the mind of people as memory, skills, experience and creativity [10]. When tacit knowledge is externalised and transformed into explicit knowledge (properly speaking it is data now), we call this implicit knowledge. Implicit knowledge is of very high value for organisations such as software projects, as it gives hints how to solve specific problems in the future.

Regardless of the ambiguous definitions of knowledge and the claims for necessity and importance for knowledge management, software engineering is a dynamic process, which is reliant on latest knowledge in the subject domain. This knowledge is dynamic and evolves with technology, organisational culture and changing needs of the organisation [9]. Knowledge management in software engineering can be improved by recognising the need for informal communication and exchange of data in order to support the exchange of implicit knowledge amongst developers. Learning and working environments thus should support awareness of developers, sharing of implicit knowledge and foster informal, ad hoc exchange of short messages [6,11] as well as facilitating inter-project social networks in form of communities of interest.

Informal learning is characterized as a process that does not follow a specified curriculum but rather happens by accident, sporadically and naturally during daily interactions and shared relationships. Experience shows that the majority of real learning is informal [4]. Informal learning is what happens when tacit knowledge of a person is communicated to another person, which internalizes and interprets the data and thus expands his own knowledge. Examples of such informal learning situations within software engineering projects are spontaneous meetings, short messages, phone calls but also asynchronous communication like entries in bulletin boards, comments in source code or comments in blogs. As hardly any formal training for developers takes place, in software engineering informal learning is the only way to stay up to date. Previous approaches for supporting ad hoc communication focus on intra-project improvements and do not include experts from outside the project. Connecting with others and using artifacts from outside the own project seem to be a crucial factor in supporting learning within a project.

Social Software Engineering

The term social software engineering denotes both the engineering process of so called social software and the software engineering within social relationship in collaborative teams. For this paper the latter denotation is the focus of interest.

Studies show, that the main part of modern software engineering is carried out in teams, requiring strong interactions between the people involved in a project [1,13,14]. Social activity thus represents a substantial part of the daily work of a developer. Social network structures in social network sites (SNSs) emerge by adding explicit friendship connections between users. By contrast, social networks in the software engineering mainly result from object-centred sociality [15]. Developers do not just communicate with each other—they connect through shared artifacts. These social connections normally exist only within a project even though many of the artifacts used come from outside of the project. The consulted domain specific experts often do not reside within the own organisation, but in other communities.

Collaborative Development Environments

BOOCH and BROWN [2] define a CDE as "a virtual space wherein all stakeholders of a project—even if distributed by time or distance—may negotiate, brainstorm, discuss, share knowledge, and generally labor together to carry out some task, most often to create an executable deliverable and its supporting artifacts." So CDEs are a virtual working environment whose key functions can be clustered in the following categories: a) coordination of developers' work, b) cooperation of developers, and c) formation of a community. CDEs shall create a working environment that tries to keep frictional losses at a minimum. Frictions are costs for setup and launch of the working environment, inefficient cooperation while artifact creation and dead time caused by mutual dependencies of tasks.

BOOCH and BROWN define five several stages of maturity of CDEs [2]; besides simple artifact storage (stage 1) and basic mechanisms for collaboration (stage 2), advanced artifact management (stage 3), advanced mechanisms for collaboration (stage 4) the main feature of CDEs on stage 5 is to "encourage a vibrant community of practice" [2].

As the current median is somewhere around stage 1 and 2 [2], it is the goal of our efforts to enhance existing CDEs for single projects with a community component that allows project-spanning collaboration. This community-embedded CDE (CCDE) shall provide the classical functions of a CDE stated above but also allow the seamlessly exchange of artifacts [12], data and expertise amongst projects and developers from multiple projects. The remainder of this paper describes specific requirements for a CCDE and presents an initial architectural design.

Solution Design

The following section introduces the requirements for a CCDE to support awareness and transparency in multi-project environments. We define functional and nonfunctional requirements for the CCDE and introduce possible data sources needed in social software engineering projects (SSEP). Finally this section provides a first architectural design of the CCDE eCopSoft.

Organisational Requirements on a CCDE

As stated in section 2.B, social software engineering is a collaborative development process performed by a team of people that often are separated by time and space [18]. A CCDE aims at closing the gap between the members of a team by providing project awareness and transparency as well as providing options to connect with other developers and teams. From an organisational point of view a CCDE splits into two parts: I) the developers community and II) the single projects hosted at the CCDE. The requirements for the first part of a CCDE requires methods, services and tools for networking, presentation of contents and exchange of opinions to foster data exchange and the emergence of a community feeling. Thus, a CCDE should be equipped with the typical community features of SNSs like groups, wikis, bulletin boards, user profiles and friend lists. On top of this basic services and tools the community component of a CCDE should offer domain specific areas like a job market for developers, an event review and a news corner for trending development topics. All services and tools of the developer community are to ensure the shared identity of developers, the sharing of news and opinions as well as the start of new projects.

The second important parts of a CCDE are the project spaces. A project space is basically the home of a hosted project on the CCDE. A project space has to support the members of the project in collaborative and coordinative tasks. With our CCDE we claim to foster transparency and awareness of collaborative projects, for what reason a project space must provide fundamental tools such as wikis, e-mails, repository, bug tracker, and roadmap planning. Further data sources for the deployment in software projects are discussed in section 3.B. Any user of the CCDE must be able to start a new project and easily select the required services and tools for his project. The instantiation of the single tools has to take place automatically and without human intervention. Adding new developers to a project must be possible in various ways: either the members of the project are selected a priori by the creator of the project or added to the project afterwards. For the latter one it is important to discern between public and private projects. It must be possible to allow anyone to contribute to a project (public) or to approve new developers for the project. The creator must be able to broadcast his search for new developers to the community (e.g. by sending a microblogging message or adding an entry in a bulletin board) and also to browse the existing developers in order to directly ask them to join the project.

Data Sources in Software Engineering Projects

The potential data sources relevant for software engineering project are manifold. This section tries to identify the most important resources to support collaborative software engineering in the project spaces of the CCDE.

The selection of data sources that are applicable in a CCDE is essentially dependent on the available interfaces of the respective backend systems. It is crucial that the applicable data sources provide interfaces (e.g. open APIs) that allow the installation, configuration and query of data without sweeping adaptations of the data sources. To integrate a new data source in the project spaces the implementation and upload to the server of a new connector module is sufficient.

Basically we need to distinct between data sources or systems that incorporate coordination activities and those that incorporate communication activities of the development team. The latter is to be distinguished between informal and formal communication [18]. Informal communication is considered as explicit communication via diverse communication channels such as telephone, video, audio conference, voice mail, e-mail or other verbal conversations. Formal conversation refers to explicit communication such as written specification documents, reports, protocols, status meetings or source code [6]. Thus essential systems and tools to support communication in software engineering projects include e-mail, wiki, version control systems, blogs, instant messaging or microblogs as well as shared bookmarks and shared RSS feeds. Also modern communication channels like VoIP or video chat could be part of the communicative toolbox of a project space. Coordination activities address system-level requirements, objectives, plans and issues. Working with the customer and end users carries them out. To support coordinative activities the following data sources and systems ought to be integrated in a project space: road map planning, issue and bug tracker, collaborative calendars, and collaborative to-do lists.

For many of the data sources mentioned well-known software systems exist that offer open APIs. Along with MediaWiki and StatusNet, several version control systems and mail servers exist that can be a possible data source for the integration in a project space. For other data sources (e.g. shared bookmarks or VoIP) these software systems applicable in a CCDE are still to be found. Besides the open APIs it is also a necessary feature of the data sources that they store their data persistently, so that another person or tool can reuse the respective artifact in another context later.

Requirements on a Sophisticated Integration Layer

The main duty of an integration layer is to process the data of all connected backend systems in a way that a central and comprehensive access to all data is possible. By integrating the different data sources into a common layer it will becomes feasible to gain additional information that could not be provided from a single backend system beforehand.

Therefore the integration layer has to be informed about changes in the different backend systems and start an analysis of the changed artifacts consequently. Changes on an artifact in a backend system have to trigger a uniform change event that can be processed and stored by the integration layer. A change event will typically deploy the analysis of the specific artifact, which requires the automatic processing of various artifact types like e-mail, wiki articles, source code and many more. Further on different analyses techniques have to be integrated pursuing different targets. These techniques ought to range from simple stuff like language detection and keyword analyses to sophisticated semantically analyses of textual artifacts and precise source code analyses. The analysis framework has to be highly extensible allowing the later addition of new techniques. All data gained throughout the analysis have to be stored in a central data structure. An efficient design of the data structure aims at fast and precise querying of the data and easy integration.

The integration layer is obliged to enhance a manually entered developer profile with automatically generated data in order to keep it up-to-date. To be able to do this and to be able to retrace the chronological sequence in the modifications of an artifact, each user interaction with one of the backend systems has to be stored as an entry in the event log of the integration layer. Additional data extracted from an event (e.g. path to a source code file, categories of a wiki entry etc.) must be stored in a global data model where artifacts are being connected system—and project spanning. With this connection it shall become possible to gain additional information about artifacts and developers and to answer specific queries like:

- Who is the main developer of a package, class or method?
- Which artifacts from other systems are highly related to the current one?
- Who is an expert in a specific development domain or technique?
- Which developers from the community could be invited to work on a new project?
- What is the expertise of a developer?

Architectural Design

The requirements stated above demand for a system that allows the connection of various data sources and that provides multiple interfaces to access the integrated data in various ways. For that reason our prototypical implementation eCopSoft (event-based cooperative software engineering platform) consists of several components on different layers (cf. fig. 1) that make use of the typical mashup design pattern: easy and fast integration of multiple data sources, done by accessing APIs to produce results that were not the original reason for producing the raw source data [16].

There is a central server component (eCopSoft core) that is responsible for harvesting and processing data from all connected data sources on the system layer. The system layer mainly consists of the data sources described in section 3.C. From a technical point of view these systems run autonomous on a server and are connected to the eCopSoft server via their respective APIs. The eCopSoftcore processes the data from all data sources, extracts event data and other metadata and stores it in an internal database. Those involved in a project can access the data stored in the backend systems and the additionally generated and aggregated metadata with various clients on the presentation layer. These tools connect to server via the eCopSoft API.

The eCopSoft application is a modular and flexible system that holds administrative and operating data, assures the connection to the backend systems and provides interfaces for accessing the operating data with various clients. Furthermore eCopSoft provides a central management for users and projects. The integration layer is the most important component in the eCopSoft architecture—all events of the backend systems are processed here. Normally an event represents a user interaction with one of the backend systems. The connector modules of the data sources act as event provider, whereas the event consumers in the integration layer process these events. Each event holds information about the user that initiated the event, the changed artifact, which kind of operation the user was carrying out (e.g. create, update, link…) as well as other event-specific information if required. On arrival of an event at the event consumers, the event and all containing information are stored in the event database. The event data is processed by the eCopSoft core and used to update the user profiles in the user profile database. Based on these comprehensive additional data about the usage of and work with artifacts in a development team the cooperative work can be explored in new ways. A visual project dashboard, artifact networks, artifact usage patterns or expert lists showing individual expertise are enhancing the individual and organizational learning process with artifact and user awareness and transparency.

To connect the several data sources with eCopSoft a connector module will be implemented for each data source. A connector module assures the creation of the project-related instances and forwards the operating data from the backend system to the integration layer. The connector modules encapsulate the specific interfaces of the backend systems represent them homogenous at server side. The creation of events can either be actively triggered by a backend system (e.g. by a SVN hook) or passively by periodically querying the data source for new data (e.g. polling a RSS feed). The automatically instantiation of the backend systems is handled via scripts as part of the eCopSoft application. We will script the instantiation of the backend systems because most systems do not provide an API for doing that out of the box. Furthermore a scripted instantiation allows various adaptations to meet the specific requirements of the eCopSoft architecture.

The clients on the presentation layer can connect to eCopSoft via a web services API. Mediated through the API queries for projects, developers, or artifacts are realisable. These queries can be qualified with additional criteria or weighted. Therewith it is possible to query the system for experts to a specific artifact or all artifacts that a specific developer contributed to. In the first instance we plan three main clients:

1. A web-based project home (cf. fig. 2, 3),
2. An Eclipse expert view plug-in and
3. An admin interface to administer the whole system.

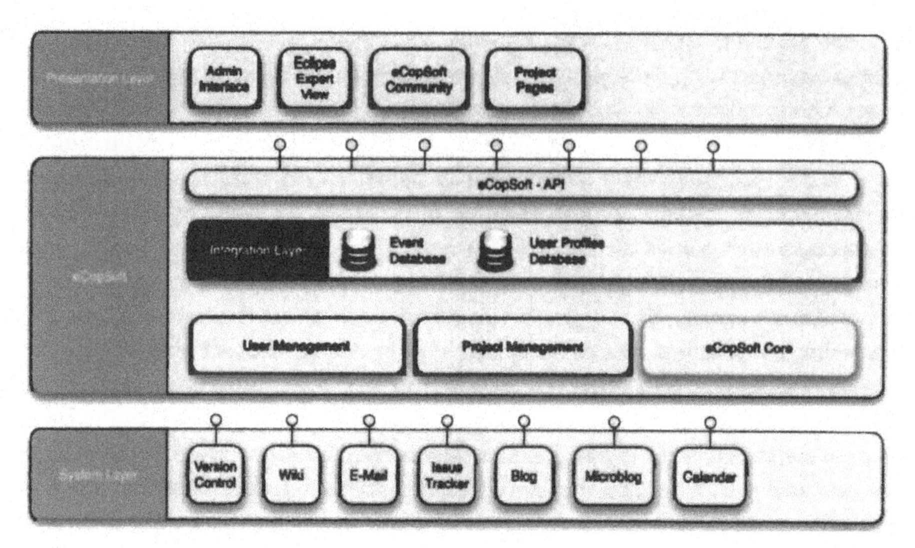

Figure 1. Schematical architecture of eCopSoft

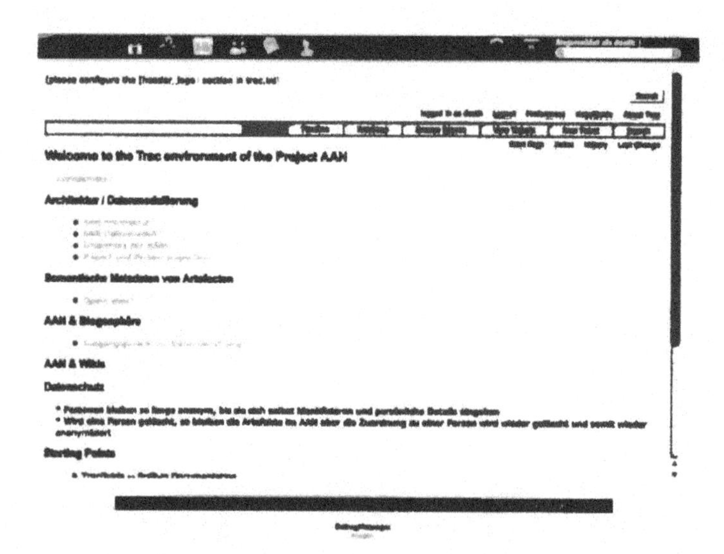

Figure 2. Screenshot of the eCopSoft web frontend showing a Trac environment for a project

Large parts of the eCopSoft system base on the Java platform, which ensures reliability, portability and scalability. Furthermore, when it comes to problem solving, there are numerous existing Java libraries that provide finished, tested and proven solutions to specific problems. This reuse of existing frameworks accelerates the whole development process a lot. To ensure future extensibility and the integration of further connector modules, eCop-Soft will be developed on an OSGi platform.

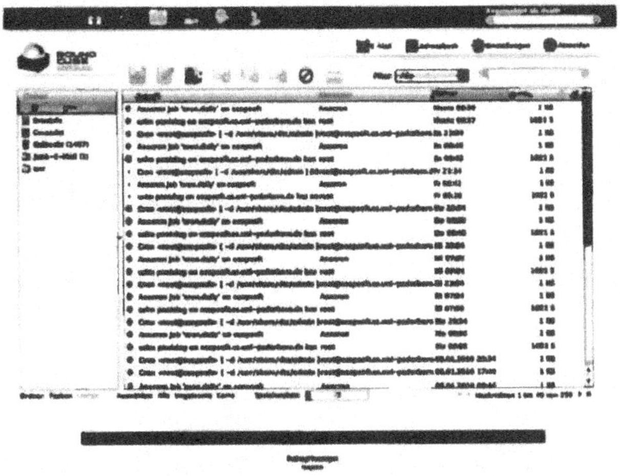

Figure 3. Screenshot of the eCopSoft web frontend showing the integrated webmail client for the project e-mail address

Conclusion and Outlook

This paper introduced the concept of a community-embedded collaborative development environment (CCDE) whose main functions are to combine classical approaches from collaborative development environments with the strengths of communities of interest. We provided requirements on functions of a community of developers as well as functional requirements for a technical integration layer to enhance awareness and transparency in social software engineering. With the help of a sophisticated integration layer the transparency of the development process can be increased as common events connect the hitherto separated backend systems. Thereby connections between artifacts (e.g. wiki articles and Java classes) manifests that have been hidden before. On the other hand an integration layer increases the personal awareness by connecting artifacts of a project directly with its contributors and thus allowing direct communication. With the help of the automatically extended developer profile the expertise and working fields of a developer become clearer. The artifact awareness will be increased by providing related artifacts, additional metadata (semantic information, classifications, used patterns...) and a lucid overview of recent changes of artifacts. Furthermore the integration layer will allow anonymously connecting to developers from other project in order to get help from them.

Although not being a classical mashup, the presented CCDE approach connects data from various sources in away that developers and users of the community could gain an advantage. In our opinion this advantage turns out to be in the assistance of individual work and the steady learning process by a more transparent process and enhanced awareness on various levels. Furthermore the possibility for a project spanning exchange of domain knowledge and artifacts enhances the data exchange and the collaboration within an organisation and thus fosters learning and interrelation. The easier data exchange, the higher awareness of the development process and contextualised data and experts creates an increased satisfaction with the whole development process and thus motivates developers.

The presented prototype eCopSoft is currently underdevelopment at the University of Paderborn and will be evaluated in software development courses. Furthermore we plan to run the CCDE as a campus-wide platform for software engineering projects, allowing the exchange of experience and data among multiple projects. The eCop-Soft platform furthermore shall reduce the administrative overhead of providing CDEs to numerous software projects by providing a one-click-deployment for new projects. The first evaluation results of eCopSoft will be part of another publication.

References

1. N. Ahmadi, M. Jazayeri, F. Lelli, and S. Nescic, "A survey of social software engineering," in 23rd IEEE/ACM International Conference on Automated Software Engineering - Workshops, pp. 1–12, 2008.
2. G. Booch, and A. W. Brown, "Collaborative development environments," in Advances in Computers, vol. 59, pp. 2–29, 2003.
3. K. Beck, Extreme Programming Explained. Embrace Change. Addison-Wesley, 1999.
4. J. Cross, Informal Learning–Rediscovering the Pathways that inspire innovation and performance. Pfeiffer, 2006.
5. P. N. Robillard, "The role of knowledge management in software development," in Communications of the ACM, vol. 42, no. 1, pp. 87–94, 1999.
6. W. Reinhardt, "Communication is the key–Support Durable Knowledge Sharing in Software Engineering by Microblogging," in Proceedings of Conference on Software Engineering 2009, Workshop Software Engineering within Social software Environments, 2009
7. I. Nonaka et al., "Emergence of "Ba,"" in Knowledge Emergence, 2001.

8. P. Schütt, "Kleine feine Unterschiede: Daten, Information und Wissen," in Wissensmanagement 02/2009, pp. 10–12, 2009.
9. A. Aurum, F. Daneshgar, J. Ward, "Investigating Knowledge Management practices in software development organisations–An Australian experience," in Information and Software Technology, vol. 50, pp. 511–533, 2008.
10. M. Polyani, The Tacit Dimension. Routledge & Kegan Paul, London, 1966.
11. P. N. Robillard, and M. P. Robillard, Types of collaborative work in software engineering. J. Syst. Softw., vol. 53, no. 3, pp. 219– 224, 2000. (doi:10.1016/S0164-1212(00)00013-3)
12. A. Sarma, "A survey of collaborative tools in software development," Technical Report at University of Irvine, Institute for Software Research, 2005.
13. T. DeMarco, and T. Lister, Peopleware: productive projects and teams. Dorset House Publishing, New York, 1987.
14. C. Jones, Programming productivity. McGraw-Hill, New York, 1986.
15. K. Knorr-Cetina, "Sociality with Objects: Social Relations in Postsocial Knowledge Societies," in Theory, Culture & Society, vol. 14, no. 4, pp. 1–30, 1997 (doi:10.1177/026327697014004001)
16. Wikipedia. Mashup (web application hybrid). (Revision as of 10:23, 27.05.2009). Available at http://en.wikipedia.org/w/index.php?title=Mashup_(web_application_hybrid)&oldid=292635186
17. J. Münch, and P. Liggesmyer (Eds.), Proceedings of the Software Engineering 2009 conference, Workshops. Social Aspects in Software Engineering, 2009.
18. J. Herbsleb, and A. Mockus, "An empirical study of speed and communication in globally distributed software development," in IEEE Transactions on Software Engineering, vol. 29, no. 6, pp. 481–494, June 2003.

Comprehending Software Architecture Using a Unified Single-View Visualization

Thomas Panas, Thomas Epperly, Daniel Quinlan, Andreas Sæbjørnsen and Richard Vuduc

ABSTRACT

Software is among the most complex human artifacts, and visualization is widely acknowledged as important to understanding software. In this paper, we consider the problem of understanding a software system's architecture through visualization. Whereas traditional visualizations use multiple stakeholder-specific views to present different kinds of task-specific information, we propose an additional visualization technique that unifies the presentation of various kinds of architecture-level information, thereby allowing a variety of stakeholders to quickly see and communicate current development, quality, and costs of a software system. For future empirical evaluation of multi-aspect, single-view architectural visualizations, we have implemented our idea in an existing visualization tool, Vizz3D. Our implementation includes techniques, such as the use of a city metaphor, that reduce visual complexity in order to support single-view visualizations of large-scale programs.

Introduction

Visualization techniques are widely considered to be important for understanding large-scale software systems [15]; yet knowing what to visualize and how to present information are themselves daunting issues. The challenges are many. First, there are several stakeholders in a software project—architects, developers, maintainers, managers—each asking different questions about the software. Answering a diverse set of questions will involve different abstraction levels, such as the architecture, the middle level structure [15], or the source code itself. Each question may require a distinct analysis; multiple analyses can generate huge volumes of data, which may be difficult to store, to manipulate, and to present. Having to manage multiple analyses, possibly through multiple visualizations (views), places a significant cognitive burden on any individual stakeholder. Multiple views also tend to make it difficult for separate stakeholders to communicate on subtle issues about a software architecture. These challenges make it hard for multiple stakeholders to reach a common understanding or consensus about the project.

Different stakeholders often interact around questions about the software system's architecture. For example, developers and project managers need to know where to make improvements, e.g., "which components do we have to modify in order to improve the performance or security of our system?" Project managers need to know how to allocate team members to each part of the system and answer questions such as, "can we meet the next deadline?" Additionally, both managers and vendors are interested in development hot spots (frequently modified components), which may indicate areas of high system maintenance cost. Designers and maintainers are more interested in the overall structure of the system; this knowledge helps them to identify reusable components, for instance.

We are investigating a visualization technique designed to help all stakeholders collectively understand and better communicate the architecture of a large-scale software system. Our particular technique represents the structure of the system using a graph-based model, following common convention [15], and we augment this model with a number of static and dynamic analyses. Our specific choices of model and analyses help stakeholders answer a wide range of questions about the overall design, quality, and costs in development and maintenance (Section 2).

The central design goal of our approach is to visualize multiple task- (or stakeholder-) specific aspects of a software architecture using just a single view (Section 3), cf. Figure 1, to communicate problems, decisions and solutions between stakeholders. In contrast, existing approaches (Section 6) primarily rely on multiple views when visualizing multiple aspects of a system [1, 17, 30, 31]. Although multi-view approaches can be very effective, users may also have considerable difficulty managing and navigating through different views [27, 28, 32]. In our current research, we are particularly interested in whether these difficulties can be overcome by judicious rendering within a single view.

Figure 1. Architectural Program Visualization of a C++ Program.

To help answer this question, we have implemented a single-view visualization tool (Section 4). We use a metaphor based on cities that provides users with an intuitive physical interpretation of the system. In addition, we use a layout algorithm that permits the system to be rendered consistently each time the visualization is performed. That is, the layout is designed to be insensitive to small structural changes, thereby helping users remember and navigate in the visualization. Our specific implementation supports C and C++ programs, extending our earlier work for Java programs [20, 34].

Our overall approach allows multiple stakeholders to see and communicate about the same view, which facilitates the collective understanding and discussion of the system's architecture (Section 5).

Visualizing Architectures

There are roughly three levels of program visualization, based on the level of abstraction [15, 26]: source code level, middle level, and architecture level. This section explains what we mean by architecture level, and describes our specific model of the program and analyses in detail.

On the source code level, typical visualization tools include program and aspect editors. Advanced tools may integrate program editors with online debuggers and profilers. Source-level visualizations are very "low-level," as they relate directly to the underlying software artifact, and are primarily of interest to developers and maintainers.

On the middle level, visualizations are problem-specific. Developers and code maintainers have specific problems to solve, and they usually apply tailored algorithms and visualizations to the program to better understand both the problem and the program. Typical middle level visualizations include sequence diagrams, abstract syntax tree (AST) representations, dominance tree visualizations, concept lattices, control and data flow graphs.

The aim of architecture-level visualization is to rapidly summarize and communicate the architecture and design decisions of the overall software system. Architectural visualization is naturally more abstract than source-or middle-level visualizations, and therefore better suited to visualizations in the large. Abstract visualizations of software architectures combined with metrics can help different stakeholders to answer many questions about a software system.

For instance, project managers might use the visualizations to understand the aspects of a project that are most expensive, code designers can communicate current implementation deviations from original design plans, and code maintainers may use the visualizations to better understand unknown software systems [7].

Common examples of architectural visualizations are function call graphs, hierarchy graphs, and directory structures. There are many ways to present these graphs, such as UML diagrams, graph browsers, and component/connector graph drawings. Many tools support these visualizations [12, 26, 31]. However, to the best of our knowledge, no tool exists that supports the visualization and communication of software architecture between different software development and maintenance stakeholders.

A Graph-Based Program Model

To support communication among various stakeholders through a single-view architecture visualization, a program model combining different aspects of a program (important to different stakeholders) is needed. We have merged several tools (Section 4) to retrieve the following architectural program models of C/C++ applications:

- A Function Call graph represents the call relationship between different (C/C++) functions.
- A Class Call graph shows the interaction structure between (C++) classes.
- A Class Contains graph holds information about classes and their functions.
- A Class Inheritance graph represents the inheritance structure of an (C++) application.
- A File Call graph shows the call structure between source files.
- A File Contains graph shows functions in relation to their files. In C/C++, related functions may be implemented in (multiple) source as well as header files. The C++ specification does not enforce a standard implementation style. As a result, File Contains and Class Contains graphs are most likely to be viewed together.
- A Directory Contains graph represents the relationship between files and their corresponding directories. Our graph-based program model is a union of the graphs implemented and described above. Table 1 shows examples of how various stake holders might communicate software architecture through a particular set of graph types.

Table 1. Correlation Graphs-Stakeholders

Graph	Program Manager	Architect/ Designer	Developer	Maintainer/ Re-engineer
Function Call		x	x	x
Class Call		x	x	x
Class Contains			x	x
Class Inheritance		x	x	x
File Call	x	x	x	x
File Contains			x	x
Directory Contains	x	x	x	x

Table 1 only suggests what information might be of interest to each stakeholder. Of course, we might modify the table to reflect the interest, expertise, or task at hand. The key point in Table1 is that a unified architectural visualization more flexibly supports and encourages interaction among stakeholders, compared to visualizing and communicating multiple stakeholder-specific representations in various notations.

Augmenting the Model with Analyses

There is more information of interest to various stakeholders than plain structural graphs can represent. For instance, when investigating a Directory Contains graph, a project manager would also like to understand the complexity of the files being viewed with respect to other metrics, such as size. The additional program information can either be shown in tables and be related to the views, or more preferably be integrated directly into the views themselves.

For our prototype, we have implemented a number of analyses, the results of which can be attached to our program model. We list examples below; though not all-inclusive, this list can be extended easily:

Run-Time Analyses

We collect run-time information about C/C++ applications using the gprof profiling tool.

- Execution Time Analysis profiles the time spent in functions, identifying performance "hot spots," or functions that execute for an exceptionally long time.
- Execution Frequency Analysis profiles the call frequency of functions to determine exceptionally frequently called functions.

Metric Analyses

We collect program information about single entities such as functions.

- Lines of Code (LOC) is measured for each function.
- Unsafe Function Calls. Certain aspects of C++ (e.g., unchecked array access, raw pointers), can lead to low level buffer overflows, page faults, and segmentation faults. In this analysis, we detect calls to "unsafe" functions, such as *sprintf, scanf, strcpy*.
- Global Variables. We traverse the program's abstract syntax tree (AST) to check for public declared variables (within the scope of classes) and global variables (outside the scope of classes). It is considered a good programming style to avoid global variables.
- New-Delete Analysis. In C++, when deleting scalar entities, the scalar C++ delete statement should be used to ensure correct program behavior. Similarly,

dynamically allocated arrays must be deleted with the delete [] statement. Our analysis is based on data-flow and control-flow information.

- Cyclomatic Complexity (CC) indicates how much effort is required to maintain a function. Our implementation of McCabe's Cyclomatic Complexity [19] counts the possible execution branches in a function for the following branching statements: if, for, while, do-while, and (switch-) case.
- Arithmetic Complexity. For each function, this analysis counts the number of arithmetic operations on float, int, float pointer, and int pointer types. Thus, functions and classes with large arithmetic operation counts can be detected. This is particularly important in scientific computing codes, since such functions should be the most robust and reliable pieces of the software.

Advanced Static Analyses

This helps to recover various hidden aspects of software. In general, these analyses can not be interpreted alone; results indicate relations between entities.

- Pattern Matching is used to locate functional and non-functional properties of a software system, e.g., MPI [22] calls in high performance computing. This approach is similar to visualizing aspects from aspect-oriented programming (AOP) [13].
- Class Membership. This analysis annotates each member function with its associated class and source file it is implemented in. It discovers fragmented member functions, i.e., member functions declared in the same class but defined in different files. This analysis may uncover "bad" coding styles or refactoring efforts that were applied to split large source files.
- Strongly Connected Components (SCC). We detect cyclic dependencies between functions, classes or files. In general, nodes in a cyclic dependency may be merged to reduce call dependencies, and hence to reduce the structural complexity of the system.

Repository Analyses

This retrieves information from a source-code repository such as CVS or SVN and attach it to the current program model. In our current implementation of C/C++ visualization, we have not attached this information yet. We base the need for such information on our previous studies visualizing Java repository information [20, 24] using VizzAnalyzer [35] and Kenyon [3].

- Work Distribution is an analysis that determines the specific developers that are working on a specific part of a software system. This information gives an idea about the progress of the maintenance or development team and can also assist to estimate time to completion.

- Frequent Change. Entities changed frequently result in higher maintenance efforts and costs. Thus, it is essential to determine frequently changed components.
- Defect Dependencies show how certain defects and their bug-fixes relate [4, 6]. Defect analyses may help to predict software project/maintenance costs.

Table 2 lists stakeholders and these analyses in which they might be particularly interested.

Table 2. Correlation Analyses-Stakeholders

Metric/Analysis	Project Manager	Architect/ Designer	Developer	Maintainer/ Re-engineer
Execution Time			x	x
Execution Frequency			x	x
Lines of Code	x	x	x	x
Unsafe Function Calls			x	x
Global Variables		x	x	x
New-Delete			x	x
Cyclomatic Complexity	x		x	x
Arithmetic Complexity			x	x
Pattern Matching		x	x	x
Class Membership			x	x
SCC		x	x	x
Work Distribution	x			
Frequent Change	x		x	x
Defect Dependencies	x			x

Table 2 only suggests what is possible. For our prototype, we have implemented all of the analyses above (except pattern matching and the repository analyses). The analysis results are attached to our model graph. With the model graph at hand, the user can now view, interactively select (reduce) and communicate information of interest.

Practical Single-View Visualization

We believe that single-view visualizations are better than multi-view visualizations when communicating about software architecture. First, single-view illustrations avoid the difficulties when managing and navigating through different views [27, 28, 32]; secondly, they rapidly summarize a system because all information is available within that view; and thirdly, having one view helps different stakeholders to easily communicate different concerns within the same familiar picture. Single-view visualizations do not replace detailed stakeholder-specific visualizations.

Rather, they unify different stakeholder-specific architectural visualizations to a common image that can easily be communicated, as is done in the engineering industry where task-specific blueprints are unified to be understood by architects, electricians, civil engineers, and others. As an example, Kruchten introduced (software) architectural blueprints in which a logical view describes classes and associations for end users and a development view describes modules and compilation dependencies for developers and managers [16]. Both views are essential. Nevertheless, we believe that a single unified view combining both the logical and development view could support the communication between end users, developers and managers. We plan to evaluate this hypothesis in future work.

To study the effects of single-view visualizations, we have developed a tool prototype, cf. Section4. For our prototype to work effectively(i.e., by reducing the cognitive overload for a viewer [28, 32]), we need to handle the inherent complexity of multiple architectural aspects within one view. Below, we describe complexity reduction techniques we have selected and implemented from literature:

- Abstraction. To cope with a flood of information, we need mechanisms for filtering, aggregating, or merging low-level details into higher-level properties [25, 36]. Abstraction techniques are partofVizz3D [20].
- Association with Source Code. The architectural visualization should be easy to relate to the underlying software artifact [36]. Therefore, we have extended Vizz3D with the capability to interactively view the source code in association with any node of the architectural view. This feature associates architecture and source levels, cf. Section 2.
- Metaphors. Many graphic designs lack an intuitive interpretation, requiring that a user be trained to understand them. We can alleviate this problem by selecting metaphors that are familiar to the user, such as those found in the real world. Tangible metaphors improve understanding and social interaction [8]. We have chosen and implemented the city metaphor [14, 24], cf. Section 4.1, to increase program understandability.
- Predictability. Two different runs of a layout algorithm, involving the same graphs, should not lead to radically different visual representations. This property is also referred to as "preserving the mental map" of the user [23]. While force-directed layouts are usually not predictable, hierarchical algorithms improve the situation, but they do not scale particularly well. For improved predictability and scalability, we have implemented a combination of force-directed and hierachical layout within our tool, cf. Section 4.2.

The items above do not constitute an all-inclusive list of complexity reduction techniques or features a tool should support. There are several other techniques discussed in the literature [31, 36]. Some techniques were not chosen because of their minor impact. For instance, focus and context [9], in which a user focuses on a visual detail without losing the visual context, also aids program comprehension. However, as Storey, et al., report in their user study, although focus and context (particularly fish-eye) views are thought to be useful, in practice they tend not to be used [32]. Moreover, it is not our primary goal to define a complete list, but rather to implement a reasonable number of techniques to support our goal to show that our single-view visualization can present all manners of architectural information.

Implementation: Vizz3D

Our prototype is based onVizz3D [20, 34], a3D information visualization system primarily developed for program visualization. Vizz3D is highly flexible and allows users to define and reassign layout algorithms and metaphors at runtime. Hence, visualizations can be configured on-line [20].

To reduce a user's cognitive load, Vizz3D has a variety of operations, such as the aggregation and filtering of nodes. In addition, users may at run-time filter all nodes and edges of certain types, thereby allowing the visualization of arbitrary combinations of the graphs defined in Section 2.1. For instance, a user may interactively select several types of edges to display; the resulting same image might then represent a Function Call graph, a Class Call graph, a File Contains graph, or all graphs simultaneously.

We have enhanced Vizz3D with the ability to view the corresponding source code when interactively selecting visual entities. In addition, to achieve predictable and consistent layouts and to provide an intuitive representation of our program model, we have configured our own layout and metaphor for Vizz3D. (Sections 4.1–4.2, below).

Our prototype works on real C and C++ applications [33]. To build our implementation, we have used, combined, and enhanced the following reverse engineering tools: ROSE [29], a C/C++ source-to-source code transformation tool used for program retrieval; ROSEVA, a program analyzer developed to create and merge the various graphs described above; VizzAnalyzer [35], a framework for reverse engineering tool integration, allowing us the rapid integration of various program analysis and visualization tools; and for language interoperability between C/C++ (ROSE and ROSEVA) and Java (VizzAnalyzer), we use Babel [2].

An Intuitive City Metaphor

Metaphors help users to better understand complex situations. For our architectural program visualization, we have chosen a city metaphor [14, 24]. Our metaphor comprises the following elements, cf. Figure12. Buildings represent functions. Building textures represent source code metrics; for instance, the (tall) blue colored buildings in Figure 1 indicate a LOC> 200. Similarly, other textures in this image indicate other thresholds. Cities (blue plates) indicate source files. Pillars (C++ only), shown perpendicular to cities in Figure 2, represent class definitions. The pillars are the foundation for water towers (spheres), representing header files. In this way, a header file can have multiple class definitions. Finally, (green) landscapes, which carry cities and water towers, represent directories. The water and sky in Figure 1 and 2 are optional aesthetic decorations. Note that the above is an example of one visual configuration; to determine the most effective configuration, cognitive studies must be conducted. Vizz3D merely allows different user defined mappings (from program model entities to visual entities). For more information see [20].

Figure 2. Architectural Visualization.

Computing Consistent Layouts

Our layout algorithm extends the force-directed algorithm of Huang and Eades [11] by combining it with a hierarchical algorithm. Because force-directed layout algorithms can in general be rather slow, we calculate the forces for coarse-grained nodes first, i.e., for header files, source files, and directories. Figure 3a) shows the layout of 36 files. The edges represent Directory Contains relationships. Files belonging to the same directory, representing components by design, are laid out close together. The size of a source file reflects the number of functions defined within it. Buildings are laid out compactly next to each other, i.e., they are not part of the force-directed layout. The second step is to apply the landscapes (directory structure) for the cities (files), cf. Figure3b).

The height of the landscape (y-axis) represents the depth of the directory path. Therefore, cities or files in a deeper directory structure are represented on a higher hierarchical level. As a result of the force-directed layout, directories containing subdirectories are laid out more closely. As subdirectories are on a higher hierarchical level, subdirectories produce "visual mountains," similar to 3D tree-maps [5], where directory structures are represented to the user in a hierarchical way.

Our layout provides predictable visualizations (see Section 3) in that different runs of the system produce fairly similar landscapes. Together with our 3D city metaphor, familiar entities can quickly be rediscovered. The current layout predictability can even be improved if the initial random seed of our layout is kept constant across multiple runs.

Representing Analysis Results

Our single-view restriction means that all analysis results are displayed in one view, raising immediate concerns about information overload. To overcome this problem, we display metrics either with 2D icons [27] within our 3D scene (i.e., 2D icons are shown on top of each building and above each city to convey information), or we use visual properties such as height, width, depth, and texture, among others. Analysis results may also be represented by color; this means, however, that only one analysis at a time can be represented. For instance, all strongly connected components are colored red (the top roof of a building).

Application Examples

We envision a variety of scenarios in which our single-view architectural visualization would be particularly useful. We outline several such scenarios in this section, emphasizing the ways in which our approach can facilitate collaboration and discussion among stakeholders.

Quality Assessment

Project managers and developers can easily assess various aspects of a software system's quality in our one-view visualization. For instance, consider Figure 1, which combines a File Call-, Class Inheritance-, Class Contains-, File Contains-and

Directory Contains graph. In addition, this figure shows software complexity information (fire texture), global variables (globe icon), oversized functions (blue buildings), unsafe functions (lock icon), and run-time information (the width and depth represents run-time information).

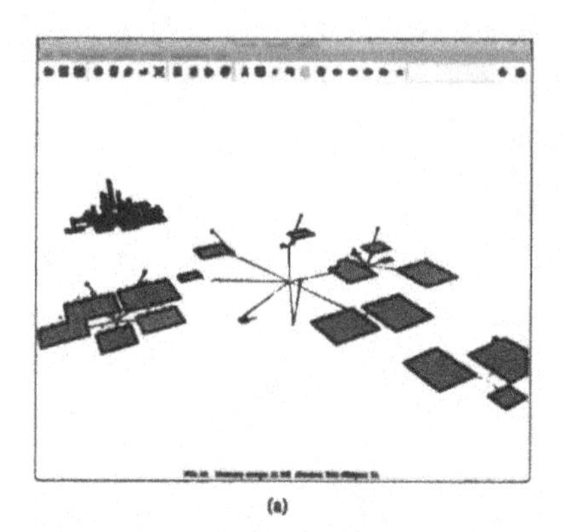

(a)

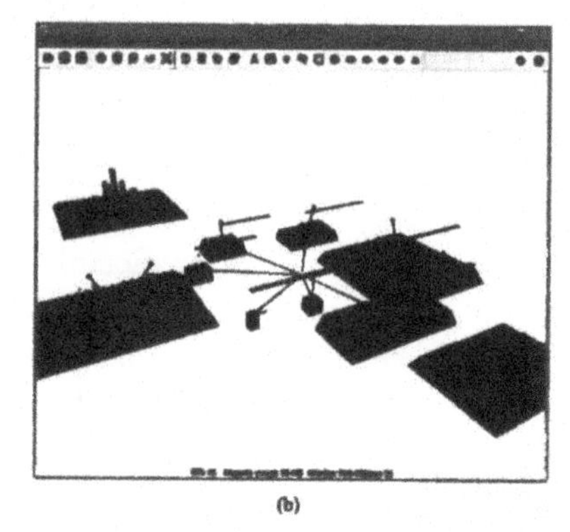

(b)

Figure 3. (a) layout algorithm between files (b) adding the directory structure

By interactively examining Figure 1, developers can communicate concerns, such as global variables, new-delete deviations, or unsafe functions, to managers, and help the managers understand where and why additional time must be spent to improve those components. Similarly, complex areas of the system, as indicated by cyclomatic or arithmetic complexity, can be easily illustrated via the common metaphor. Developers of all skill levels can use this kind of visualization to detect and communicate concerns, and they may do so over a variety of communication media, such as teleconferencing or virtual reality displays.

We anticipate that single-view collaboration will help stakeholders detect code quality problems earlier, make meetings more effective, and reduce project costs in general.

Componentization

Once a software system becomes large, it is essential to decomponentize it, i.e., to split it into smaller reusable components that eventually can be maintained or sold separately. A single-view architectural visualization may again help to communicate componentization issues and costs among management, developers, and re-engineers, using analyses such as pattern matching, class membership, and SCC.

For example, consider the analysis of class membership violations. As described in Section 2.2, a class membership analysis determines fragmented member functions, i.e., member functions declared in the same C++ class (usually header file) but defined indifferent source files. Suppose a file represents a component. From a reusability perspective, a "good" coding style might prefer that all member functions of one class be implemented in the same source file. A re-engineer might write a simple analysis to check this condition and print the results as text to the screen.

Though this screen dump provides useful information, this information might be of limited use if the original developers are not present to guide changes. An architectural visualization, on the other hand, might supply some of the developers' expertise through additional metrics visualized at the same time. For instance, consider Figure 4, where the class membership relationship is indicated by assigning the same color to each member function of a class. In Figure 4, the directory "/wpp3D/" contains four source files and one header file. The header file contains one class (pillar) that contains eight member functions, indicated by their red coloration.

We see that one member function is defined in a different source file, i.e., a fragmentation is present. However, due to the additional LOC information, even a non-expert could conclude that the exceptional member function has been refactored due to its size.

Our single-view architectural visualization helps us to understand the specific analysis problem at hand. Moreover, it shows the entire context of all additional entities and relationships involved. For example, one can determine whether the exceptional method described above occurs in the same directory as the other methods in its class or not. Furthermore, call edges may be activated turning the Class Contains graph into a merge of the Class Contains and Function Call graphs. The additional information may help to investigate the purpose of certain functions in the context of the whole program.

Another example is the detection of strongly connected components. Again, if only printed to the screen as a list, the context might be lost. For instance, Figure 5 shows an architectural visualization of a file call graph. The files themselves are color coded; a green color indicates a cycle. Seeing the analysis result in the context of the directory structure, one could assume that all files in the cycle belong to the lower right directory. There are however two exceptions. It is now up to the developer or re-engineer to determine whether the exceptional files should be moved.

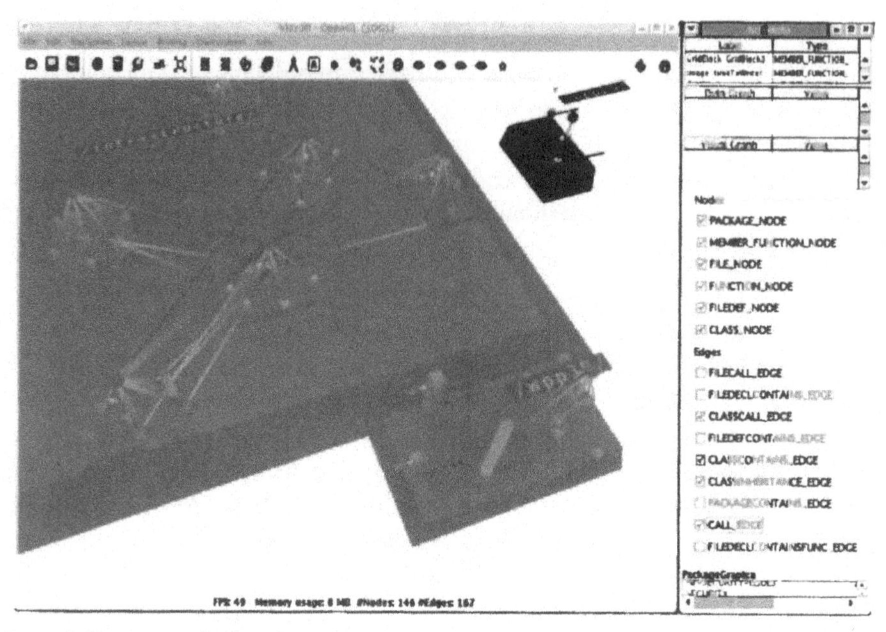

Figure 4. Class Membership Visualization.

Comparison to Related Work

There is a large literature on existing architectural visualization tools. We can classify these approaches accordingto the number of aspects theyvisualize using how many views.

- Single-aspect, single-view. Illustrate only one aspect of a software architecture, e.g., CrocoCosmos [18], sv3D [21].
- Multi-aspect, single-view. Our work is an example of this class. Another example is SHriMP [31], allowing the interactive, single-view navigation between architecture and source code. SHriMP is a great tool for architectural browsing and understanding of source code. However, SHriMP was not developed with stakeholder communication in mind; it does not support a natural metaphor or the ability to add multiple analysis and metric results within the view.
- Multi-aspect, multi-view. Illustrate multiple views of the architecture level, e.g., CodeCrawler [17], BLOOM [30]3. Other approaches, such as ArgoUML [1], exist. However, these applications visualize not just architectural level information, but rather a combination of architecture and middle level. Further recommendations and cognitive studies on (architectural) multi-views can be found in [10, 16, 32].

We believe that multiple views are important for detecting and answering many problems, especially when depicting low-level analysis results in middle-level visualizations. In such cases, it is impossible to view all the possible aspects of software in one image. However, when illustrating the architecture of a software system to various stakeholders, we believe that showing the different architectural aspects in the same view with the same metaphor and layout is desirable. In this paper, we have suggested how it is possible to merge and filter the essential information for different stakeholders. This helps stakeholders get precise answers to their questions and, moreover, enables them to communicate the answers among themselves and others.

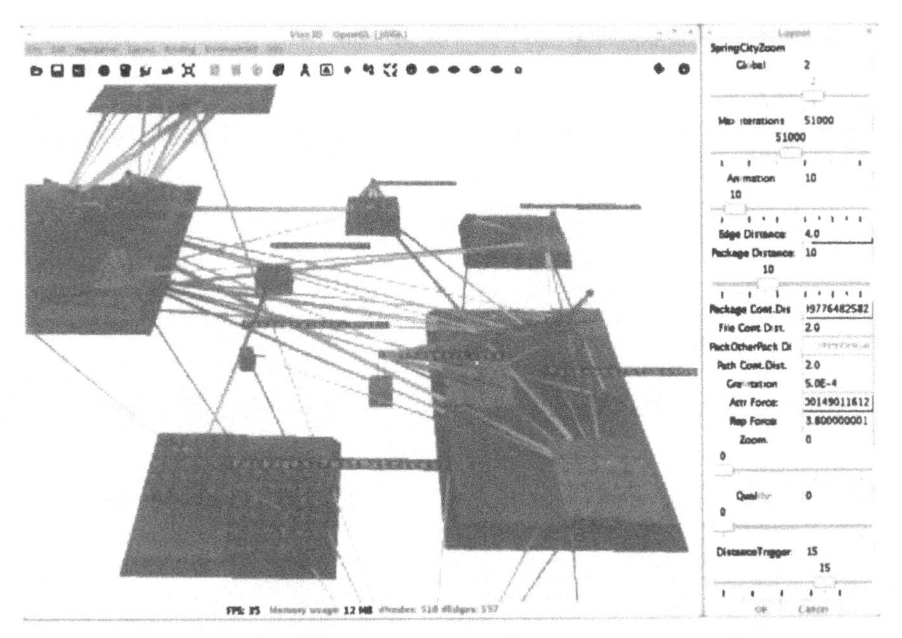

Figure 5. Strongly Connected Components Visualization.

Conclusion and Future Work

Our tool is a proof-of-concept design for a multi-aspect, architecture-level, single-view visualizer. This paper reviews our philosophy and implementation, with particular emphasis on how different stakeholders can use such visualizations as an aid in collaborative understanding, development, maintenance, and re-engineering of a large-scale software system. The key features of our approach are the use of an intuitive city metaphor for representing the structure of the system architecture, a single view for visualizing multiple aspects and analysis results, and powerful filtering and focusing techniques built into the tool implementation.

We have integrated and extended a variety of analysis and visualization tools, allowing us in future work to evaluate the trade-offs of using a unified single-view for architecture-level program visualization and communication. In preparation for such future experiments, we have classified and implemented program analyses, identified and implemented complexity reduction techniques for large scale visualizations, and implemented a layout algorithm and metaphor in our visualization tool, Vizz3D.

Acknowledgements

This work was performed under the auspices of the U.S. Department of Energy by University of California, Lawrence Livermore National Laboratory under Contract W-7405Eng-48, UCRL-CONF-227293.

We thank Jeremiah Willcock and Christian Iwainsky for numerous helpful discussions.

References

1. ArgoUML. Available at: http://argouml.tigris. org/, 2004.
2. Babel. Available at: http://www.llnl.gov/CASC/ components/, July 2006.
3. J.Bevan,S. Kim, and L. Zou. Kenyon: A common software stratigraphy system. Available at: http://www.soe. ucsc.edu/research/labs/grase/kenyon/, 2005.
4. D. Beyer. Co-change visualization. In Proceedings of the 21st IEEE International Conference on Software Maintenance (ICSM 2005, Budapest, September 25-30), Industrial and Tool volume, pages 89–92, Budapest, 2005.
5. T. Bladh, D. Carr, and J. Scholl. Extending tree-maps to three dimensions: a comparative study. In M. Masoodian, S. Jones, and B. Rogers, editors, 6th Asia-Pacific Conference on Computer-Human Interaction (APCHI 2004), New Zealand, June 2004.
6. M. Burch, S. Diehl, and P. Weissgerber. Visual data mining in software archives. In SoftVis '05: Proceedings of the 2005ACM symposium on Software visualization, pages 37– 46, New York, NY, USA, 2005, ACM Press.
7. S. Demeyer, S. Ducasse, and O. Nierstrasz. Object-Oriented Reengineering Patterns. Morgan Kaufmann Publishers, 2003.

8. C. R. dos Santos, P. Gros, P. Abel, D. Loisel, N. Trichaud, and J. Paris. Metaphor-aware 3d navigation. In IEEE Symposium on Information Visualization, pages 155–65. Los Alamitos, CA, USA, IEEE Comput. Soc., 2000.
9. G.W. Furnas. The FISHEYE view: A new look at structured files. Technical Report 81-11221-9, Murray Hill, New Jersey 07974, U.S.A., 12 1981.
10. J.C. Grundy, R. Mugridge, and J. Hosking. Visual specification of multi-view visual environments. In IEEE Symposium on Visual Languages, Halifax, Nova Scotia, Canada. IEEE CS Press, September 1998.
11. M.L. Huang and P. Eades. A fully animated interactive system for clustering and navigating huge graphs. In 6th Int. Symposium on Graph Drawing, pages 374–383. Springer LNCS 1547, 1998.
12. M. Junger and P. Mutzel, editors. Graph Drawing Software. Springer, 2004.
13. G. Kiczales, K. Lieberherr, H. Ossher, M. Aksit, and T. Elrad. Discussing Aspects of AOP. Communications of the ACM, 44(10), October 2001.
14. C. Knight and M.C. Munro. Virtual but visible software. In IV00, pages 198–205, 2000.
15. R. Koschke. Software Visualization in Software Maintenance, Reverse Engineering, and Reengineering: A Research Survey. Journal on Software Maintenance and Evolution, 15(2):87–109, March 2003.
16. P. Kruchten. The "4+1" view model of architecture. IEEE Software, 12(6):42–50, November 1995.
17. M. Lanza. Codecrawler a lightweight software visualization tool. In VisSoft 2003 (2nd International Workshop on Visualizing Software for Understanding and Analysis). IEEE Computer Society Press, 2003.
18. C. Lewerentz and F. Simon. Metrics-based3DVisualization of Large Object-Oriented Programs. In 1st International Workshop on Visualizing Software for Understanding and Analysis, June 2002.
19. W. Li and S. Henry. Maintenance Metrics for the Object Oriented Paradigm. In IEEE Proceedings of the 1st International Software Metrics Symposium, May 1993.
20. W. Lowe and T. Panas. Rapid Construction of Software Comprehension Tools. International Journal of Software Engineering and Knowledge Engineering, December 2005.
21. A. Marcus, L. Feng, and J. I. Maletic. 3D Representations for Software Visualization. In Proceedings of ACM Symposium on Software Visualization, 2003.

22. Message Passing Interface Forum (MPIF). MPI: A Message-Passing Interface Standard. Technical Report, University of Tennessee, Knoxville, June 1995. http://www.mpi-forum.org/.
23. K. Misue, P. Eades, W. Lai, and K. Sugiyama. Tree visualisation and navigation clues for information visualisation. J. of Visual Languages and Computing, 6:183–210, 1995.
24. T. Panas, R. Berrigan, and J.C. Grundy.A3d metaphor for software production visualization. In IV03, London, UK, June 2003. IEEE.
25. T. Panas, W. Lowe, and U. Aßmann. Towards the unified recovery architecture for reverse engineering. In International Conf. on Software Engineering Research and Practice, Las Vegas, USA, June 2003.
26. T. Panas, J. Lundberg, and W. Lowe Reuse in reverse engineering. In International Workshop on Program Comprehension, Bari, Italy, June 2004.
27. G. Parker, G. Franck, and C. Ware. Visualization of large nested graphs in 3d: Navigation and interaction. Journal of Visual Languages and Computing, 9(3):299–317, 1998.
28. M. Petre, A. Blackwell, and T. Green. Cognitive questions in software visualization. Software Visualization: Programming as a Multimedia Experience, pages 453–480, January 1998.
29. D. Quinlan, S. Ur, and R. Vuduc. An extensible open-source compiler infrastructure for testing. In Proc. IBM Haifa Verification Conference, volume LNCS 3875, pages 116–133, Haifa, Israel, November 2005.
30. S.P. Reiss and M. Renieris. The BLOOM Software Visualization System. In Software Visualization –From Theory to Practice, MIT Press, 2003.
31. M.-A.D. Storey, F.D. Fracchia, and H.A. Mueller. Cognitive design elements to support the construction of a mental model during software visualization. In Proc. of the 5th Int. Workshop on Program Comprehension (WPC '97), Washington, DC, USA, 1997. IEEE Computer Society.
32. M.-A. D. Storey, K. Wong, and H. A. Muller How do program understanding tools affect how programmers understand programs? Science of Computer Programming, 36(2–3):183–207, 2000.
33. T. Panas. Quality Analysis of SMG2000.Technical report, CASC, Lawrence Livermore National Laboratory, November 2006.
34. Vizz3D. Available at: http://vizz3d. sourceforge.net, July 2006.
35. Vizz Analyzer. Available at: http://www.arisa.se/, 2006.

36. P. Young and M. Munro. Visualising software in virtual reality. In Proc. IEEE6th Int. Workshop on Program Comprehension, June 24-26, pp19-26. Ischia, Italy, IEEE Computer Society Press., 1998.

Determination of Neural Fiber Connections Based on Data Structure Algorithm

Dilek Göksel Duru and Mehmed Özkan

ABSTRACT

The brain activity during perception or cognition is mostly examined by functional magnetic resonance imaging (fMRI). However, the cause of the detected activity relies on the anatomy. Diffusion tensor magnetic resonance imaging (DTMRI) as a noninvasive modality providing in vivo anatomical information allows determining neural fiber connections which leads to brain mapping. Still a complete map of fiber paths representing the human brain is missing in literature. One of the main drawbacks of reliable fiber mapping is the correct detection of the orientation of multiple fibers within a single imaging voxel. In this study a method based on linear data structures is proposed to define the fiber paths regarding their diffusivity. Another advantage of the proposed method is that the analysis is applied on entire brain diffusion tensor data. The implementation results are promising, so that the method will be developed as a rapid fiber tractography algorithm for the clinical use as future study.

Introduction

Functional magnetic resonance imaging (fMRI) serves to determine the brain activity during perception or cognition. BOLD contrast for fMRI is remarkable in cognitive neuroscience, surgical treatment planning, and preclinical studies in examining the main parameters such as the blood flow, blood volume, resting state connectivity, and anatomical connectivity within the brain [1]. To define the cause of the detected activity, the anatomy of the underlying tissue must be analyzed. The functional properties of the region of interests (ROIs) in the brain can be investigated by combination of different modalities such as diffusion tensor magnetic resonance imaging (DTMRI or DTI), ADC fMRI, and BOLD fMRI [2]. As a noninvasive imaging modality DTMRI helps identification and visualization of the fiber connections in the anatomy [3–5]. DTMRI is unique in its ability providing in-vivo anatomical information noninvasively. The potential of DTI is to make the determination of anatomical connectivity in the investigated brain regions by mapping the axonal pathways in white matter noninvasively [6].

The lack of a complete neural fiber map in literature makes the post processing of the data very important. Methods and updates are to be researched to define the fiber trajectories in the uncertainty regions where multiple fiber orientations cross within a single imaging voxel [7, 8]. Our proposed technique aims to track the white matter fibers according to data structure algorithm noniteratively and depending on the structural information of the underlying tissue. The proposed algorithm is based on two major processes. One is decision making and the other one is storing process. Decision making process is basically an operation based on comparison between the orientations of diffusivities of adjacent voxel pairs. In other words, it is the determination of the path to be traced for computing the neural pathways. The decision making involves setting a similarity measure having a constant scalar value for a subject. The voxels which succeeded to pass the threshold is stored in a data structure. This process is performed for all the adjacent voxel pairs in the examined brain MR images. So the study applies the method to the entire human brain DT images to construct maps of neural fibers in uncertainty regions.

Material and Methods

Principles of Diffusion Tensor Analysis

The Stejskal-Tanner imaging sequence is used to measure diffusion weighted images [3, 4, 9]. The diffusion tensor D is calculated from this raw data source at each point in the tissue formulated by the Stejskal-Tanner equation as [10, 11]

$$S_i = S_0 e^{-b g_i^T D g_i}, \tag{1}$$

where S_i is the signal received with diffusion gradient pulses, S_0 is the RF signal received for a measurement without diffusion gradient pulses, b is the diffusion weighting factor, and $|g|$ is the strength of the diffusion gradient pulses.

The diffusion tensor D is a real, symmetric second-order tensor, represented in matrix form as a real, symmetric 3×3 matrix [3, 4]. The six unique elements of the diffusion tensor D are calculated according to the three-dimensional Gaussian Stejskal-Tanner model as (2) by acquiring at least six diffusion-weighted measurements in noncollinear measurement directions g along with a nondiffusion-weighted measurement S_0 [3, 4, 7, 12, 13]. On regular DTMR scans more than six diffusion-weighted measurements are taken which creates an over constrained system of equations solved using least square methods [9, 12, 14, 15]:

$$\begin{bmatrix} x_1^2 & y_1^2 & z_1^2 & 2x_1y_1 & 2y_1z_1 & 2x_1z_1 \\ x_2^2 & y_2^2 & z_2^2 & 2x_1y_1 & 2y_2z_2 & 2x_2z_2 \\ \vdots & \vdots & \vdots & \vdots & \vdots & \vdots \\ x_n^2 & y_n^2 & z_n^2 & 2x_ny_n & 2y_nz_n & 2x_nz_n \end{bmatrix} \begin{bmatrix} D_{xx} \\ D_{yy} \\ D_{zz} \\ D_{xy} \\ D_{xz} \\ D_{yz} \end{bmatrix} = \begin{bmatrix} -\frac{1}{b}\ln\frac{S_1}{S_0} \\ -\frac{1}{b}\ln\frac{S_2}{S_0} \\ \vdots \\ -\frac{1}{b}\ln\frac{S_n}{S_0} \end{bmatrix}. \tag{2}$$

Equation (2) equals a vector containing natural logarithmic scaled RF signal loss resulting from the Brownian motion of spins, and x_i, y_i, z_i denote the n gradient measurement directions. An orthogonal basis is the eigensystem of the symmetric matrix D by finding its eigenvalues and eigenvectors are calculated [16]. Principal component analysis (PCA) is used to perform the diffusion tensor analysis and compression. The diagonalization of the diffusion tensor as (3) results in a set of three eigenvalues $\lambda_1 > \lambda_2 > \lambda_3$ representing the principal diffusion orientation in an investigated pixel [5, 8]. The eigensystem is defined by the eigenvectors e_i and the corresponding eigenvalues λ_i (4). The eigenvectors e_i represent the principal diffusion directions:

$$D_x e_i = \lambda_i e_i \qquad (i = 1, 2, 3), \tag{3}$$

$$|D_x - \lambda I| = 0. \tag{4}$$

Examining the raw data for every pixel, the eigensystem of D is calculated in each pixel. The eigensystem calculation for analyzed image data provides information about the diffusion distribution throughout the investigated image data. The first principal component λ_1 shows the dominant diffusivity direction. The second and third principal components λ_2 and λ_3 provide information of the intermediate and the smallest principal diffusivity, respectively [17].

List Data Structure Implementation

The linear data structure used here helps to create a list of investigated region of interest eigenvectors where data item insertions and retrievals/deletions are made at one end, namely, the top of the list. A data item insertion is called pushing and removing is called popping the list. The created list can be called a linked list in which all insertions and deletions are performed at the list head (top) [18]. For each data item push, the previous top data item and all lower data items move farther down. When the time arrives to pop a data item from the list, the top data item is retrieved and deleted from the list. To clarify the implementation routine, application steps are explained on the synthetic data as in Figure 1.

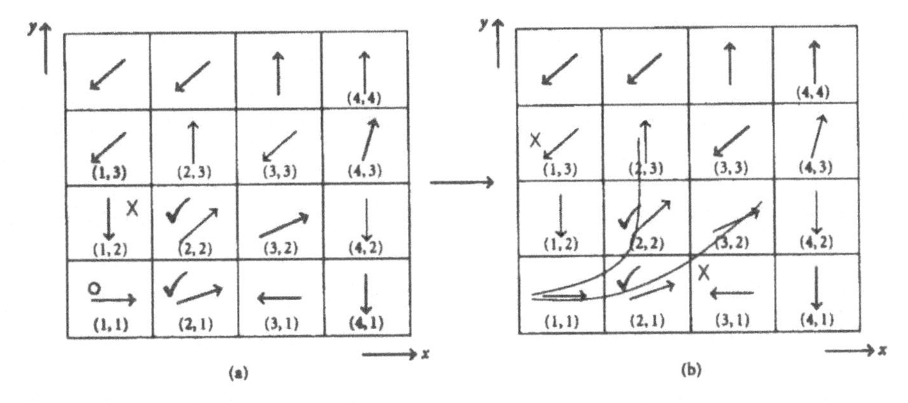

Figure 1. Sample synthetic eigenvector pattern. (a) (1,1) is the starting node, where green checks represent the neighbors within the similarity measure.

The starting point is selected as x=1 and y=1 as shown in Figure 1(a). This selected coordinate having the eigenvector [1,0] is the bottom of the linked list. The predefined similarity measure is a set of angular thresholds π/j (j= 4, 6, 12, 18, 20). Pixel (1, 2) is not within the limits of similarity measure $\pi/4$ (see Figure 1(a)). Pixel (2, 1) is stored in the stack on the top again in compliance with similarity. Top is now assigned to the new node. Next, pixel (2, 2) fulfilling the selected similarity measure is stored on the top of the list. The eigenvector [0.7 .07] with its neighboring pixels' eigenvectors is being compared for similarity. As a result, neighbors with coordinates (1, 3) and (3, 3) with both having the eigenvector [-0.7 -0.7] are eliminated (see Figure 1(b)). The implementation follows by pushing the coordinates (2, 3) and (3, 2) to the list. Pixel (3, 2) is popped. Then its neighbors are examined as in Figure 2(a). The routine follows by determining pixels matching with the predefined similarity rule $\pi/4$. The synthetic fiber path (represented in blue) is defined as a result as in Figure 2(b).

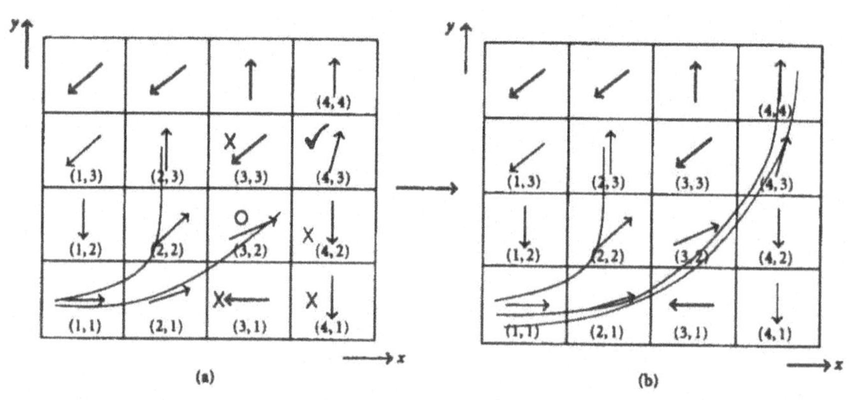

Figure 2. Listed data structure analysis results shown on sample pattern with its principal eigenvectors. Two possible resulting fiber paths are represented.

Selecting the similarity measure as $\pi/4$ allows the pixel (2, 2) to be on the list as described above. But examining the pattern by a different try for a varying angular threshold such as $\pi/6$ or $\pi/12$, this pixel is not being assigned for the neighboring pixel list. As a result the track represented in red on Figure 2(b) is the outcome of the computational routine. The decision making here about to select a track follows regarding to the underlying tissue's structural information.

The proposed approach relies on the assumption of the unique path description of an axon. Each element in the implementation represents a voxel in the ROI, and each voxel is related with its neighboring voxels. Regarding the neighboring voxel knowledge, the computation sorts the elements in the list for tracking, where the elements which do not fulfill the criteria are kept in a secondary matrix. While examining the investigated pattern pixelwise, the elements in the secondary matrix come up as potential neighboring pixels in question. The repeated check for if they are within the similarity criteria and if they belong to the fiber track gives the chance of a double check in the system. By that way, the neighboring is updated and a more secure resulting track is being defined and followed. The routine updates itself so that for the one selected starting node the first and second neighboring pixels are investigated and the computational routine is stretched to a wide range via this increased neighborhood.

Results

The proposed method is implemented on simulated fiber eigensystem to determine the predefined synthetic trajectories in Section 2.2. The output of the algorithm is in agreement with the visual inspection results as shown in Figure 2. Variation of the similarity measure causes major differences in the calculated neural path as seen in Figure 2(b). Small values of the similarity measure decreases the number of voxels in the solution which are defined by the decision making as neighboring voxels while increased similarity measure selections generate more well-defined and close results to the underlying tissue structure.

Following the promising results of the synthetic data implementations, the method is applied on real DT brain images. As explained in detail in Section 2.1 ((3) and (4)), the eigensystem of D is determined by PCA [19] and interpreted graphically as seen in Figure 3.

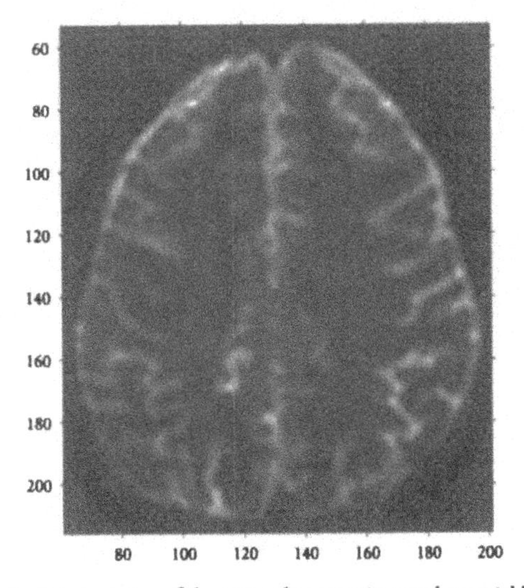

Figure 3. Calculated principal eigenvectors of the entire slice superimposed on axial brain MR image.

It is obvious that visual detection of any fiber path on the 2D axial MR image representing the eigenvectors is pretty hard unlike the simulated case. Therefore the developed linear list data structure algorithm is applied to the entire brain for neural fiber mapping. The search process of the pattern in the selected limits is completed in examining the eigenvectors of each pixel based on the predefined similarity measure. This examined data set sample might be a whole image data or a single ROI as in Figure 4.

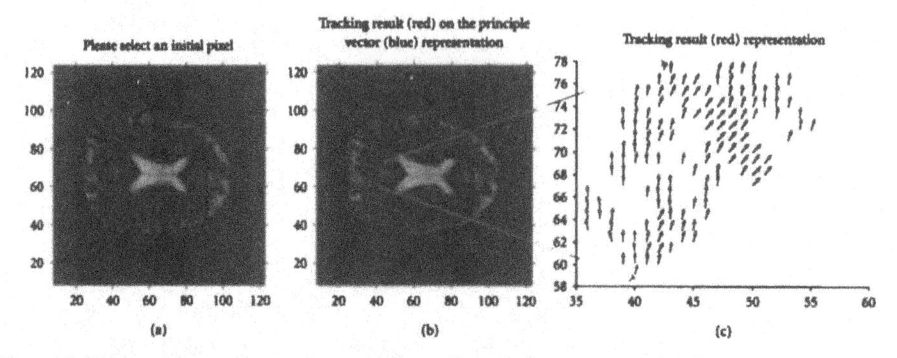

Figure 4. Fiber tracking results traced on axial slice with a similarity measure of $\pi/20$. (a) Starting point at [44,70]. (b) Calculated neighboring pixels with related diffusivity mapped on entire eigenvector map. (c) Zoomed region of interest.

The selection of the investigated brain region's size is directly related with the elapsed time of the computation. To be able to visualize the results of the algorithm, not the whole brain volume but only a selected and easily recognized region is computed. The results of such an example are represented in Figures 5 and 6 from different view angles in 3D.

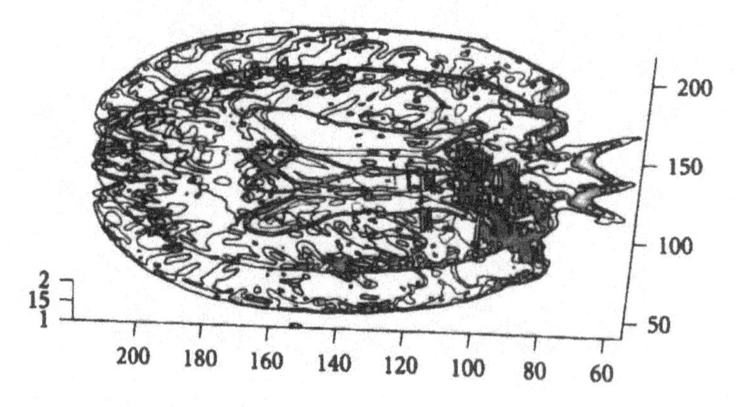

Figure 5. Tracking results of the implementation are represented on 2 consecutive slices.

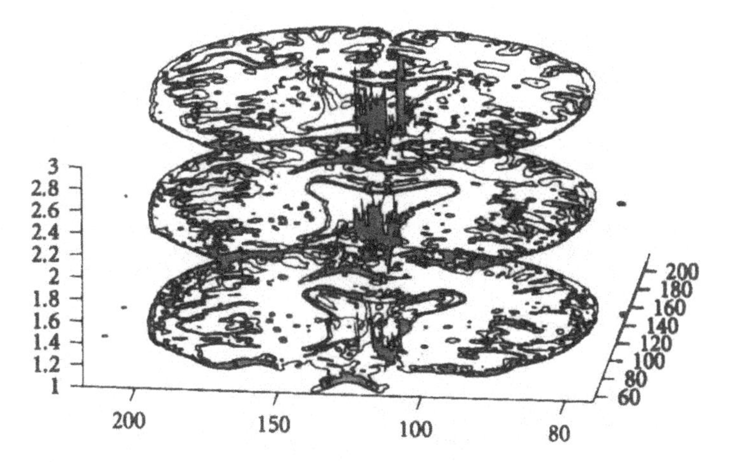

Figure 6. Fiber tracking results of the ROI close to inferior frontal lobe registered with the anatomic MR images.

Discussion

Some modalities such as PET and fMRI makes it possible to map the brain functions noninvasively. A parallel fMRI experiment with DTI is promising for understanding the brain function in both neuroimaging and neuroanatomical techniques' sense [2]. The knowledge derived from the DTI make it possible to map the in vivo information of the human neural fiber pathways noninvasively. This is an important motivation in diffusion tensor analysis research. The postprocessing of DTI analyzing tools plays great role in determination of the anatomical structural maps of fiber tracts. To follow a fiber tract and to build a neural map, each voxel's trajectory is approximated by a set of computed lines in each voxel regarding their major diffusivity. Each resulting tract defines a curvature representing a small bundle of axons in the pathway.

In the existence of fiber crossings and branches in an investigated ROI, the accuracy of the computed neural paths by DTI analyzing tools is unclear. One of the main limitations of diffusion tensor analysis relies on providing a solution for identification of the orientations of the brain fibers in uncertainty regions which is of great importance [3, 4, 8]. Therefore this problem arising in these so-called uncertainty regions is tried to be eliminated by different research groups [12, 20–23].

The aim of this study is to propose a rapid and reliable tracking algorithm which may eliminate the uncertainty region problem in DTI analysis. As seen in results, the synthetic fiber tracking implementation succeeded for predefined neural pathways. This motivated us to implement the algorithm on real diffusion tensor brain images. The computed tracts are found in agreement with the spatial visual inspection. Detailed anatomic information can be gathered via the computed tractography based on the Talairach atlas to become a gold standard, which is still missing.

Future work relies on eliminating the tracking problems in the uncertainty areas by upgrading the proposed method so that the calculation will be implemented on neural system basis and physiological background. The results will provide the base to reliable brain mapping.

Conclusion

This work aims to develop a promising approach which may eliminate the uncertainties in DT-MRI fiber tractography reconstructions and enhance a neural mapping. The degree of uncertainty in fiber orientation is subject to change by the selection made for similarity measure to detect neighboring voxel pairs. The fiber tracking tools are limited to trajectory-based representations. Therefore the detection of the anatomical connectivity and reliable computation of the neural map should be applied carefully being aware of any mistaken result.

It has been shown that linear list data structure gives promising analysis results in diffusion tensor fiber tract estimation. The identifying similarity measure varies in a range which is accepted in the means of anatomical fiber structure knowledge. Comparing the resulting tracts in synthetic eigenvector pattern with the known predefined pathways, the algorithm gives promising results and works well for the tracking purposes. The computed neural pathways varying with the change of the similarity measure cause to decrease or increase the number of the neighboring voxels for a selected starting voxel. The differing resulting pathways can be thought as an error of the method where it might be also in some cases the possible orientation of a fiber bundle in a wide range, which may be determined by an anatomical brain atlas, that is, Talairach atlas.

Besides the existing algorithms the proposed technique provides the possibility to compute the whole eigensystem of the investigated brain volume. The neighboring voxel pair calculation compares the investigated node in every step of the algorithm within the entire image volume. Each voxel is checked for more than one trial in the total analysis. In that way the decision making of the algorithm becomes more precise.

Acknowledgement

This work was supported in part by Bogazici University Scientific Research Project 07HX104D.

Academic Editor

Fabio Babiloni

References

1. D. G. Norris, "Principles of magnetic resonance assessment of brain function," Journal of Magnetic Resonance Imaging, vol. 23, no. 6, pp. 794–807, 2006.
2. S. A. Huettel, A. W. Song, and G. McCarthy, Functional Magnetic Resonance Imaging, Sinauer Associates, Sunderland, Mass, USA, 2004.
3. P. J. Basser, J. Mattiello, and D. LeBihan, "MR diffusion tensor spectroscopy and imaging," Biophysical Journal, vol. 66, no. 1, pp. 259–267, 1994.
4. P. J. Basser, J. Mattiello, and D. LeBihan, "Estimation of the effective self-diffusion tensor from the NMR spin echo," Journal of Magnetic Resonance, vol. 103, no. 3, pp. 247–254, 1994.
5. P. J. Basser, S. Pajevic, C. Pierpaoli, J. Duda, and A. Aldroubi, "In vivo fiber tractography using DT-MRI data," Magnetic Resonance in Medicine, vol. 44, no. 4, pp. 625–632, 2000.
6. S. Mori, B. J. Crain, V. P. Chacko, and P. C. M. Van Zijl, "Three-dimensional tracking of axonal projections in the brain by magnetic resonance imaging," Annals of Neurology, vol. 45, no. 2, pp. 265–269, 1999.
7. P. J. Basser and D. K. Jones, "Diffusion-tensor MRI: theory, experimental design and data analysis—a technical review," NMR in Biomedicine, vol. 15, no. 7-8, pp. 456–467, 2002.
8. D. Le Bihan, C. Poupon, A. Amadon, and F. Lethimonnier, "Artifacts and pitfalls in diffusion MRI," Journal of Magnetic Resonance Imaging, vol. 24, no. 3, pp. 478–488, 2006.
9. P. J. Basser, "Inferring microstructural features and the physiological state of tissues from diffusion-weighted images," NMR in Biomedicine, vol. 8, no. 7-8, pp. 333–344, 1995.
10. E. O. Stejskal, "Use of spin echoes in a pulsed magnetic-field gradient to study anisotropic, restricted diffusion and flow," The Journal of Chemical Physics, vol. 43, no. 10, pp. 3597–3603, 1965.
11. E. O. Stejskal and J. E. Tanner, "Spin diffusion measurements: spin echoes in the presence of a time-dependent field gradient," The Journal of Chemical Physics, vol. 42, no. 1, pp. 288–292, 1965.

12. C. F. Westin, S. E. Maier, H. Mamata, A. Nabavi, F. A. Jolesz, and R. Kikinis, "Processing and visualization for diffusion tensor MRI," Medical Image Analysis, vol. 6, no. 2, pp. 93–108, 2002.
13. P. B. Kingsley, "Introduction to diffusion tensor imaging mathematics—part I. Tensors, rotations, and eigenvectors," Concepts in Magnetic Resonance, vol. 28, no. 2, pp. 101–122, 2006.
14. S. Pajevic and C. Pierpaoli, "Color schemes to represent the orientation of anisotropic tissues from diffusion tensor data: application to white matter fiber tract mapping in the human brain," Magnetic Resonance in Medicine, vol. 42, no. 3, pp. 526–540, 1999.
15. C. H. Sotak, "The role of diffusion tensor imaging in the evaluation of ischemic brain injury—a review," NMR in Biomedicine, vol. 15, no. 7-8, pp. 561–569, 2002.
16. A. I. Borisenko and I. E. Tarapov, Vector and Tensor Analysis with Applications, Dover, New York, NY, USA, 1979.
17. C. Pierpaoli, A. Barnett, S. Pajevic, et al., "Water diffusion changes in wallerian degeneration and their dependence on white matter architecture," NeuroImage, vol. 13, no. 6, pp. 1174–1185, 2001.
18. N. Wirth, Algorithms and Data Structures, Prentice-Hall, Englewood Cliffs, NJ, USA, 1986.
19. D. Goksel and M. Ozkan, "Towards rapid analysis of diffusion tensor MR imaging," ESR Supplements, vol. 16, no. 1, p. 286, 2006.
20. S. Pajevic and P. J. Basser, "Parametric and non-parametric statistical analysis of DT-MRI data," Journal of Magnetic Resonance, vol. 161, no. 1, pp. 1–14, 2003.
21. D. K. Jones, "Determining and visualizing uncertainty in estimates of fiber orientation from diffusion tensor MRI," Magnetic Resonance in Medicine, vol. 49, no. 1, pp. 7–12, 2003.
22. D. K. Jones and C. Pierpaoli, "Towards a marriage of deterministic and probabilistic tractography methods: bootstrap analysis of fiber trajectories in the human brain," in Proceedings of the 12th International Society for Magnetic Resonance in Medicine (ISMRM '04), p. 1276, Kyoto, Japan, May 2004.
23. D. K. Jones, A. R. Travis, G. Eden, C. Pierpaoli, and P. J. Basser, "PASTA: pointwise assessment of streamline tractography attributes," Magnetic Resonance in Medicine, vol. 53, no. 6, pp. 1462–1467, 2005.

A Preliminary Analysis of Software Engineering Metrics-based Criteria for the Evaluation of Learning Objects Reusability

J. Sanz Rodríguez, J. M. Dodero and S. Sanchez-Alonso

ABSTRACT

Reusability of learning objects is evaluated on the basis of a priori software reusability analysis, which are related to cohesion and coupling aspects. A number of reusability metrics extracted from metadata records are defined and analyzed to provide an aggregate reusability evaluation for learning objects in a repository. The evaluation is validated and compared with an expert-based a posteriori evaluation method.

Keywords: Learning objects, metadata, reusability.

Introduction

Reusability is a key issue on e-learning contents and systems. Providing reusable learning objects can facilitate its further development and adaptation, augment learning object development productivity, reduce development costs and improve quality. Although reusability is an intrinsic characteristic of the learning object that can provide a priori a measure of its quality, reusing learning objects is an empiric and observable fact that can be compared with such measures by means of a posteriori data compiled from their actual use. Nevertheless, studies on reusability indicators and design criteria that guarantee reusability are scarce [1].

The objective of this work is to identify concrete metrics that can be used to qualify learning objects with aspects related to the capability of being reusable. Such measures can be useful to learning object producers, who can have quantitative data on the reusability of the designed objects, as well as to learning object consumers, who can search in repositories for objects that can be more easily adapted to their specific needs.

Learning Object Evaluation

Several initiatives have approached the evaluation of learning objects to provide an estimation of the guaranteed quality. MERLOT (http://www.merlot.org) classify objects in seven discipline categories (i.e. Arts, Economy, Education, Humanities, Mathematics, Science and Technology) and compile experts and users' evaluations on three dimensions (i.e. content quality, usability and effectiveness as a learning tool) on a 1-5 numeric scale [2].

ELera (http://www.elera.net) extends this evaluation scheme by the LORI (Learning Object Review Instrument) tool [3], which evaluates aspects such as content quality, objective fulfillment, feedback and adaptation capability, motivation, presentation, usability, accessibility, reusability and standards compliance. Each aspect is assessed on a 1-5 scale, based on a Delphi-style collaborative evaluation scheme with the participation of groups of experts, in which objects are first evaluated on an asynchronous, individual basis; afterwards individual evaluations are discussed to agree on eventual assessments.

Usual learning object evaluation methods are based upon compiling opinions from users and experts about different aspects of a learning object. In contrast to these, the learning object reusability evaluation model proposed here is aprioristic and is based upon the learning object structure and common metadata that describe it. However, harnessing learning object metadata for that aim depicts some issues, particularly related to information fragmentation and the potential lack of integrity on the harvested metadata. Therefore, we need to augment metadata with extended information that enacts reusability.

Evaluation Methodology

We have based on an evaluation methodology used to measure reusability of object-oriented software [4], based on the following steps:

1. Study and identify the learning object aspects and factors having influence on the capability of reusing
2. Define metrics to measure reusability factors that have been identified, based upon analysis of IEEE Learning Object Metadata (LOM) standards [5] and the learning object structure
3. Formulate an aprioristic evaluation model formed by the aggregation of the metrics according to their significance for evaluating reusability
4. Evaluate the model though application and comparison with the reusability data obtained by LORI for a significant set of learning objects of the eLera repository.

Reusability Factors

The factors that determine the ability of a learning object to be reused [6][7][8] can be classified as structural or contextual issues. From a structural viewpoint, reusable learning objects must be:

- Self-contained: a learning object should have sense by itself; references to other resources will decrease reusability; the more pre-requisites it needs, the more difficult will be adapting it to other contexts.
- Modular: a learning object must be combinable with other objects to form composite structures as lessons and courses.
- Properly grained: proper size and a proper learning objective for a learning object will facilitate reusing it.
- Traceable: a learning object should be easily identifiable and traceable through the correct metadata.
- Modifiable: a learning object should be modifiable to reformulate it under a given context different to the originally designed.
- Usable: a reusable learning object must be easy to use and interactive interface elements it contains should be intuitive.

- Standardized: a reusable learning object must be compliant to a shared specification or standard.

From a contextual viewpoint, the more context-dependent and context-specific a learning object is, the more limited its reusability will be. We can deal with contextual factors in the following dimensions: technological, educational and social.

- The technological dimension of context includes platform dependencies and software needed to run the learning object, as well as representation issues (reusable learning objects should separate contents and format issues).
- The social and educational contexts require the following features: learning objects must be generic, i.e. independent from a given subject or discipline; they have to be prepared for using on different education and assessment levels; they must be pedagogically neutral, i.e. do not involve a specific pedagogical method; they must lack institutional, legal, social and cultural dependencies; they are independent of time and location in which they are run.

We have to mention that some factors described above cannot be actually considered up to its extreme in order to achieve the greatest reusability; for instance, a generic, discipline-independent learning object is more reusable than a discipline-specific one, but clearly it is not useable, since it has to commit the learning objectives for which it is intended, and such objectives are always subject-specific. A different thing is that, for instance, a learning object dealing with Statistics is more reusable if it does not involves examples that deal with a given discipline (e.g. mechanical engineering) that hinders to include it in another object (e.g. a biology course). Similar issues can be discussed about the pedagogical neutrality or time-independence features, to say only some of them.

Designers tend to produce objects with multiple dependences to enrich the learning process, in contrast to independent and self-contained objects that contribute with not much significant knowledge. This situation is a challenge to design cohesive, uncoupled objects containing both structural and contextual aspects that do not jeopardize reusability [9].

Learning Object Reusability Metrics

We have analyzed common software metrics in order to provide reusability metrics for learning objects, based upon the reusability factors discussed above. Traditionally, software engineering based upon an old design principle to strive for strong cohesion and loose coupling [9]. These two principles head for building maintainable software that easily adapt to new requirements. Since learning objects are designed for reuse, we analyzed how these principles apply to determine learning object reusability. Although reusability metrics are mainly related to cohesion and coupling, we have also analyzed metadata elements to evaluate other reusability factors, such as portability, size and complexity and difficulty of comprehension. Clearly, these are not completely independent factors, but they depict clear intersections up to some extent. We will describe further on how this issue can be managed.

In order to evaluate and compare our aprioristic model with a posteriori values, we have normalized metrics values in the [1,5] interval, which is the same scale of readily available evaluation models such as MERLOT and LORI.

Cohesion

Cohesion analyzes the kind of relationships among different modules. A module must realize a single task to be maximally cohesive [11]. Greater cohesion implies greater reusability [12]. Cohesion is a software quality indicator that, applied to learning objects, is fulfilled by the following elements:

- A learning object involves a number of concepts (LOM 9 Classification category). The lesser number of concepts, the greater cohesion it will depict [13].
- A learning object must have an only and clear learning objective [1]. The more learning objectives it has, the lesser cohesive it will be.
- The semantic density of a learning object (LOM 5.4 Educational category) indicates how concise it is. The more conciseness, the more cohesion for the learning object.
- A learning object must be self-contained to be highly cohesive [13]. LOM 7 Relation category is used to define as many instances as relationships the learning object has (notably is-version-of, has-version, is-format-of, has-format, references, is referenced-by, is-based-on, is-basis-for, requires, is required-by, is-part-of and has-part). The more relationship instances a learning object has, the less self-contained and, therefore, less cohesive. Moreover, LOM 1.8 Aggregation level element summarizes the level of aggregation of a learning object as ranging from 1 for single resources to 4 for a set of related courses. The lower level of aggregation, the more cohesion.

We can conclude that learning object cohesion is directly proportional to semantic density and inversely proportional to the number of relationships, aggregation level, number of concepts dealt with, and number of learning objectives covered. These metadata elements can be source for a valid estimation of the reusability of a learning object. This way, we can classify learning objects cohesion values as depicted on Table I.

Table 1. Cohesion values to measure learning object reusability.

Cohesion	Capability of reuse	Value
Very high	Independent and fully self-contained objects. Adaptations are rarely required	5
High	Self-contained objects including some dependencies. Reusable after simple adaptations.	4
Medium	Objects with multiple dependencies. Reusable after a considerable number of adaptations.	3
Low	Objects with multiple dependencies. Reusable after many adaptations.	2
Very low	Completely dependent objects. Reusable after completely changing the object.	1

Coupling

Coupling measures interdependencies among software modules and must be minimized [12]. A module must communicate with the minimum number of modules and must exchange as minimal information as possible, in order to minimize the impact provoked from changes on other modules. Learning object coupling describes interrelationships among distinguishable objects, so the lesser coupling, the greater reusability [13].

LOM 9 Relation category indicates the number of objects related to a given learning object, so we conclude that coupling is directly proportional to the number of relationships present in that category.

Size and Complexity

Software size and complexity can be measured through several methods, e.g. lines of code, McCabe's software complexity, Halstead's difficulty, etc. In general terms, granularity provides clear information on learning object reusability, since fine-grained objects are more easily reusable. Learning object granularity is directly proportional to the following LOM elements:

- Size: the number of bytes of a learning object. These data should be weighted depending on the learning object format, since there are different interpretations of size for texts, images and videos, for instance.
- Duration: the estimated time to run the learning object. This is specifically useful for videos or animations.

- Typical Learning Time: the estimated time required to complete the learning object. This is a reliable source of information to estimate the size and complexity of a learning object

Learning object size and complexity can be classified according to values of Table II.

Portability

Portability metrics measures the ability to transfer software from one system to another and is based on analyzing modularity and hardware/software context independence [14]. Learning objects portability can be measured as the context dependence at technological and socio-educational levels. The few dependendencies found, the more portable the learning object.

Technical Portability

The following LOM values can be analyzed when considering portability at a technical level:

- Format: determines the learning object components' delivery format, such as video/mpeg, application/xtoolbook, text/html, etc. Some formats are more readily portable (e.g. text/html is more widespread than application/x-toolbook.
- Requirements: involves the hardware and software required to run the object. As the number of requirements increase and these are more complex, less portable is the object.

Learning objects' technical portability can be qualified by means of the values shown in Table III [15].

Table 2. Values to measure learning object size

Size	Description	Value
Very small	Atomic resources	5
Small	Small-sized resources	4
Medium	Medium-size lessons	3
Big	Big-sized aggregated courses	2
Very big	Very big-sized courses	1

Table 3. Values to measure learning object technical portability.

Technical portability	Description	Value
Very high	The object is based on an technology available on practically all systems and platforms (e.g. html).	5
High	The object is based on an technology available on many systems and platforms (e.g. pdf).	4
Medium	The object is based on an technology that is not available on many systems (i.e. common platform-specific file format).	3
Low	The object is based on an technology that is hardly available on different systems (i.e. uncommon proprietary file formats).	2
Very low	The object is based on a proprietary technology that is not available on many systems (i.e. a specific server technology)	1

Educational Portability

When moving at the educational level, we can deal with vertical or horizontal portability [15]. Vertical portability means the possibility for a learning object to be used and reused on different educational levels; in contrast, horizontal portability determines the inter-disciplinarity of the object. We have considered the following LOM values:

- Context: potential educational contexts in which an object can be used (i.e. school, high school, higher education, professional training, etc.) Educational portability is greater for learning objects that can be used and reused on more different educational contexts.
- Typical age range: potential age ranges in which an object can be used. Educational portability increases as the number of ranges grows.
- Language: the human languages supported by the object. An object is more reusable if it is available on more languages.
- Classification: information used to classify a learning object within the discipline it belongs or is related to. The more specific the classification scheme, the lesser reusable the learning object can be.

Difficulty of Comprehension

Software difficulty measures the cognitive effort to understand a software component. It is based on analyzing the component complexity, how self-descriptive and well documented it is [14].

Table 4. Values to measure learning object educational portability

Educational portability	Description	Value
Very high	The object is generic, pedagogically neutral and can be used on different educational levels	5
High	The object can be used for several disciplines and educational levels	4
Medium	The object can be used without modifications on a specific area and educational level	3
Low	The object depicts educational dependencies and can be reused with several modifications on a different educational context and level	2
Very low	The object depicts many educational dependencies and can be hardly reused on different educational contexts and levels	1

We can state that the difficulty to comprehend a learning object directly influences the capability of a designer to reformulate and reuse it on another aggregated object. We can consider here the LOM 5.8 Difficulty category, although other LOM elements can be clearly correlated (e.g. LOM 5.4 Semantic density or LOM 5.9 Typical learning time). Even LOM 7 Relation category or LOM 1.8 Aggregation level elements can be heavily correlated to the difficulty. For this reason, and since these correlated elements have been considered for inclusion in other reusability factors above, we do not consider this factor separately.

Reusability Evaluation Model

Learning object reusability depends on cohesion, coupling, portability and difficulty category elements. Several LOM values can be aggregated to build an a priori evaluation model. We discard the difficulty of comprehension factor due to the great number of dependencies it shows with elements from all other categories. Moreover, we assume that coherent metadata values are available for all considered LOM elements on analyzed objects. Let be the set of evaluation criteria as extracted from LOM records. To estimate the reusability of a learning object we require an aggregation process. For that aim we used first an ordered weighted averaging operator:

$$M_w(x) = \sum_{n}^{i=1} w_i x_i$$

where a learning object x is characterized as the vector $(x_1,...,x_n)$ with $x_i \in \{1, 2, 3, 4, 5\}$, $\sum_i w_i = 1$ and $w_i \geq 0 \forall i \in C$.

Weight values w_i are provided by the evaluation as parameters that can be estimated and agreed to enhance or soften the contribution of a given factor to the aggregated reusability evaluation. For instance, Table V provides an estimated, primitive set of values that depend on the number of available evidences extracted from LOM. This must be calibrated and validated if needed.

Table 5. Weights for reusability model validation

Metric	Weight
Cohesion	0.3
Coupling	0.0
Technical portability	0.3
Educational portability	0.3
Size and complexity	0.1

Reusability Model Validation

Validation of our model has been carried out by a detailed analysis of eight learning objects from the eLera repository. After that, we compared the aggregated reusability metric with LORI reusability evaluations as done by experts. The learning objects have been selected because they received the highest number of expert evaluations in the repository, so it guarantees the reliability of such evaluations. However, we found that a lot of metadata information was missing to compute our aprioristic reusability value, so we had to complete that information.

Table VI shows reusability values obtained and compared with LORI evaluations. They are graphically depicted on Figure 1.

If we consider a 0.5 permissible difference, we have that the aprioristic reusability evaluation model fits 62.5%of cases with experts' opinions. If we consider a permissible difference of 1.0, the model fits 87.5% of cases with experts' evaluations. There exists a significant 95%-confidence correlation between size and educational portability, and a 90%confidence correlation between cohesion and size and between cohesion and educational portability. Therefore, we can assume that there is a degree of interdependence between the selected metrics of cohesion, size and educational portability.

Table 6. Aprioristic reusability values compared with a posteriori LORI reusability values

eLera Object	Cohesion	Educational portability	Technical Portability	Size	Aprioristic Reusability	LORI Reusability
Newton's First Law	3	3	4	3	3.3	3.2
Pythagoras' Theorem	4	3	4	4	3.7	2.6
6 Billion Human Beings	3	2	4	2	3	3.5
Population Growth and Balance	3	2	4	3	3	3.3
Map of the Human Heart	4	3	5	3	3.9	3.4
Element Hangman	4	4	4	4	4	4.6
Clinical Pharmacology	4	3	4	3	3.6	4
Elizabethan Times	4	3	5	4	4	4.8

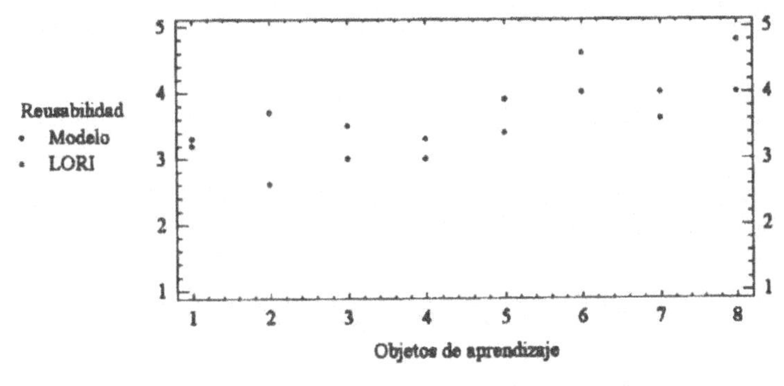

Figure 1. Aprioristic reusability values compared with a posteriori LORI reusability values

Conclusions

We can conclude that the aprioristic reusability estimations provided by the model approximate to those provided a posteriori by expert evaluation. Although some aspects of the model must be improved, it provides an approach to develop a formal, aprioristic reusability model. Therefore we can conclude that reusability metrics adapted from traditional software engineering reusability factors can provide a clear measurement of learning objects reusability. Including such computed reusability values as metadata records allows to enhance indexing and searching capabilities [16] as well as developing new reusable learning objects, so improving productivity and quality in learning object-based systems.

Aspects to be improved include the ordered weights estimation, and the treatment of interdependencies among analyzed reusability factors. The latter can be managed by utilizing more powerful aggregation operators, such as the Choquet integral, which takes into account existing interdependencies and reduces their influence on the aggregate evaluation value [17]. The former can be managed through studying the correlation on a learning object-basis among LORI reusability value and concrete metadata on a significant amount of repository objects. However, metadata records must be filled-in and readily available for that aim.

References

1. Sánchez-Alonso, S, Sicilia, M.A. Normative specifications of learning objects and processes. International Journal of Instructional Technology and Distance Learning, 2(3), 3–12. 2005.
2. Vargo, J., Nesbit J.C., Belfer, K., Archambault A.: Learning Object Evaluation: Computer-Mediated Collaboration and Interrated Reliability. International Journal of Computers and Applications, Vol. 25, No. 3, 2003.
3. Nesbit J., Belfer K., Leacock T.: Learning Object Review Instrument (LORI) User Manual. Available at www.elera.net.
4. Etzkorn, L. H., Hughes, W. E., Davis C. G.: Automated reusability quality analysis. In Information and Software Technology, 43 (2001), pp 295–308. (doi:10.1016/S0950-5849(00)00169-5)
5. IEEE Learning Technology Standards Committee (LTSC): "Learning Object Metadata (LOM)," Final Draft Standard, IEEE 1484.12.1–2002.
6. Daniel, B, Mohan, P.: A Model for evaluating learning objects. In Proceeding of the IEEE International Conference on Advanced Learning Technologies (ICALT 2004). 30 Aug-1 Sep 2004. pp 50–60.
7. Huddlestone J, Pike, J.: Learning object reuse–A four tier model. People and systems - who are we designing for. Nov. 2005.
8. Palmer K., Richardson, P. Learning Object Reusability–Motivation, Production and Use. In 11th International Conference of the Association for Learning Technology (ALT). University of Exeter, Devon, England, 14–16 september 2004.
9. Boyle, T: Design principles for authoring dynamic, reusable learning objects. Australian Journal of Educational Technology, 2003, pp. 46–58.
10. Selby, R.W., Basili, V.R. Analyzing error-prone system structure, IEEE Transactions on Software Engineering, 17 (2) (1991) 141–152. (doi:10.1109/32.67595)

11. Sommerville, I.. Software engineering, 6th Ed. Addison-Wesley. 2000.
12. Vinoski, S.: Old Measures for New Services. IEEE Internet Computing. November-December 2005, pp. 72–74. (doi:10.1109/MIC.2005.131)
13. Yang, D, Yang, Q: Customizable Distance Learning: Criteria for Developing Learning Objects and Learning Model Templates. In Proceedings of the 7th international conference on Electronic commerce (ICEC'05), ACM International Conference Proceeding Series, August, Xi'an, (China). ACM. (New York), 2005, pp. 765–770.
14. Poulin, J.: Measuring Software Reusability. In Third International Conference on Software Reuse, Rio de Janeiro, Brasil, 1-4 November 1994, pp. 126–138.
15. Currier, S, Campbell, L: Evaluating learning resources for reusability: the "dner & learning objects" study. Proceeding of The Australasian Society for Computers in Learning in Tertiary Education (ASCILITE 2002) Auckland, New Zealand. 2002.
16. Williams, D. Evaluation of learning objects and instruction using learning objects. Available at reusability.org.
17. Dodero, J. M., Sicilia, M. A., Fernández, C. On the use of the Choquet integral for the collaborative creation of learning objects, Computing and Informatics, 23(2), 2004, pp. 101–113.

Extending Conceptual Schemas with Business Process Information

Marco Brambilla, Jordi Cabot and Sara Comai

ABSTRACT

The specification of business processes is becoming a more and more critical aspect for organizations. Such processes are specified as workflow models expressing the logical precedence among the different business activities (i.e., the units of work). Typically, workflow models are managed through specific subsystems, called workflow management systems, to ensure a consistent behavior of the applications with respect to the organization business process. However, for small organizations and/or simple business processes, the complexity and capabilities of these dedicated workflow engines may be overwhelming. In this paper, we therefore, advocate for a different and lightweight approach, consisting in the integration of the business process specification within the system conceptual schema. We show how a workflow-extended conceptual schema can be automatically obtained, which serves both to enforce the organization business process and to manage all its relevant domain data in a unified way. This extended model can be directly processed with current CASE tools, for instance, to generate an implementation of the system (including its business process) in any technological platform.

Introduction

All software systems must include a formal representation of the knowledge of the domain. In conceptual modeling, this representation is known as the conceptual schema of the software system [1]. However, software development processes for complex business applications usually require the additional definition of a workflow model to express logical precedence and process constraints among the different business activities (i.e., the units of work).

Workflow models are usually specified through dedicated languages (e.g., Business Process Management Notation–BPMN [2]) and implemented with the help of specialized workflow management systems (WFMSs), for example, see [3, 4], which are heavy-weight applications focused on the control aspects of the business process enactment. This is clearly the best option to manage large workflow models. However, in some cases organizations may prefer a more lightweight approach that does not require acquiring a specific workflow subsystem.

This paper tackles the problem of defining a light-weight approach to the implementation of business processes within software applications, without the use of specialized WFMSs, which represents a relevant issue in several application scenarios. Indeed. alternative solutions to complete WFMSs can be preferred in case of simple business requirements, small organizations, or when the business process needs are going to be drowned into a larger system that is being implemented ad hoc for the organization. In these cases, designing and implementing the workflow using the same methods, notations and tools used to develop the rest of the system can be convenient and cost effective for the organization.

Along these development lines, some approaches have focused on the implementation of workflow models in specific technology platforms, as relational databases (generally in the form of triggers [5]), Web applications (by means of hypertextual links and buttons properly placed in Web pages, thus restricting the user navigation [6]), or Web services (through transformation into Business Process Execution Language for Web Services–BPEL4WS [7] specifications). This way, the workflow definition becomes part of the system implementation and no specific workflow engine is required. However, these approaches can be hardly generalized to technologies different from the ones for which they have been conceived (e.g., to new technology platforms), make difficult a wider adoption of business processes within the organizations, and present some limitations regarding the supported expressivity for the initial workflow model and/or its integration with the conceptual schema.

As an alternative, in this paper we propose a formalized model-driven development (MDD) approach for developing workflow-based applications and advocate for the automatic integration of the workflow model within the (platform-independent) conceptual schema. The resulting workflow-extended conceptual schema includes in a single schema both the business process specifications and the domain knowledge, providing a unified view of the system and allowing treating both dimensions in a homogeneous way when implementing, verifying, and evolving the system. The integration is done at the model level. Therefore, current modeling tools can be used to manage our workflow extended schema, no matter the target technology platform or the purpose of the tool (e.g., verification, code-generation, etc.).

The rest of the paper is structured as follows: Section 2 summarizes and motivates our approach and its advantages. In Sections 3 and 4 the conceptual schema, the workflow concepts, and our case study are illustrated. Section 5 introduces the normalization phase. In Sections 6 and 7 we provide the definition of the workflow-extended conceptual schema and of the (Object Constraint Language OCL [8]) process constraints, respectively. Section 8 sketches possible implementation strategies for this extended model. Section 9 portrays our prototype tool implementation. Then, Section 10 compares our approach with related work and in Section 11 we draw our conclusions and discuss future work.

Overview of the Proposed Approach and of its Benefits

Our MDD approach for developing workflow-based applications is sketched in Figure 1: the designer specifies the conceptual schema (e.g., in UML) and the workflow model of the application (e.g., in BPMN), using the appropriate design tools. At this stage some links between the workflow model and the conceptual schema can be already identified. Typically, they represent the usage relationship that associates objects of the application domain to activities in the workflow model.

The workflow model may need a normalization transformation for homogenizing the notation and making it fit for the next (automatic) steps.

The conceptual schema and the workflow model undergo to the integration transformation phase that produces the workflow-extended conceptual schema. More specifically, given a conceptual schema c and a workflow model w, it is possible to automatically derive a full fledged conceptual schema c' enriched with the types needed to record the required workflow information in w (mainly its activities and the enactment of these activities in the different workflow executions) and with a set of process constraints over such types to control the correct workflow execution. Several workflow models can be integrated with the same conceptual schema since the process constraints of each workflow model do not interfere. This is guaranteed by the construction process of the workflow-extended model. This extended schema can then be managed using any commercial UML CASE tool.

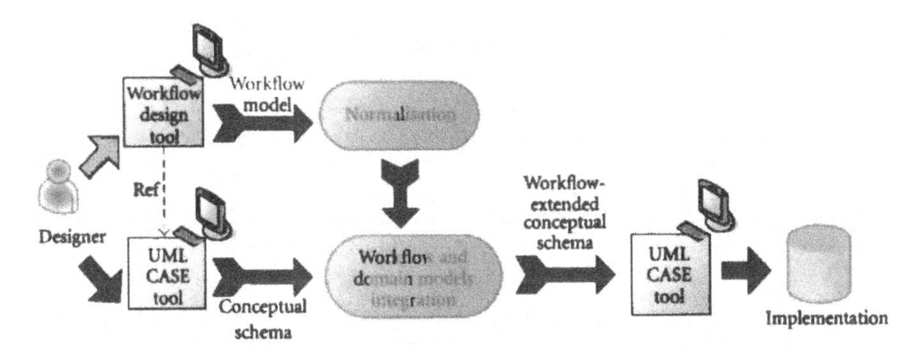

Figure 1. MDD process for workflow-based applications.

The whole approach has been implemented in a prototype tool that automatically translates the workflow specifications into a set of types and constraints on the conceptual schema, according to a set of translation patterns described in the paper.

The focus of the paper will be on the platform-independent transformations of the conceptual models; however, some ideas on how to implement the workflow-extended conceptual schema into target platforms will be provided. As reference models, throughout the paper we will use UML class diagrams for the representation of conceptual schemas and OCL constraints to represent the process constraints. For the workflow, we will adopt a particular business process notation, namely BPMN [2], for sake of readability and concreteness. Indeed, business analysts are well aware of business process modeling techniques but are not so familiar with software engineering notations and practices. Recently, BPMN and other domain-specific notation have been increasingly accepted in the field, thus we based our examples on a notation that business roles in the enterprises are familiar with.

Our model transformations are based on the concepts and definitions specified by (Business Process Definition Metamodel BPDM [9]), a platform- and notation-independent metamodel for defining business processes. Since BPDM is a common metamodel for all business process notations (e.g., it includes all concepts of BPMN and UML activity diagrams), our approach can be used exactly in the same way when using activity diagrams or any other BPDM-compliant notation to model the workflows. The proposed approach is therefore general-purpose and is valid regardless of the adopted business process notation.

Motivation and Discussion

The main advantage of the proposed approach is that the workflow-extended conceptual schema includes in a single conceptual schema both the business process specifications and the domain knowledge. Since the workflow-extended model is automatically generated from the workflow model and the conceptual schema, a unified view of the system is hence available without any additional effort by the designer. This allows treating both dimensions in the resulting model in a homogeneous and consistent way when implementing, verifying, and evolving the system. Thanks to this unified view, our workflow-extended schemas enable the definition of more expressive business constraints, generally not allowed by common business process definition languages such as timing conditions [10] or conditions involving both business process and domain information.

Moreover, since the integration of the workflow and conceptual schemas is done at the model level, the resulting workflow-extended conceptual schema is a platform-independent model. Thanks to the current state of the art of model-to-model and model-to-text transformation tools, integrating different notations in the same approach (e.g., UML class diagrams, OCL, and BPMN) does not make a difference. Indeed, the extraction and integration process will simply consider models conforming to different metamodels (e.g., UML and BPDM). Anyway, the model transformations involved are straightforward and compliant with the MDD approach.

Once the final workflow-extended schema is produced, it can benefit from any method or tool designed for managing a generic conceptual schema, no matter the target technology platform or the purpose of the tool, spawning from direct application execution, to verification/validation analysis, to metrics measurement, and to automatic code-generation in any final technology platform. Those methods do not need to be extended to cope with the workflow information in our workflow-extended schema, since it is a completely standard UML model [11]. In this sense, with our approach we do not need to develop specific techniques for workflow models nor to use specific tools for managing them.

Finally, once (automatically) implemented (with the help of any of the current UML/OCL CASE tools offering code-generation capabilities), the workflow-extended conceptual schema ensures a consistent behavior of all enterprise applications with respect to the business process specification. As long as the applications properly update the workflow information in the extended model, the generated process constraints enforce that the different tasks are executed according to the initial business process specification.

Original Contributions of the Paper

To our knowledge, ours is the first approach that automatically derives a platform-independent conceptual schema integrating both domain and business process information in a unified view. A first version of this proposal has been published in [12]; however, this paper extends [12] in several directions. In particular, the main original contributions of this paper include the following.

(i) The introduction of a normalization phase to simplify the initial workflow models and extend the set of workflow patterns we can directly cover with our method.

(ii) A complete description of the process that allows obtaining the workflow-extended conceptual schema starting from the domain model and the workflow model.

(iii) An extended and refined version of the translation of process constraints, including also the management of the start, end, and intermediate events in the business process specification. Such events can represent different event types (message, exception, rule, timer, etc.).

(iv) The specification of different integration scenarios that can be used in the transformation process and a discussion on their trade-offs in terms of the complexity of the resulting extended schema and of the process constraints.

(v) The description of different implementation alternatives for the workflow extended schema towards target platforms.

(vi) The description of our tool implementation, supporting all the (automatic) model transformations.

Conceptual Schemas

A conceptual schema (also known as domain model) defines the knowledge about the domain that an information system must have to perform its business functions. Without loss of generality, we will represent conceptual schemas using UML [11].

The most basic constructors in conceptual schemas are entity types (i.e., classes in the UML terminology), relationship types (i.e., associations) and generalizations.

An entity type E describes the common characteristics of a set of entities (i.e., objects) of the domain. Each entity type E may contain a set of attributes.

A binary relationship type R has a name and two participants. A participant is an entity type that plays a certain role in the relationship type. Each relationship (i.e., link) between the two participants represents a semantic connection between the entities. A participant in R may have a minimum and maximum cardinality. The minimum cardinality min between participants p1 and p2 in R indicates that all entities of E1 (type of the participant p1) must be related at least with min entities of E2 (type of the participant p2). A maximum cardinality max between p1 and p2 in R defines that entities of E1 cannot be related with more than max entities of E2.

A generalization is a taxonomic relationship between a more general entity type E (supertype) and a set of more specific entity types $E_1,\ldots,E_n$ (subtypes).

As an example, Figure 2 shows a conceptual schema, represented in UML, meant to (partially) model a simple e-commerce application. It consists of the entity types Product, Quotation, QuotationLine (to record the details of the products included in the quotation), and Order (an order is generated by each quotation accepted by the customer, and then, its quotation lines are referred to as order lines). According to the cardinality constraints in the relationship types, all quotation must include at least one product and orders must be of a single quotation.

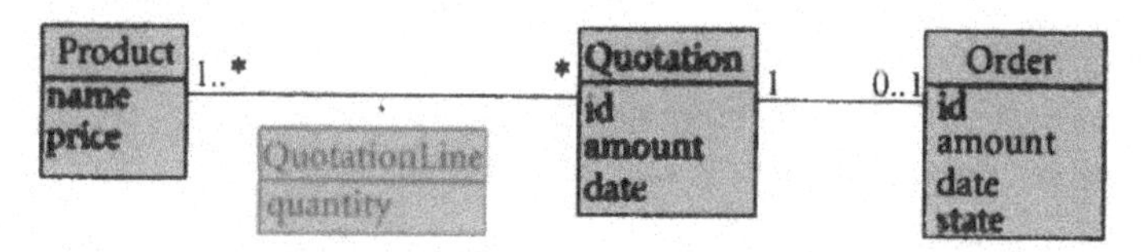

Figure 2. A partial conceptual schema for an e-commerce application.

Business Processes Concepts

Several visual notations and languages have been proposed to specify workflow models, with different expressive power, syntax, and semantics. Without loss of generality, in our work we have adopted the Workflow Management Coalition terminology, the Business Process Definition Metamodel [9] (BPDM), and the Business Process Management Notation [2] (BPMN).

BPDM is a standard proposed by OMG for representing and modeling business processes independent of any notation or methodology. This is done by proposing a unified metamodel that captures the common meaning behind the different notations and technologies. The metamodel is a MOF-compliant [13] metamodel. As such, BPDM also defines a XML syntax for storing and transferring business process models between tools and infrastructures. BPDM has been evaluated in [14] as the best business process interchange format in terms of expressivity.

BPMN perfectly fits with the BPDM metamodel and provides a graphical notation to express BPDM business processes. However, the specification of the business process can be provided with any other notation or language, including UML Activity Diagrams [11]. Several works evaluated and compared the different notations for specifying business processes (e.g., see [14–17]), highlighting strengths and weaknesses of every proposal. The results of our approach using one of these alternative notations would be quite similar. Indeed, our approach can be directly applied to any specification compliant with BPDM.

In our work, we focus on the core part of the BPDM metamodel. The workflow model is hence based on the concepts of Process (the description of the business process), Case (a process instance, that is, a particular workflow execution), Activity (the elementary unit of work composing a process), Activity instance (an instantiation of an activity within a case), Actor (a user role intervening in the process), Event (some punctual situation that happens in a case), and Constraint (logical precedence among activities and rules enabling activities execution). Processes can be internally structured using a variety of constructs: sequences of activities; gateways implementing AND, OR, XOR splits, respectively, realizing splits into independent, alternative and exclusive threads; gateways implementing joins, that is, convergence point of two or more activity flows; conditional flows between two activities; loops among activities or repetitions of single activities. Each construct may involve several constraints over the activities.

In the sequel, we will exemplify the proposed approach on a case study consisting of a simplified purchase process, illustrated using the BPMN notation in Figure 3.

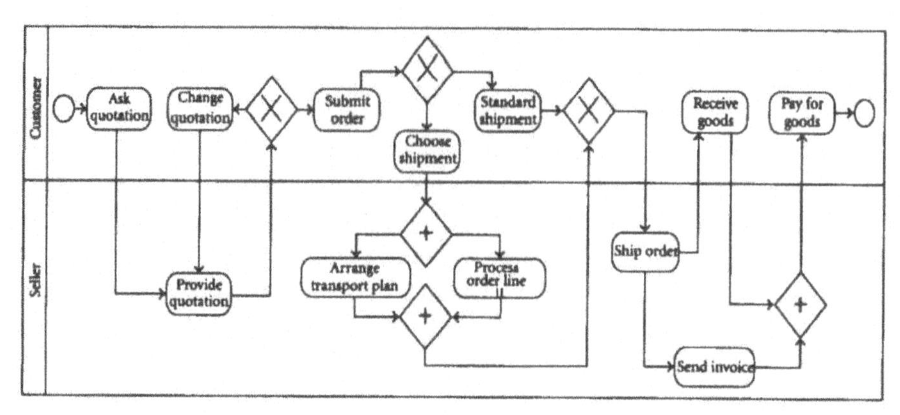

Figure 3. Example of a workflow model.

According to the BPDM semantics, the depicted diagram specifies a process involving two actors (represented by the two swimlanes): a customer and a seller. The customer starts the process by asking for a quotation about a set of products (Ask quotation activity). The seller provides the quotation (Provide quotation activity) and the customer may decide (exclusive choice) to modify the request (Change quotation activity, followed by the repetition of the Provide quotation activity) or to accept it (then the order is submitted). For simplicity, it is not modeled what happens if they never reach an agreement. Depending on the complexity of the order, the process can follow two alternative paths: the first consists only of a Standard Shipment activity, while the second requires the customer to specify the kind of shipment he prefers (Choose shipment). After the choice, the Seller takes the order in charge and performs two parallel activities: the arrangement of the transport plan and the processing of each order line. The latter is represented by the multi-instance activity called Process order line: a different instance is started for each order line included in the order. Once all order lines have been processed and the shipment has been defined (i.e., after the AND merge synchronization), the path reaches the join point with the alternative path of the standard shipment. Independently on the kind of shipment, the Ship order activity is performed, and then two uncontrolled branches take place: the customer receives the goods and the seller issues and sends the invoice. When both activities have completed (synchronization AND gateway), the user pays for the goods, and thus closes the process.

Normalization Phase

Before addressing the actual integration of the workflow model and the conceptual schema, the business process specification usually needs to be normalized.

This step simplifies the processing of the workflow model later on without losing generality in the coverage of the business process specification admitted in our method.

Workflow languages allow different equivalent representations of the same business semantics (see [2] for details) and define several complex constructs that can be derived from more basic ones. The normalization phase tackles these problems by applying a set of model to model transformations that ensure a coherent representation and render all the complex concepts in terms of simple ones. Notice that this phase does not aim at the reconciliation of different business processes. Instead, it aims at unifoming the notation of different design styles that could be adopted even within the same notation. The main issues addressed in this phase are the following:

(i) *Nested Structures*. If the business process is specified by means of nested subprocesses, they are flattened into a single-level business process that includes all tasks that were included in the subprocesses. If the subprocess contained only one lane, all the activities are moved inside the current lane of the main process; if more lanes were contained in the subprocess, they are transferred to the current pool of the main process, together with their respective activities, thus introducing new lanes in the flattened process.

(ii) *Different Notation Styles*. All different notations with the same BPDM semantics are homogenized in a single BPMN notation style (some examples are shown next). Thanks to this step, the business process will use only a single representation for each modeled behavior.

(iii) *Concatenation of Gateways*. If two or more gateways are directly connected by a control flow, the transformation adds a fake intermediate activity in the middle of every gateway pair. This simplifies the integration job, since it permits to work in a modular fashion when generating the constraints and the rules imposed by the gateways. Fake activities can be treated as activities that can be immediately enacted as soon as their process constraints are satisfied, and then automatically completed without neither user interaction nor business action execution.

Figure 4 shows the result of applying the normalization phase on the workflow model specified in Figure 3. The elements added because of the normalization are highlighted in color and bold line face. Only the last two transformations apply to this example. To avoid the alternative notation for XOR-merge (consisting in directly connecting two incoming arrows to an activity), the XOR-merge after the Ask quotation activity is made explicit and added to the model; analogously, to avoid two outgoing arrows from the Ship order activity, an AND-spit gateway is added. To remove the configurations of two connected gateways, a fake activity is added.

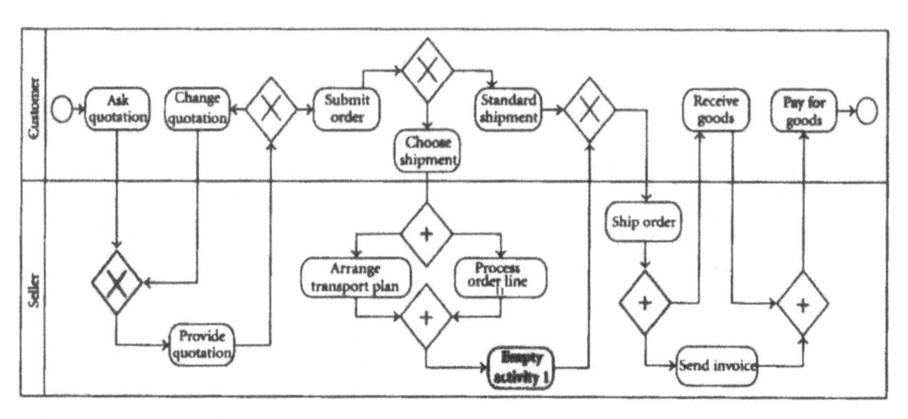

Figure 4. Example of a normalized workflow schema.

Extending Conceptual Schemas with Business Process Information

Given an initial conceptual schema c, the workflow-extended conceptual schema c¢ of the workflow-based application w is obtained by extending c with some additional elements derived from the business process specification w. We will focus on the case of a single business process; however, our extensions to the conceptual schema suffice when considering different business processes on the same domain. Indeed, in our approach several workflow models can be integrated with the same conceptual schema since the process constraints of each workflow model do not interfere due to the construction process of the workflow-extended model.

Generation of the Workflow-Extended Conceptual Schema

The workflow-extended conceptual schema must include: (i) the original conceptual schema, (ii) user-related information, (iii) workflow-related information, (iv) a set of possible relationships between the conceptual schema, the workflow information and the user information, and (v) a set of process constraints guaranteeing a consistent state of the whole model with respect to the workflow definition (see Section 7). To illustrate the generation of these different parts of the workflow-extended conceptual schema we will use the workflow model of Figure 4 and we will assume that the initial conceptual schema is the one shown in Figure 2.

More formally, we define a workflow-extended conceptual schema as follows. Given an initial conceptual schema with entity types (i.e., classes) E={e1,...,en}, representing the knowledge about the domain, and a workflow model w with activities A={a1,...,am}, the workflow-extended conceptual schema is obtained in the following way.

Domain Subschema

All entity types in E and their relationships (i.e., associations) and generalizations remain unchanged in the workflow-extended model (bottom part of Figure 5).

User Subschema

User-related information is added to the extended model by means of two entity types (see the top-left part of Figure 5): entity type User represents individual workflow actors; entity type Role represents groups of users, having access to the same set of tasks. A user may belong to different roles.

Workflow Subschema

Workflow-related information (top-right part of Figure 5) includes several fixed types (i.e., independent of the particular workflow model):

Entity type Process represents the supported workflows. As an example, an instance of the Process type would be our Purchase workflow. Other instances would represent additional workflows over the same domain subschema.

Entity type Case denotes an instance of a process, which has a name, a start time, an end time, and a status, which can be: ready, active, cancelled, aborted, or completed [2]. Every execution of a process results in a new instance of this type. This new instance is related with the appropriate process instance.

Entity type ActivityType represents the different activities that compose a process. Activity types are assigned to roles, which are responsible for executing them. In our case study, AskQuotation, ProvideQuotation, and so forth, would be instances of ActivityType.

Entity type Activity denotes the occurrence of a particular activity within a Case, described by the start time, the end time, and the current status, which can be: ready, active, cancelled, aborted, or completed. Only one user can execute a particular activity instance, and this is recorded by the relationship type Performs. The Precedes relationship keeps track of the execution order between occurred activities.

Entity type EventType represents the events that may affect the sequence or timing of activities of a process (e.g., temporal events, messages etc.). There are three different kinds of events (eventKind attribute): start, intermediate, and end. For start and intermediate events we may define the triggering mechanism (eventTrigger). For end events, we may define how they affect the case execution (eventResult).

Entity type Event denotes the occurrence of a particular type of event in the system, and a set of workflow-dependent subtypes.

For each activity aŒA, a new subtype sa is added to the entity type Activity (ActivityType is a powertype [11] for this set of generalizations). The name of the subtype is the name of a (e.g., in Figure 5 we introduced ProcessOrderLine, AskQuotation, ShipOrder, and so on). These subtypes record the information about the specific activities executed during a workflow case. For instance, the action of asking a quotation for the purchase X in a case C of a workflow W would be recorded in the system as an instance in the AskQuotation subtype related with the corresponding instance "C" in the Case type (in its turn related with the "W" instance in the Process type).

Relationships between Workflow, Subschema, and Domain Subschema

Each subtype sa is related with a (possibly empty) set of entity types EaŐE. These new relationship types are useful to record the objects modified/managed during the execution of a certain activity. Also, they are required to evaluate conditions appearing in some process constraints. In the case study (see Figure 5), a set of relationship types are established: Quotations are associated with the activities Ask Quotation and Provide Quotation; QuotationLines are associated with the ProcessOrderLine activity; and Orders are associated with the activities Submit Order, Process OrderLine, Ship Order, Pay for goods and so forth. When necessary, these associations between the domain and the workflow subschemata may be automatically generated if the workflow specification includes auxiliary primitives for describing the data flow between activities and/or when the designer defines some pattern-matching among the names of the activities and of the entity types. Otherwise, they must be manually specified.

Complexity of the Workflow-Extended Conceptual Schema

Clearly, the workflow-extended schema is more complex than the original conceptual schema. However, we believe that this increased complexity is compensated by the fact that it may be automatically generated (with our method) and processed (for instance, with code-generation tools) and thus, the designer does not need to directly manipulate it. Moreover, the size of the extension is (1) constant regardless the size of the initial conceptual schema and (2) linear with respect to the number of activities in the workflow. Therefore, in most cases, the extension will be small when compared with the size of the initial conceptual schema.

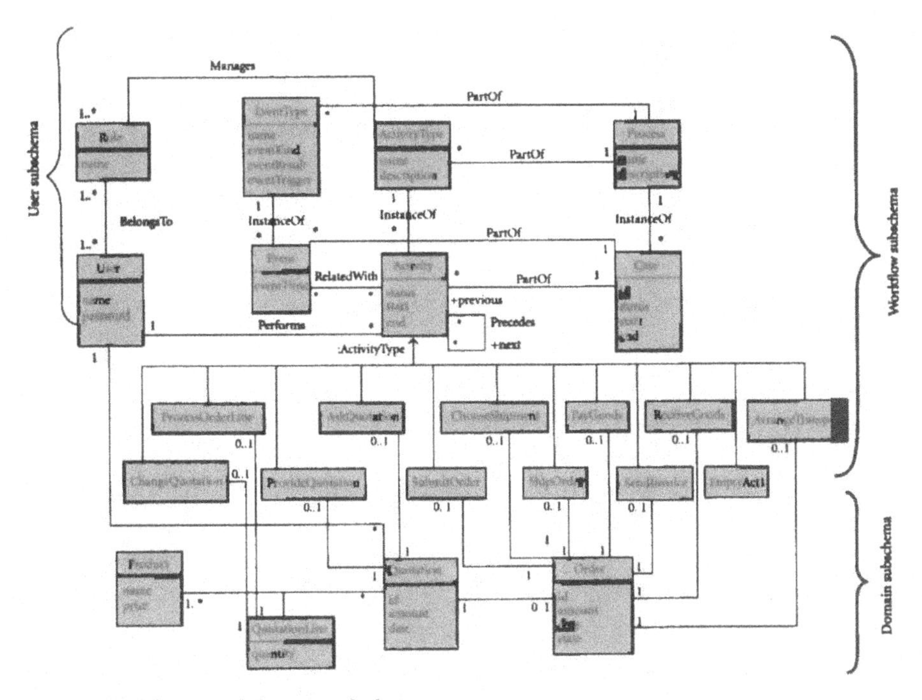

Figure 5. Workflow-extended conceptual schema.

We would like to remark that when proposing our workflow-extended schema we opted for balancing the size of the workflow subschema with the complexity of the process constraints. Richer schemas with further relationship types and/or attributes could be defined, according to the requirements of the specific workflow application (for example, we could have used a more complex pattern for the specification of the role-user relationship [18]). Similarly, simpler extensions could be used instead but then, as a trade-off, the process constraints would become much more complex. To better illustrate this discussion, two other alternative workflow-extended models are provided in the Appendix A. All three alternatives share the same philosophy and provide the same kind of benefits, and thus, designer may choose any of them when applying our method.

Translation of Process Constraints

The structure of a workflow model implies a set of constraints regarding the execution order of the different activities, the number of possible instances of each activity in a given case, the conditions that must be satisfied in order to start a new activity, and so forth. These constraints are usually referred to as process constraints. The behavior of all enterprise applications must always satisfy these constraints. Thus, the generation of the workflow-extended model must consider all process constraints.

Process constraints are translated into constraints over the population of the $s_{a1},\ldots,s_{am}$ subtypes of Activity (see previous section). The generated constraints guarantee that any update event over the population of one of these subtypes (for instance, the creation of a new activity instance or the modification of its status) will be consistent with the process constraints defined in the workflow model.

We specify process constraints by means of invariants written in the OCL language [8]. Invariants in OCL are defined in the context of a specific type, the context type. The actual OCL expression stating the constraint condition is called the body of the constraint. The body is always a boolean expression and must be satisfied by all instances of the context type, that is, the evaluation of the body expression over every instance of the context type must return a true value. For instance, a constraint like: *contextAinv: condition* implies that all instances of the type A must verify condition.

Constraints are defined to restrict only the execution of the workflow they are created for (the context type of the constraints is always a specific activity and not an entity type of the domain subschema). Therefore, no interferences among different workflows occur, even if they are defined over an overlapping subset of the conceptual schema.

Even though some of the constraints may seem quite complex, we would like to remark that all of them are automatically generated from the workflow model, and thus, they do not need to be manipulated (nor even necessarily understood) by the designer but for other tools. However, to simplify its presentation in the extended model, we could easily define a stereotype for each constraint type, as done in [19].

Next subsections define a set of patterns for the generation of the process constraints corresponding to the different typical constructs appearing in workflow models (sequences, split gateways, merge gateways, conditions, loops, etc.). The patterns can be combined to produce the full translation of the workflow model.

Sequences of Activities

A sequence flow between two activities (Figure 6) indicates that the first activity (A) must be completed before starting the second one (B). Moreover, if A is completed within a given case, B must be eventually started before ending the case (we do not require B to be completed since, for instances, it could be interrupted by the trigger of an intermediate exception event). This behavior can be enforced by means of the definition of three OCL constraints.

Figure 6. Sequence flow.

The first constraint (seq1 constraint) is defined over the entity type corresponding to the destination activity (B in the example) stating that for all activity instances of type B the preceding activity instance must be of type A and that it must have been already completed. Its specification in OCL is the following: *context B inv seq1: previous->size()=1* and *previous->exists(a a.oclIsTypeOf(A)* and *a.status='completed')*.

This OCL definition enforces that B instances (since B is the context type of the constraint) have a previous activity (because of the size operator over the value of the navigation through the role previous) and that such activity is of type A (enforced by the exists operator). B and A are Activity subtypes as defined in Section 6.

The other two required constraints are the following:

(i) A constraint seq_2 over the second activity to prevent the creation of two different B instances related with the same A activity instance *context B inv seq2: B.allInstances()-> isUnique(previous)*

(ii) A constraint seq_3 over the Case entity type verifying that when the case is completed there exists a B activity instance for each completed A activity instance. This B instance must be the only instance immediately following the A activity instance *context Case inv seq_3: status='completed' implies self.activity-> select(a a.oclIsTypeOf(A) and a.status='completed')->forAll(a a.next->exists(b b.oclIsTypeOf(B)) and a.next->size()=1)*

Split Gateways

A split gateway is a location within a workflow where the sequence flow can take two or more alternative paths. The different split gateways differ on the number of possible paths that can be taken during the execution of the workflow. For XOR-split gateways only a single path can be selected. In OR-splits several of the outgoing flows may be chosen. For AND-splits all outgoing flows must be followed.

For each kind of split gateway, Table 1 shows the process constraints required to enforce the corresponding behavior.

Table 1. Constraints for split gateways.

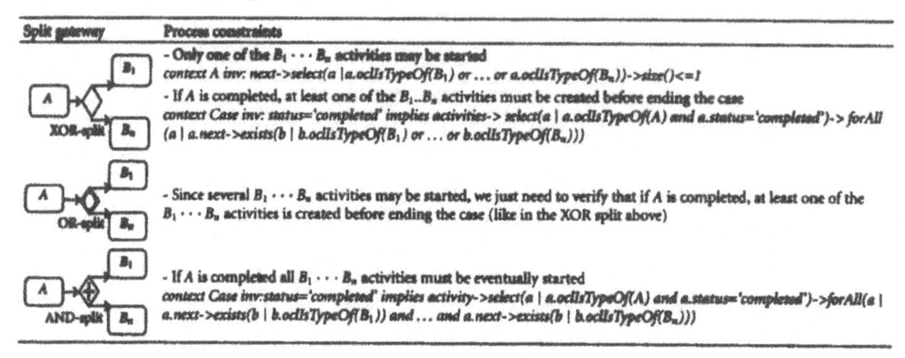

<table>
<tr><th>Split gateway</th><th>Process constraints</th></tr>
<tr><td>XOR-split</td><td>- Only one of the $B_1 \cdots B_n$ activities may be started
context A inv: next->select(a |a.oclIsTypeOf(B₁) or … or a.oclIsTypeOf(B_n))->size()<=1
- If *A* is completed, at least one of the $B_1..B_n$ activities must be created before ending the case
context Case inv: status='completed' implies activities-> select(a | a.oclIsTypeOf(A) and a.status='completed')-> forAll (a | a.next->exists(b | b.oclIsTypeOf(B₁) or … or b.oclIsTypeOf(B_n)))</td></tr>
<tr><td>OR-split</td><td>- Since several $B_1 \cdots B_n$ activities may be started, we just need to verify that if *A* is completed, at least one of the $B_1 \cdots B_n$ activities is created before ending the case (like in the XOR split above)</td></tr>
<tr><td>AND-split</td><td>- If *A* is completed all $B_1 \cdots B_n$ activities must be eventually started
context Case inv:status='completed' implies activity->select(a | a.oclIsTypeOf(A) and a.status='completed')->forAll(a | a.next->exists(b | b.oclIsTypeOf(B₁)) and … and a.next->exists(b | b.oclIsTypeOf(B_n)))</td></tr>
</table>

Besides the process constraints appearing in the table, we must also add to all the activities $B_1 \dots B_n$ the previous constraints seq_1 and seq_2 to verify that the preceding activity A has been completed and that no two activity instances of the same activity B_i are related with the same preceding activity A. We also require that the activity instance/s following A is of type B_1 or … or B_n.

Merge Gateways

Merge gateways are useful to join or synchronize alternative sequence flows. Depending on the kind of merge gateway, the outgoing activity may start every time a single incoming flow is completed (XOR-Merge) or must wait until all incoming flows have finished in order to synchronize them (AND-Merge gateways). The semantics of the OR-Merge gateways is not so clear. If there is a matching OR-split, the OR-Merge should wait for the completion of all flows activated by the split. If no matching split exists, several interpretations are possible, being the simplest one to wait just till the first incoming flow. This is the interpretation adopted in this paper. For a complete treatment of this construct see [20].

Table 2 presents the different translation patterns required for each kind of merge gateway. Besides the constraints included in the table, a constraint over A should be added to all the gateways to verify that two A instances are not created for the same incoming set of activities (i.e., the intersection between the previous instance/s of all A instances must be empty).

Table 2. Constraints for merge gateways.

<table>
<tr><th>Merge gateway</th><th>Process constraints</th></tr>
<tr><td>B_1
A
B_n XOR-merge</td><td>- All A activity instances have as a previous activity instance a completed activity instance of type B_1 or…or B_n
context A inv: previous->size()=1 and previous->exists(b | (b.oclIsTypeOf(B_1) or … or b.oclIsTypeOf(B_n)) and b.status='completed')
- Each $B_1 \cdots B_n$ activity instance is followed by an A activity
context Case inv: status='completed' implies activity->select(b | b.oclIsTypeOf(B_1) or … or b.oclIsTypeOf(B_n))->forAll(b | b.next->exists(a |a.oclIsTypeOf(A)))</td></tr>
<tr><td>B_1
A
B_n OR-merge</td><td>- An A activity instance must wait for at least an incoming flow
context A inv: previous->select(b | (b.oclIsTypeOf(B_1) or … or b.oclIsTypeOf(B_n)) and b.status='completed')->size()≥ 1</td></tr>
<tr><td>B_1
A
B_n AND-merge</td><td>- An activity instance of type A must wait for a set of activities $B_1 \cdots B_n$ to be completed
context A inv: previous->exists(b | b.oclIsTypeOf(B_1) and b.status='completed') and … and previous->exists(b | b.oclIsTypeOf(B_n) and b.status='completed')
- Each set of completed $B_1 \cdots B_n$ activity instances must be related with an A activity instance
context Case inv: status='completed' implies not (activity->exists(b | b.oclIsTypeOf(B_1) and b.status='completed' and not b.next->exists(a | a.oclIsTypeOf(A)) and …and activity ->exists(b | b.oclIsTypeOf(B_n) and b.status='completed' and not b.next->exists(a | a.oclIsTypeOf(A)))</td></tr>
</table>

Condition Constraints

The sequence flow and the OR-split and XOR-split gateways may contain condition expressions to control the flow execution at run-time. As an example, Figure 7 shows a conditional sequence flow. In the example, the activity B cannot start until A is completed and the condition cond is satisfied. The condition expression may require accessing the entity types of the domain subschema related to B in the workflow-extended model. Through the Precedes relationship type, we can also define conditions involving the previous A activity instance and/or its related domain information.

Figure 7. A conditional sequence flow.

To handle these condition expressions we must add, for each condition defined in a sequence flow or in an outgoing link of OR and XOR gateways, a new constraint over the destination activity. The constraint ensures that the preceding activity satisfies the specified condition, according to the following pattern: *?context B inv: previous->forAll(a | a.cond)*

Note that these additional constraints only need to hold when the destination activity is created, and thus, they must be defined as creation-time constraints [21].

Loops

A workflow may contain loops among a group of different activities or within a single activity. In this latter case we distinguish between standard loops (where the activity is executed as long as the loop condition holds) and multi-instance loops (where the activity is executed a predefined number of times). Every time a loop is iterated a new instance of the activity is created. Figure 8 shows an example of each loop type.

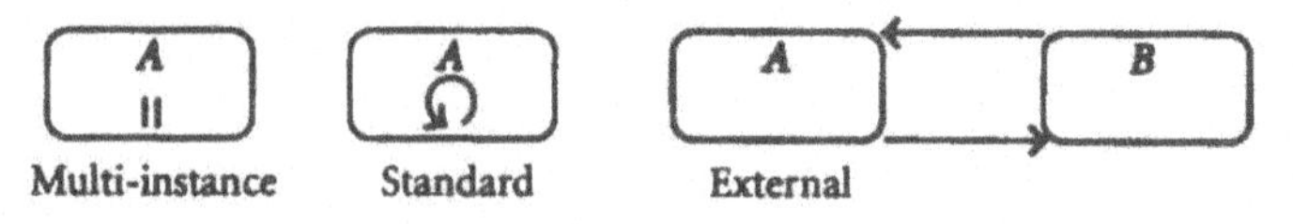

Figure 8. Loop examples.

Management of external loops does not require new constraints but the addition of a temporal condition in all constraints stating a condition like "an instance of type B must be eventually created if an instance of type A is completed." The new temporal condition on those constraints ensures that the B instance is created after the A instance is completed (earlier B instances may exist due to previous loop iterations).

Standard loops may be regarded as an alternative representation for conditional sequence flows having the same activity as a source and destination. Therefore, the constraints needed to handle standard loop activities are similar to those required for conditional sequence flows. We need a constraint checking that the previous loop instance has finished and another one stating that the loop condition is still true when starting the new iteration (again, this is a creation-time constraint). The loop condition is taken from the properties of the activity as defined in the workflow model. Moreover, we need also to check that the activity/ies at the end of the outcoming flows of the loop activity are not started until the loop condition becomes false. To prevent this wrong behavior we should treat all outgoing flows from the loop activity as conditional flows with the condition "not loopCondition." Then, constraints generated to control the conditional flow will prevent next activity/ies to start until the condition "not loopCondition" becomes true.

Multi-instance loop activities are repeated a fixed number of times, as defined by the loop condition, which now is evaluated only once during the execution of the case and returns a natural value instead of a boolean value. At the end of the case, the number of instances of the multi-instance activity must be an exact multiple of this value. Assuming that the multi-instance activity is called A, the OCL formalization of this constraint would be: *context Case inv: (activity->select(a | a.oclIsTypeOf(A)) ->size() mod loopCondition)=0.*

For multi-instance loops the different instances may be created sequentially or in parallel. Besides, we can define when the workflow shall continue. It can be either after each single activity instance is executed (as in a normal sequence flow), after all iterations have been completed (similar to the AND-merge gateways), or as soon as a single iteration is completed (similar to the basic OR-merge gateway).

Event Management

An event is something that "happens" during the course of the workflow execution. There are three main types of events: Start, Intermediate and End (see Figure 9). A workflow schema may contain several start, intermediate, and end events.

Figure 9. Examples of events.

Start events initiate a new flow, while end events indicate the termination of a flow. Intermediate events are instead used to change the normal flow (for instance, to handle exceptions or to start additional activities). Intermediate events can be attached to an activity (the triggering of the event aborts the activity execution) or can be placed in the middle of a sequence flow between two activities (the flow does not continue until the event is issued).

When a start event is issued, an instance of each activity connected to the event has to start afterwards. Conversely, no activity instance is created in a case before the occurrence of at least a start event. In particular, activity instances for activities connected only to flows coming from one or more start events (as activity A in the previous figure) cannot be created until one of those start events is issued. The formalization of these constraints is the following:

(i) context Event inv: eventType.name='StartEvent1' and case.status='completed' implies case.activity-> select (a | a.oclIsTypeOf(A) and a.event=self)->size()=1

(ii) context Case inv: activity->notEmpty() implies event->exists(e |e.eventType.eventKind='StartEvent')

(iii) context A inv: self.event->exists(ev | ev.eventType.name ='StartEvent1').

No activity instances can be created in the case after the event has been issued. Assuming that EndEvent1 (Figure 9) is defined as a terminate event, the following constraint must be added to the workflow-extended model: *context Event inv: eventType.name='EndEvent1' implies case.activity->forAll (a | a.start< eventTime)*

For intermediate events, the target activity of the event must be executed after the triggering of the event (and it cannot be executed otherwise). Depending on the kind of intermediate event, the interrupted activity will change its status to cancelled or aborted (which, for instance, may prevent the next activity in the normal sequence flow to be started).

The following process constraints are generated for the IntermediateEvent1 example in Figure 9:

(i) *context Event inv: eventType.name= 'IntermediateEvent1' and case.status='completed' implies case.activity-> exists(a | a.oclIsTypeOf(B))*

(ii) *context Case inv: activity-> exists(a | a.oclIsTypeOf(B)) implies event->exists(e | e.eventType.name="IntermediateEvent1")*

Obviously, this last constraint is true as long as B has no other incoming flows. Otherwise, all incoming flows form an implicit XOR-Merge over B and we should generate the constraints according to the pattern for the XOR-Merge gateway.

Applying the Translation Patterns

As an example, Table 3 summarizes the process constraints resulting from applying the translation patterns over the workflow running example (Figures 4 and 5).

For the sake of brevity, in this section constraints are described in an informal and compact way. The complete set of constraints and their OCL specification is exemplified in Table 4 only for the Provide Quotation activity. The translation of all the other constraints is provided in the Appendix B.

Table 3. Process constraints for the workflow running example.

Activity	Constraints
Ask Quotation	- A new *Ask Quotation* activity must be created every time a start event occurs.
Provide Quotation	- A quotation cannot be provided until an *Ask Quotation* or a *Change Quotation* finishes. A single new *Provide Quotation* instance must exist for each completed *Ask Quotation* or *Change Quotation* activity. - After providing a quotation we can either ask for a change in the quotation or submit the order, but not both (at least one of them must be executed).
Change Quotation	- The previous *Provide Quotation* activity must have been completed (a single new ask quotation activity can be generated). Otherwise, it must have been created in response to the occurrence of a start event (due to the implicit XOR merge gateway corresponding to the two incoming arrows).
Submit Order	- The previous *Provide Quotation* activity must be completed. Besides, only a single *Submit Order* instance must be created for the same *Provided Quotation* instance. - After submitting an order, a *Choose Shipment* or a *Standard Shipment* activities must be executed (but not both).
Standard Shipment	- The previous *Submit Order* activity must be completed. Besides, only a single *Standard Shipment* instance must be created for the same *Submit Order* instance. - Once the standard shipment is completed, a new *Ship Order* activity must be created.
Choose Shipment	- The previous *Submit Order* activity must be completed. Besides, only a single *Choose Shipment* instance must be created for the same *Submit Order* instance. - After choosing the shipment, both the *Arrange Transport* and *Process Order Line* activities must be executed.
Arrange Transport	- The preceding *Choose Shipment* activity instance must be completed. Besides, a single *Arrange Transport* activity instance must be executed for each *Choose Shipment* activity instance.
Process OrderLine	- The preceding *Choose Shipment* activity must be completed. - The system must exactly execute as many *Process OrderLine* activity instances as the number of order (quotation) lines for the related order.
Empty Activity1	- The new *Empty activity* instance can be created (as completed) when the transport has been arranged and all order lines have been processed. - Then, a new *Ship Order* instance must be executed before ending the case.
Ship Order	- Once a *Standard Shipment* xor an *Empty Activity1* instance has been completed, the order can be shipped. - For each order shipped, an invoice must be sent and the reception of the goods must be acknowledged by the customer.
Send Invoice	- The preceding *Ship Order* activity instance must be completed. Besides, a single *Send Invoice* activity instance must be executed for each *Ship Order* activity instance.
Receive Goods	- The preceding *Ship Order* activity instance must be completed. Besides, a single *Receive Goods* activity instance must be executed for each *Ship Order* activity instance.
Pay Goods	- An order cannot be paid until the invoice has been send and the good have been received. When both previous activities have been done, a single *pay goods* activity shall be created in response.

Table 4. Constraint definitions for the Provide Quotation activity.

<table>
<tr><td>Constraints due to incoming XOR-merge</td><td>–The preceding activity must be of type Ask Quotation or Change Quotation and must be completed
context ProvideQuotation inv: previous->size()=1 and previous->exists(a (a.oclIsTypeOf(AskQuotation) or a.oclIsTypeOf(Change Quotation)) and a.status='completed')
–No two instances may be related with the same Ask Quotation or Change Quotation instance
context ProvideQuotation inv: ProvideQuotation.allInstances()-> isUnique(previous)
–A Provide Quotation instance follows each completed Ask Quotation or Change Quotation activity
context Case inv: status='completed' implies activity->select(b | b.oclIsTypeOf(AskQuotation) or … or b.oclIsTypeOf(ChangeQuotation))-> forAll(b |b.next->exists(a | a.oclIsTypeOf(ProvideQuotation)))</td></tr>
<tr><td>Constraints due to the outgoing XOR- split</td><td>–The next activity must be either another Change Quotation instance or a Submit Order instance, but not both
context ProvideQuotation inv: next->select (a| a.oclIsTypeOf(ChangeQuotation) or a.oclIsTypeOf(SubmitOrder))->size()<=1
–If the Provide Quotation instance is completed, a Change Quotation or a Submit Order must necessarily be created before ending the case.
context Case inv: status='completed' implies activity->select (a | a.oclIsTypeOf (ProvideQuotation) and a.status='completed')-> forAll (a | a.next-> exists (b | b.oclIs Type Of(Change Quotation) or b.oclIsTypeOf(SubmitOrder)))
–Only Change Quotation activity instances or Submit Order instances may follow a Provide Quotation instance
context ProvideQuotation inv: next->forAll (b| b.oclIsTypeOf(ChangeQuotation) or b.oclIsTypeOf(SubmitOrder)</td></tr>
</table>

The Provide Quotation activity involves a set of constraints due to the incoming XOR-merge from Ask Quotation and Change Quotation activities and a set of constraints due to the subsequent XOR split with Submit Order and Change Quotation.

Implementation of the Workflow-Extended Conceptual Schema

Once the workflow-extended schema is available, we may automatically generate an implementation of the system that ensures a consistent behavior of all enterprise applications with respect to the business process specification.

Since a workflow-extended conceptual schema is a completely standard UML model (i.e., no new modeling primitives have been created to express the extension of the original model with the required workflow information) any method or tool able to provide an automatic model-to-code generation from UML models to a final technology platform P can also cope with the automatic generation of our workflow-extended schema in the same platform P, using general-purpose MDD techniques and frameworks.

For instance, a tool able to generate a database schema from an UML/OCL model can follow exactly the same procedure to generate a database implementation for our extended schema that guarantees the satisfaction of all workflow constraints. As usual, classes (including also the classes in the workflow subschema) will be transformed into tables, while OCL constraints (either domain or workflow constraints) will be expressed as triggers (this is not the only option, see [22] for a discussion of the different mechanisms to implement OCL constraints in databases). Similarly, a tool able to generate Java schemas from UML/OCL models could be directly used to generate a Java-based implementation of the workflow-extended schema. In this case, classes will be expressed as Java classes while constraints could be implemented as method preconditions that prevent the execution of the method if the system is not in the right workflow state.

As an example, Figure 10 shows a possible (i) database implementation and (ii) Java-based implementation for two sequential activities A and B (Figure 6), performed by the same user. In the former, the constraint is implemented as a trigger over the table AtoB representing the Precedes relationship type (see Figure 5) between both activities (AtoB table has only two columns, a_id and b_id and records the information about which A activities precede each B activity; this is the typical database implementation for many-to-many associations in conceptual schemas). In the latter, the constraint is verified as part of the method AssignPreviousActivity redefined in the B class (corresponding to the B activity in the workflow-extended model). In both situations, when the user tries to create a new B activity and the previous A activity is not completed, an exception is raised since the B activity cannot start yet. The tables, triggers and/or Java classes and method bodies to implement the workflow-extended model translation (including the OCL constraints) can be automatically generated using current code-generation tools such as [23–25] among others.

```
create trigger AtoBSeqConstraint
before insert on AtoB
for each row
Declare v_Status Varchar(10);
        EInvalidActivity Exception;
Begin
   SELECT status into v_Status
   FROM A a
    WHERE a.id =:new.a_id;
    If (v_Status<>'completed')
      then raise EInvalidActivity; end if;
End;
```

(a)

```
class B
{
 ...
 void AssignPreviousActivity(A a)
 throws Exception
 {
    if (! a.status.equals("completed"))
         throw new Exception("Invalid Activity");
    else previous.add(a);
 }
}
```

(b)

Figure 10. Examples of a sequence constraint implemented in particular technologies.

Note that the previous strategies to implement the sequence constraint between A and B activities are efficient ones since the constraint is only checked when linking a B activity instance with an A activity instance, regardless how many activities are part of the workflow (and the checking just compares that exact pair of instances, instead of checking all existing A and B instances). We can benefit from the fact that in our workflow-extended conceptual schema the process constraints are expressed in OCL and rely on existing methods for analyzing OCL expressions (as in [26]) to automatically compute the information about when and how checking the constraints in order to get an efficient implementation for all process constraints.

For Web applications, an interesting alternative is to fully exploit MDD approaches, such as [23, 27]: an initial hypertext model can be derived from the workflow-extended conceptual schema so that the hypertext structure enforces some of the process constraints among activities assigned to the same user by means of driving the user navigation through the Web site. Process constraints involving activities belonging to different users must be enforced using one of the previous techniques, they cannot be controlled at the hypertext level (or group of users). This can be done by designing in the proper way the set of pages and links that can be browsed. For instance, Figure 11 shows a hypertext model that from the home page forces the user to go through the Web pages implementing A before starting B.The hypertext model is defined in WebML [23], a conceptual language for the specification of Web applications, already extended with workflow-specific primitives [28]. The operation units StartActivity and EndActivity are in charge of recording the information about the activities' progress in the corresponding entity types of the conceptual schema. More complicated constraints appearing in the workflow-extended model can be enforced by means of appropriate branching and task assignment primitives.

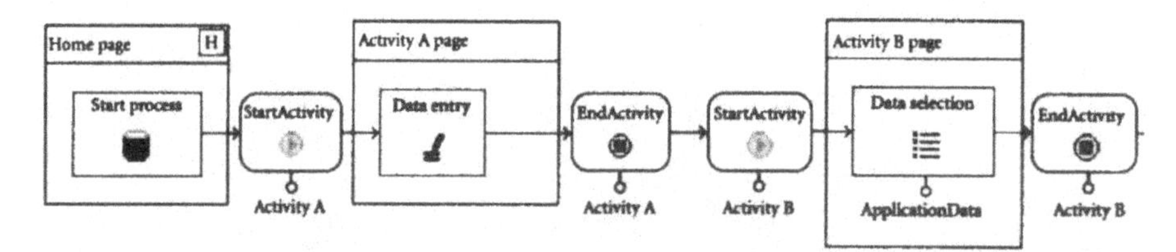

Figure 11. Example of a sequence constraint implemented within the hypertext model of a Web application.

Usually, designers will be able to choose among different strategies/platforms when implementing the workflow-extended conceptual schema. For instance, assuming a typical three tier (or n-tier) architecture, designers can decide whether to check the process constraints in the presentation layer (for example, as shown in Figure 11), in the business layer (as in Figure 10(b)) or in the persistence layer (as in Figure 10(a)). Each alternative may imply a slightly different behavior of the system at run-time in terms of its consistency, user experience, flexibility, reliability, and so forth. For instance, a database-based implementation represents a safer alternative in terms of the data consistency (regardless how users interact with the system to update the data, the data will be always consistent). Instead, enforcing the constraints at the hypertext level provides a better user experience since it reduces the probability that the user actions end up in an error due to a wrong activity selection.

Tool Support

To show the viability of the approach, we describe the tool framework we used for realizing the whole development process presented in Figure 1. Our framework tries to reuse as many existing tools as possible. We directly developed only the pieces that were missing for covering all the design phases. Once the designer provides the initial models, the rest of the process is performed automatically. Figure 12 shows the tools we used for each step of the design.

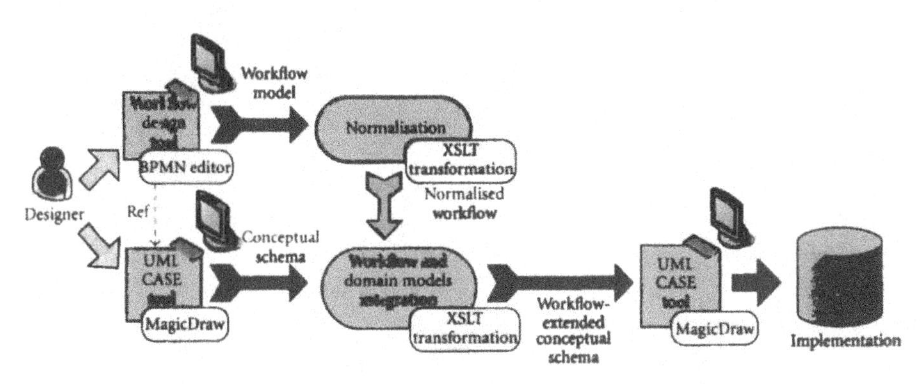

Figure 12. Tools used for the MDD generation of Workflow-extended conceptual schemas.

The design of the conceptual schema has been done using MagicDraw [29] that exports it as an XMI file [30].

For workflow design and transformation, we have developed a visual editor prototype [31] that supports the design of BPMN diagrams and their automatic model transformations. This BPMN editor has been implemented as an Eclipse plugin and it is flexible and extensible. It covers the whole BPMN notation (including subprocesses) and can manage user-defined properties of objects and new transformations of the workflow models. The workflow schema is stored as an XML document according to an internal format, but proper transformations are available for importing and exporting in standard notations (e.g., to BPDM, XPDL, BPEL, etc.). The tool includes the normalization transformation, implemented as an XSLT transformation over the workflow XML representation.

Given the XML representation of the normalized workflow model and the XMI representation of the initial conceptual schema, our main transformation generates a new XMI file containing the workflow-extended model and the process constraints, according to the guidelines presented in this paper. This XMI file can be imported back and used within the MagicDraw tool.

Related Work

With respect to traditional approaches to workflow management, implemented in a plethora of commercial WFMSs, our work takes a radically different point of view, by focusing on the business process modeling and on its transformation to a software engineering specification that integrates the domain information and that can be refined and exploited by automatic code generation tools.

This approach allows for more control and personalization of the system implementation and presents a number of additional benefits as commented in Section 2. As a downside, some aspects such as integration with legacy systems, monitoring and supervision, fault management and so forth, if needed, must be provided by the application that embeds the business specification itself, instead of relying on a WFMS for providing them.

In the software engineering field, research on business process has mainly addressed the correctness of the design of the workflow model (see [32] as a representative example). Other works address the direct implementation of business process models in specific final technology platforms: for instance, [5] proposes an implementation of process constraints over relational database structures, by exploiting event-condition-action rules; [6] implements workflow models using Web technologies by mapping the workflow specification to a DSL for Web design called WebML; and [7] exploits BPEL4WS for implementing them. All these approaches are hardly reusable when trying to implement our workflow schema in different technologies or when we want to migrate our current implementation to an alternative platform. Instead, since our method works at a platform-independent level, we are able to generate an implementation of the workflow-extended method in any final technology platform using current model-driven development (MDD) approaches, as seen in Section 8. Integrating the workflow and the domain information in a single schema also allows us to treat both dimensions in a homogeneous and consistent way (for instance, this enables the possibility of defining more complex business rules mixing domain and workflow information). This is not contemplated in the previous approaches.

Up to now, integration of workflows and MDD approaches has only been explored from a general framework perspective. For instance, [33] proposes to transform the workflow model to a DSL specification. However, it only provides some general guidelines for the transformation and a comprehensive framework specifying the different components that lead from the design-time specification to the runtime execution of the workflow model. However, no details are provided on the transformation rules that map a workflow model to a specific DSL. The work in [34] proposes an approach for configuring generic process models depending on the domain information provided by the stakeholders by mean of filling questionnaires developed ad-hoc for that specific process. Questionnaires are created from the information in the initial domain model defined by the designers. Their goal is to generate, as a result, an adapted and individualized business process but not to integrate in a single conceptual schema both the process and domain information.

Some proposals (as in [35, 36], or [37]) tried to extend and adapt the UML notation for workflow modeling purposes but they did not address the unification of the business process and the conceptual schema's views of the system. As far as we know, ours is the first proposal where both workflow information and process constraints are automatically derived from a workflow model and integrated within the platform-independent conceptual schema.

Moreover, ours is also the first translation of a workflow model into a set of equivalent OCL declarative constraints. An explicit definition of all the workflow constraints induced by the different workflow constructs is necessary regardless how these constraints are to be enforced and managed in the final workflow implementation. Very few examples of translations from process constraints to other declarative languages exist (e.g., see [38] for a translation to LTL temporal logics). In literature, OCL has only been used in relation with workflow models as an auxiliary tool for a better specification of the business process. For instance, in [39] OCL is used to manually specify workflow access control constraints and derive authorization rules, in [40] to express constraints with respect to the distribution of work to teams, in ArgoUWE [41] to check for well-formedness in the design of process models, in [42] to manually specify business models with UML, and in [7] to specify the contracts for the transformation of activity diagrams into BPEL4WS.

Conclusions

In this paper, we presented an automatic approach to integrate the semantics of business process specifications within conceptual schemas. The main advantage of using conceptual schemas to handle the workflow information is that we can develop workflow-based applications without requiring the use of a specific workflow management subsystem.

Once the designer has specified both the workflow and the conceptual schemas, we build an integrated workflow-extended conceptual schema by adding to the conceptual schema (i) the definition of a set of new entity and relationship types for workflow status tracking and (ii) the rules for generating the integrity constraints on such types, needed for enforcing the business process specification. The integration of both the domain and the workflow aspects in a single extended conceptual schema permits a homogeneous treatment of both dimensions of the workflow-based application.

The workflow-extended conceptual schema is a completely standard UML model. This provides additional benefits. For instance, we can apply the usual model-driven development methods over our extended model to generate its automatic implementation in any technology platform. As long as these methods are able to deal with UML/OCL models, they will be able to directly manage our workflow-extended schema. In the same way, we could reuse verification and validation tools for UML models and apply them to check our extended schema.

As a further work, we would like to explore the possibility of using our extended schema as a bridge to facilitate reverse-engineering of existing applications into their original workflow models and to ease keeping them aligned. We also plan to develop a method that, from the generated process constraints, is able to compute the list of activities that can be enacted by a user in a given case (i.e., those activities that can be created without violating any of the workflow process constraints according to the case state at that specific time) to provide a better user-experience when executing the workflow-based applications. Instead of letting the user choose the desired activity and then check whether the activity can be started, we would directly provide the list of secure activities avoiding possible errors in the activity selection. Along this line, we also plan to investigate the different application layers (data layer, business logic layer, presentation layer) where the process constraints can be implemented, and define some recommendation framework for the developers (and the automatically generated code) for the best implementation strategy of constraints depending on the kind of experience the application is supposed to provide to the users.

Future investigations will also address the empirical evaluation of our approach. In particular, we would like to compare the quality of manually developed applications with respect to the ones produced with our approach. For instance, we would like to compare the percentage of workflow constraints detected and included by programmers when manually developing the applications with the coverage of workflow constraints obtained when using our approach. We are confident that a manual application development will miss many workflow constraints since a manual detection of all relevant constraints and possible inconsistencies is an error-prone activity. We also plan to evaluate the effort required to develop this kind of applications with and without our approach.

Appendix A

As we have seen in Section 6, our workflow-extended schema is the result of a trade-off between the size of the model and the complexity of the OCL expression needed to represent the process constraints. However, that is not the only feasible alternative. In this Appendix we present two different alternatives: the first one aims at minimizing the size of the workflow-extended schema, while the second one tries to reduce the complexity of the required OCL expressions. The designer may choose among the three alternatives (these two plus the one presented in the main part of the paper) when following our approach for the integration of business processes and conceptual schemas.

A Minimal Workflow-Extended Conceptual Schema

In a "minimal" workflow extended schema (Figure 13), to reduce the size of the model, no workflow-dependent subtypes (i.e., the Activity subtypes recording the information about specific activities executed during the workflow case) are created. This implies that the Activity entity type must be extended with an additional attribute type to distinguish and classify the enacted activities (before we could directly determine that by examining the specific subclass of each activity instance). In our running example, the possible values for this type attribute are: AskQuotation, ProvideQuotation, SubmitOrder, and so forth.

Moreover, with this model, all relationships between the workflow subschema and the domain subschema must be done now at the Activity type level, instead of linking the domain classes with their specific related activities. As a result, there must exist a different relationship type between the Activity type and each domain class in the model (except for domain classes not related with any activity). An additional set of integrity constraints must be defined to ensure a correct instantiation of these new relationship types. For instance, in our example, an AskQuotation activity is only related to a Quotation instance. In our original workflow extended schema this was already enforced by the model itself (AskQuotation was only linked to the Quotation type) but in this minimal model, we need to add the following constraint:

contextActivityinv:

self.type="AskQuotation"implies

self.product->isEmpty()andself.order->isEmpty()and

self.quotationLine->isEmpty()

To ensure that activity instances of type AskQuotation are only associated with quotation instances. A similar constraint must be added for each workflow activity related to business data objects.

The definition of process constraints also becomes more complex. Now the activity type must act as context type for all the process constraints. Therefore, the first part of all constraints must be devoted to select from all activities those affected by the constraint. For instance, the first sequence constraint presented in Section 7.1:

context B inv seq1: previous->size()=1 and previous ->exists(a|a.oclIsTypeOf(A) and a.status='completed')

is now expressed as:

context Activity inv seq1: Activity::allInstances()-> select(type="B")->forAll(b |b.previous->size()=1 and previous->exists(a|a.type="A" and a.status='completed'))

Note that the constraints we obtain are more complex, and also the model becomes much less readable since now it is not trivial to detect all constraints affecting a particular activity type; indeed all constraints are attached to the Activity concept and are therefore mixed.

A Maximal Workflow-Extended Conceptual Schema

As an opposite approach, we could prefer to sacrifice the size of model in order to get a simpler translation of the process constraints.

In this maximal workflow-extended conceptual schema (Figure 14), the set of workflow-dependent subtypes includes the definition of an entity type for each activity and, additionally, a different entity type for each gateway. Each gateway type is related to the activity types corresponding to the activities linked by the gateway in the workflow model. All gateway types are defined as subtypes of the Gateway supertype, which includes a type attribute with information on the gateway kind: AND-merge, AND-split, OR-merge, and so forth.

On the one hand, this gateway subtypes increase the size of the workflow-extended schema and complicate its management since now the system must take care of creating at run-time the appropriate instances of the gateway types whenever one of their incoming activities are completed (split gateways are automatically created as completed gateways; merge gateways are declared completed when all required incoming activities have finished).

On the other hand, process constraints can be now defined in terms of the gateways, which results in a more clear and readable definition of the constraints. That is, if an activity is affected by several gateways (for instance, an activity may be the outgoing activity of an AND-merge and the initial activity for an AND-split), the set of constraints of each gateway is attached to the corresponding gateway type instead of being mixed altogether in the activity type.

Additionally, some of the OCL constraints can be avoided because they are already enforced by the model definition itself. For instance, one of the common constraints for all split gateways among an activity A and a set of B1 ···Bn activities states that the previous activity for a Bi activity must be unique, of type A, and completed. The first two conditions are ensured in this maximal model due to the associations (and multiplicities) between the B1 ···Bn activities and the split gateway type and between the gateway and the A type. The condition that the A activity must be completed still needs to be defined as an OCL constraint, which could be expressed as simply as follows (Split1 is assumed to be the type corresponding to the split gateway):

contextSplit1 inv: self.previous.completed

This situation is illustrated in the example of Figure 14 showing the AND-Split between ShipOrder, ReceiveGoods and SendInvoice. Note that Received-Goods and SendInvoice instances must be related with an instance of the gateway, which, in turn, must be related with an instance of ShipOrder. This guarantees that ReceiveGoodsand SendInvoice instances cannot be executed without creating first a ShipOrder instance.

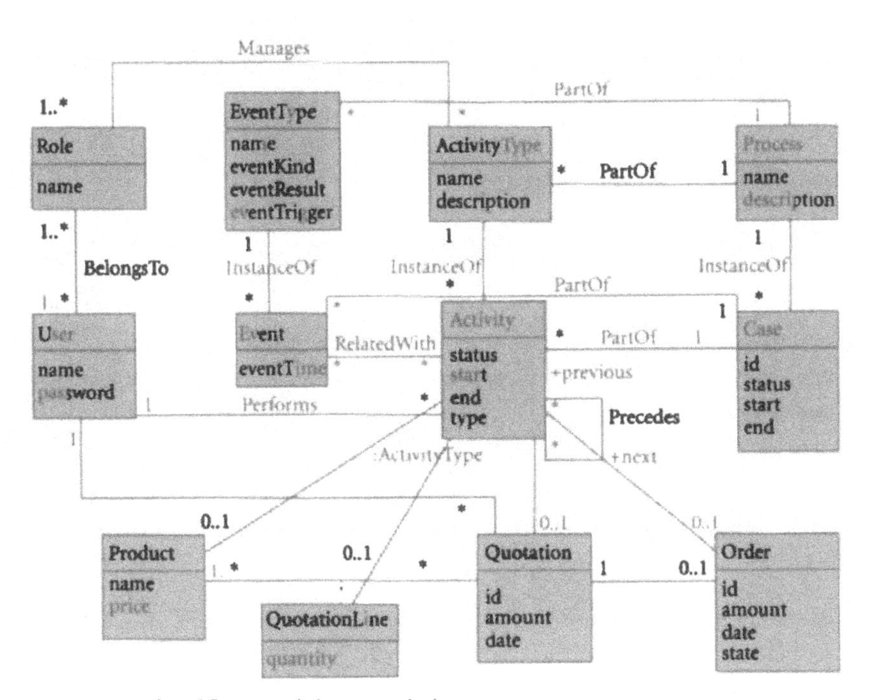

Figure 13. A minimal workflow-extended conceptual schema.

Appendix B

The application of the translation patterns over the workflow schema of Figure 4 produces the workflow-extended conceptual schema of Figure 5 plus the following set of process constraints expressed in OCL.

To simplify its presentation, constraints are grouped according to the main activity they affect. For each constraint we also indicate the workflow construct generating the constraint. Apart from the constraints specific for each activity, all activity instances must not start before the occurrence of a start event or after the occurrence of a terminate end event, as already seen in Section 7.6.

Ask Quotation Activity

(i) Constraints due to the start event

(a) A single AskQuotation activity instance must eventually exist for each issued Start event:

context Event inv: eventType.name ='Start' and case.status='completed' implies case.activity->select(a | a.oclIsTypeOf(AskQuotation) and a.event= self) ->size()=1

Provide Quotation Activity

(i) Constraints due to the XOR-Merge

(a) The preceding activity must be of type AskQuotation or Change Quotation and must be completed:

>exists(a|(a.oclIsTypeOf(AskQuotation) or a.oclIsTypeOf(ChangeQuotation)) and a.status= 'completed')

(b) No two instances may be related with the same Ask Quotation or Change Quotation instance. Note that when we iterate over the loop between Change Quotation and Provide Quotation activities, new activity instances are generated in each iteration:

context ProvideQuotation inv: ProvideQuotation.all-Instances()-> isUnique(previous)

(c) A Provide Quotationinstance follows each completed AskQuotation or Change Quotation activity:

context Case inv: status='completed' implies activity ->select(b | b.oclIsTypeOf(AskQuotation) or ... or b.oclIsTypeOf(ChangeQuotation))-> forAll (b |b.next ->exists(a |a.oclIsTypeOf(ProvideQuotation)))

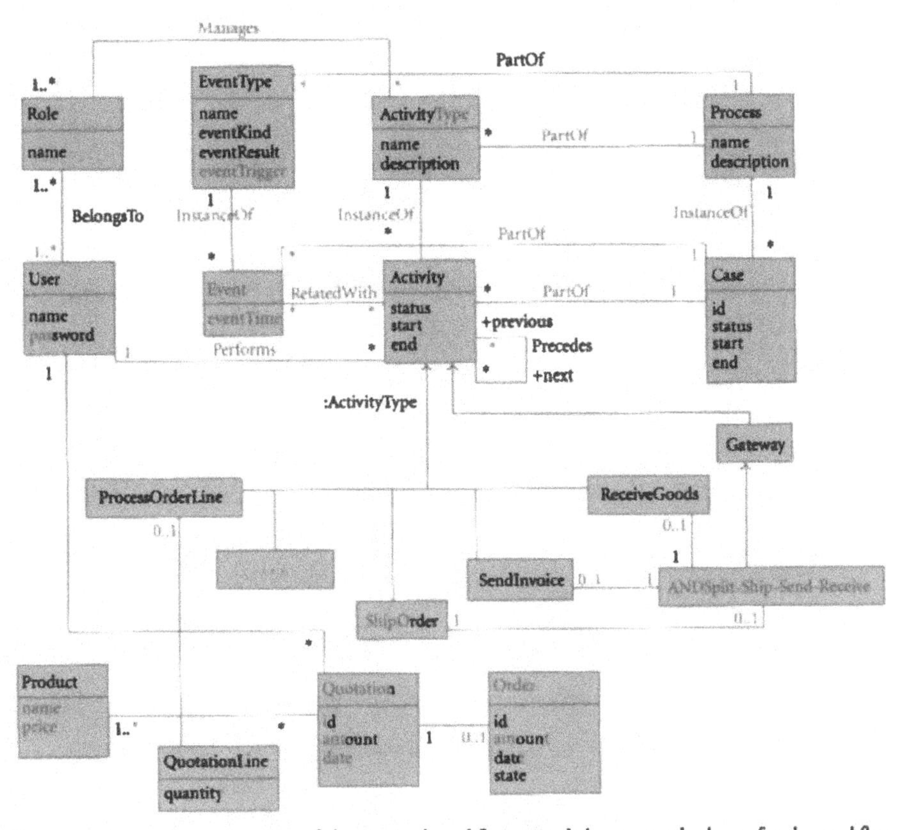

Figure 14. A partial representation of the maximal workflow-extended conceptual schema for the workflow model of Figure 3, showing the new entity type for one of the workflow gateways.

(ii) Constraints due to the XOR-split

(a) The next activity must be either another Change Quotation instance or a Submit Order instance, but not both

contextProvideQuotationinv: next->select(a |a.oclIsTypeOf(ChangeQuotation) or a.oclIsTypeOf(Submit-Order))-> size ()<=1

(b) If the Provide Quotation instance is completed, a Change Quotation or a SubmitOrder must necessarily be created before ending the case *contextCaseinv:status='completed'impliesactivity->select(a|a.status='completed' and a.oclIsTypeOf(ProvideQuotation))-> forAll (a | a.next -> exists (b | (b.oclIsTypeOf(ChangeQuotation) or b .oclIsTypeOf(SubmitOrder)) andb.start ≥a.end))*

(c) Only Change Quotation activity instances or Submit Order instances may follow a Provide Quotation instance

contextProvideQuotationinv: next->forAll(b|b.oclIsTypeOf(ChangeQuotation) orb.oclIsTypeOf(SubmitOrder))

Change Quotation Activity

(i) Constraints due to outgoing flow from the ProvideQuo-tationXOR-split

(a) The previous activity must be of type ProvideQuotation and must be completed

context ChangeQuotation inv: previous->size()=1 and previous->exists(a | a.status= 'completed' and a.oclIsTypeOf(ProvideQuotation))

(b) No two instances of Change Quotation may be related with the same Provide Quotation instance

context ChangeQuotation inv: ChangeQuotation.all-Instances()-> is Unique(previous)

(ii) Constraints due to the subsequent XOR-merge

(a) The next activity must be of type ProvideQuotation

contextChangeQuotationinv:next-> forAll(a |a.oclIsTypeOf(ProvideQuotation))

Submit Order Activity

(i) Constraints due to outgoing flow from the Provide QuotationXOR-split

(a) The previous activity must be of type Provide Quotation and must be completed

context SubmitOrder inv: previous->size()=1 and previous->exists(a | a.status='completed' and a.oclIsTypeOf(ProvideQuotation))

(b) No two instances of SubmitOrder may be related with the same Provide Quotationinstance

context SubmitOrder inv: SubmitOrder.allInstances() -> isUnique (previous)

(ii) Constraints due to the XOR-split between ChooseShipment and Standard-Shipment

(a) The next activity must be either of type Choose-Shipment or StandardShipment,but notboth

context SubmitOrder inv: next->select(a | a.oclIsTypeOf(ChooseShipment)or a.oclIsType Of(Standard-Shipment))-> size ()<=1

(b) If the Submit Order instance is completed, a Choose Shipment or a Standard Shipment activity must be created before ending the case

contextCaseinv: status='completed'impliesactivity->select(a|a.status='completed' and a.oclIsTypeOf(SubmitOrder))-> forAll (a |a.next-> exists(b |(b.oclIs-TypeOf(Choose Shipment) or b.oclIs-TypeOf(StandardShipment)) andb. start>=a.end))

(c) Only ChooseShipment or StandardShipment activity instances may follow a SubmitOrderinstance

context SubmitOrder inv: next->forAll(b | b.oclIsTypeOf(ChooseShipment) or b.oclIsTypeOf(Standard-Shipment))

Standard Shipment Activity

(i) Constraints due to outgoing flow from the SubmitOrder XOR-split

(a) The previous activity must be of type Submit Order and must be completed

context StandardShipment inv: previous->size()= 1 and previous->exists(a | a.status='completed' and a.oclIsTypeOf(SubmitOrder))

(b) No two instances of Standard Shipment may be related with the same SubmitOrder instance

contextStandardShipmentinv:StandardShipment.all-Instances()-> isUnique(previous)

(ii) Constraints due to the subsequent XOR-merge

(a) The next activity must be of type ShipOrder

context StandardShipment inv: next-> forAll(a | a .oclIsTypeOf(ShipOrder))

Choose Shipment Activity

(i) Constraints due to outgoing flow from the SubmitOrder XOR-split

(a) The previous activity must be of type Submit Order and must be completed *context ChooseShipment inv: previous->size()=1 and previous->exists(a | a.status='completed' and a.oclIsTypeOf(SubmitOrder))*

(b) No two instances of Standard Shipment may be related with the same SubmitOrder instance

context ChooseShipment inv: Choose Shipment.all-Instances()-> isUnique(previous)

(ii) Constraints due to the AND-split between Arrange Transport and Process OrderLine

(a) For each Choose Shipment activity, the Arrange Transport and the Process OrderLineactivities must be executed

context Case inv: status='completed' implies activity ->select(a | a.status='completed' and a.oclIsTypeOf(ChooseShipment))-> forAll(a | a.next-> exists(b |b.oclIsTypeOf(ArrangeTransport)) and a.next-> exists(b |b.oclIsTypeO f(ProcessOrderLine)))

(b) Only Arrange Transport activity instances or Process OrderLine instances may follow a Choose Shipment instance

context ChooseShipment inv: next->forAll(b | b.oclIsTypeOf(ArrangeTransport) or b.oclIsTypeOf(Process-OrderLine))

Arrange Transport Activity

(i) Constraints due to the outgoing flow of the Choose Shipment AND-split

(a) The previous activity must be of type Choose Shipment and must be completed

context ArrangeTransport inv: previous->size()= 1 and previous->exists(a | a.oclIsTypeOf(ChooseShipment) and a.status='completed')

(b) No two instances of ArrangeTransport may be related with the same ChooseShipment

context Arrange Transport inv: ArrangeTransport.all-Instances()-> isUnique(previous)

(ii) Constraints due to the subsequent AND-merge

(a) The next activity must be of type EmptyActivity1

context ArrangeTransport inv: next-> forAll(a | a .oclIsTypeOf(EmptyActivity1))

Process Order Line Activity

(i) Constraints due to the outgoing flow of the ChooseShipment AND-split

(a) The previous activity must be of type Choose Shipment and must be completed

contextProcessOrderLine inv:previous->size()=1 and previous->exists(a | a.oclIsTypeOf(ChooseShipment) and a.status='completed')

(b) No two instances of Process OrderLinemay be related with the same ChooseShipment instance

context ProcessOrderline invProcessOrderline.allInstances()->isUnique (previous)

(ii) Constraints due to the multi-instance loop

(a) There must exist a Process OrderLineinstance for each OrderLine of the order related with the activity

context Case inv: (activity->select(a |a.oclIsTypeOf(ProcessOrderLine))-> size()) mod (ProcessOrderLine.allInstances()-> any(p |p.case=self).order.quota-tion. quotationLine ->size()) =0

Empty Activity 1 Activity

(i) Constraints due to the AND-Merge

(a) We cannot start (and complete) an Empty Activity1 instance until the Arrange Transport activity and all required Process OrderLine instances have been executed

context EmptyActivity1 inv: previous->exists(b |b.oclIsTypeOf(Arra ngeTransport) (and b.status = 'completed') and previous->select(b | b.oclIsTypeOf(ProcessOrderLine) and b.status='completed')-> size()= self.order. quotation.orderLines->size()

(b) An Empty Activity1 instance must eventually exist if the ArrangeTransportand Process OrderLineactivities have been issued

contextCaseinv:status='completed'implies not(activity->exists (b | b.oclIsTypeOf(ArrangeTransport) and b.status='completed'and notb. next->exists(a|a.oclIsTypeOf(EmptyActivity1))) and activity-> exists(b | b .oclIsTypeOf(ProcessOrderLine) and b.status='completed'and notb.next- >exists(a |a.oclIsType-Of(EmptyActivity1))))

(c) The previous instances of two different Empty Activity1 instances must have an empty intersection *contextEmptyActivity1inv:EmptyActivity1.allInstances()->forAll(s1,s2 | s1 <> s2 implies s1.previous ->intersection(s2.previous)->isEmpty())*

(ii) Constraints due to the subsequent XOR-merge

(a) The next activity must be of type ShipOrder

context EmptyActivity1 inv: next-> forAll(a | a.oclIsTypeOf(ShipOrder))

Ship Order Activity

(i) Constraints due to the XOR-Merge

(a) The preceding activity must be of type Standard Shipment or EmptyActivity1and must be completed

context ShipOrder inv: previous->size()=1 and previous->exists(a | (a.oclIsTypeOf(StandardShipment) or a.oclIsTypeOf(EmptyActivity1)) and a.status='completed')

(b) No two instances may be related with the same previous StandardShipment or EmptyActivity1 instances

context ShipOrder inv: ShipOrder.allInstances()-> isUnique(previous)

(c) A Ship Order instance follows completed Standard Shipment orEmptyQuotation1activities

context Case inv: status='completed' implies activity>select(b | b.oclIsTypeOf(StandardShipment) or...or b.oclIsTypeOf(EmptyActivity1)) -> forAll(b | b.next>exists (a |a.oclIsTypeOf(ShipOrder)))

(ii) Constraints due to following AND-split

(a) For each ShipOrder activity, the Send invoice and the ReceiveGoodsactivities must be executed

contextCaseinv: status='completed'implies activity-> select(a |a.status='completed'and a.oclIsTypeOf(Ship-Order))-> forAll(a |a.next ->exists (b | b.oclIsTypeOf(SendInvoice)) and a.next->exists(b | b.oclIsTypeOf(ReceiveGoods)))

(b) Only SendInvoice activity instances or ReceiveGoods instances may follow a ShipOrder instance context ShipOrder inv: next->forAll(b | b.oclIsTypeOf(SendInvoice) orb.oclIsTypeOf(ReceiveGoods))

Send Invoice Activity

(i) Constraints due to the outgoing flow of the Ship Order AND-split

(a) The previous activity must be of type ShipOrder and must be completed

context SendInvoice inv: previous->size()=1 and previous->exists(a |a.oclIsTypeOf(ShipOrder) and a.status='completed')

(b) No two instances of SendInvoice may be related with the same ShipOrder

context SendInvoice inv:SendInvoice.allInstances()-> isUnique(previous)

(ii) Constraints due to the subsequent AND-merge

(a) The next activity must be of type PayGoods

contextSendInvoice inv: next-> forAll(a |a.oclIsTypeOf(PayGoods))

Receive Goods Activity

(i) Constraints due to the outgoing flow of the Ship Order AND-split

(a) The previous activity must be of type ShipOrder and must be completed

contextReceiveGoodsinv:previous->size()=1 and previous->exists(a |a.oclIsTypeOf(ShipOrder) and a.status='completed')

(b) No two instances of ReceiveGoods may be related with the same Ship Order

context ReceiveGoods inv:ReceiveGoods.allInstances() ->isUnique(previous)

(ii) Constraints due to the subsequent AND-merge

(a) The next activity must be of type PayGoods

contextReceiveGoodsinv:next->forAll(a |a.oclIsTypeOf(PayGoods))

Pay Goods Activity

(i) Constraints due to the AND-Merge

(a) We cannot start a PayGoods instance until the Send Invoice and the ReceiveGoods activities have been executed

context PayGoods inv: previous->exists(b | b.oclIsTypeOf(SendInvoice) and b.status='completed') and previous->exists(b | b.oclIsTypeOf(ReceiveGoods) and (b.status='completed')

(b) APay Goods instance must eventually exist if the Send Invoice and the Receive Goods activities have been issued *context Case inv: status='completed' implies not (activity->exists(b | b.oclIsTypeOf(SendInvoice) and b.status='completed' and not b.next->exists(a | a .oclIsTypeOf(PayGoods))) and activity->exists(b |*

b.oclIsTypeOf(ReceiveGoods) andb.status='completed' and notb.next->exists(a |a.oclIsTypeOf(PayGoods))))

(c) The previous instances of two different Pay Good activities must have an empty intersection.

context PayGoods inv: PayGoods.allInstances()-> forAll(s1,s2 |s1 <> s2 implies s1.previous-> intersection(s2.previous)->isEmpty())

Acknowledgements

This work has been partially supported by the Italian grant FAR N. 4412/ICT, the Spanish-Italian integrated action HI2006-0208, and the Spanish Research Project TIN2008-00444.

References

1. ISO/TC97/SC5/WG3, "Concepts and Terminology for the Conceptual Schema and Information Base," ISO, 1982.
2. OMG/BPMI, "Business Process Management Notation v.1," OMG Adopted Specification, 2006.
3. Oracle, "Workflow," www.oracle.com/technology/products/ias/workflow/index.html.
4. IBM, "WebSphere MQ Workflow," http://www-01.ibm.com/software/integration/wmqwf/.
5. J. Bae, H. Bae, S.-H. Kang, and Y. Kim, "Automatic control of workflow processes using ECA rules," IEEE Transactions on Knowledge and Data Engineering, vol. 16, no. 8, pp. 1010–1023, 2004.
6. M. Brambilla, S. Ceri, P. Fraternali, and I. Manolescu, "Process modeling in web applications," ACM Transactions on Software Engineering and Methodology, vol. 15, no. 4, pp. 360–409, 2006.
7. J. Koehler, R. Hauser, S. Sendall, and M. Wahler, "Declarative techniques for model-driven business process integration," IBM Systems Journal, vol. 44, no. 1, pp. 47–65, 2005.

8. OMG, "UML 2.0 OCL Specification," OMG Adopted Specification (ptc/03-10-14), 2003.
9. OMG, "Business Process Definition Metamodel (BPDM)," OMG Standard, dtc/2007-07-01, 2007.
10. C. Combi and G. Pozzi, "Temporal conceptual modelling of workflows," in Proceedings of the 22nd International Conference on Conceptual Modeling (ER '03), vol. 2813 of Lecture Notes in Computer Science, pp. 59–76, 2003.
11. OMG, "UML 2.0 Superstructure Specification," OMG Adopted Specification (ptc/03-08-02), 2003.
12. M. Brambilla, J. Cabot, and S. Comai, "Automatic generation of workflow-extended domain models," in Proceedings of the 10th International Conference on Model Driven Engineering Languages and Systems (MoDELS '07), vol. 4735 of Lecture Notes in Computer Science, pp. 375–389, 2007.
13. OMG, "MOF Core Specification," OMG Available Specification (formal/06-01-01), 2006.
14. J. Mendling, G. Neumann, and M. Nüttgens, "A comparison of XML interchange formats for business process modelling," in Workflow Handbook, Workflow Management Coalition, 2005.
15. B. List and B. Korherr, "An evaluation of conceptual business process modelling languages," in Proceedings of the ACM Symposium on Applied Computing, vol. 2, pp. 1532–1539, 2006.
16. S. A. White, Process Modeling Notations and Workflow Patterns, BPTrends, 2004.
17. W. M. P. van der Aalst, M. Weske, and G. Wirtz, "Advanced topics in workflow management: issues, requirements and solutions," Journal of Integrated Design and Process Science, vol. 7, pp. 49–77, 2003.
18. J. Cabot and R. Raventós, "Conceptual modelling patterns for roles," Journal on Data Semantics V, pp. 158–184, 2006.
19. D. Costal, C. Gómez, A. Queralt, R. Raventós, and E. Teniente, "Facilitating the definition of general constraints in UML," in Proceedings of the 9th International Conference on Model Driven Engineering Languages and Systems (MoDELS '06), vol. 4199 of Lecture Notes in Computer Science, pp. 260–274, 2006.
20. M. T. Wynn, D. Edmond, W. M. P. van der Aalst, and A. H. M. ter Hofstede, "Achieving a general, formal and decidable approach to the OR-join in workflow using Reset nets," in Proceedings of the 26th International Conference

on Application and Theory of Petri Nets (ICATPN '06), vol. 3536 of Lecture Notes in Computer Science, pp. 423–443, 2006.

21. A. Olivé, "A method for the definition of integrity constraints in object-oriented conceptual modeling languages," Data and Knowledge Engineering, vol. 59, no. 3, pp. 559–575, 2006.
22. M. Brambilla and J. Cabot, "Constraint tuning and management for web applications," in Proceedings of the Tool Presentation at 6th International Conference on Web Engineering (ICWE '06), pp. 345–352, 2006.
23. S. Ceri, P. Fraternali, A. Bongio, M. Brambilla, S. Comai, and M. Matera, Designing Data-Intensive Web Applications, Morgan Kaufmann, 2002.
24. B. Demuth, H. Hussmann, and S. Loecher, "OCL as a specification language for business rules in database applications," in Proceedings of the 4th International Conference on the Unified Modeling Language (UML '01), vol. 2185 of Lecture Notes in Computer Science, pp. 104–117, 2001.
25. KlasseObjecten, "Octopus: OCL Tool for Precise Uml Specifications," http://www.klasse.nl/octopus/index.html.
26. J. Cabot and E. Teniente, "Incremental evaluation of OCL constraints," in Proceedings of the 18th International Conference on Advanced Information Systems Engineering (CAiSE '06), vol. 4001 of Lecture Notes in Computer Science, pp. 81–95, 2006.
27. O. Pastor, J. Fons, V. Pelechano, and S. Abrahão, "Conceptual modelling of web applications: the OOWS approach," in Web Engineering, pp. 277–302, Springer, New York, NY, USA, 2006.
28. M. Brambilla, "Extending hypertext conceptual models with process-oriented primitives," in Proceedings of the 22nd International Conference on Conceptual Modeling (ER '03), vol. 2813 of Lecture Notes in Computer Science, pp. 246–262, 2003.
29. NoMagicInc., "MagicDraw UML v. 10.5," http://www.magicdraw.com/.
30. OMG, "XML Metadata Interchange (XMI) Specification v.2.0," OMG Adopted Specification (formal/03-05-02), 2003.
31. M. Brambilla, "Generation of WebML web application models from business process specification," in Proceedings of the Tool Presentation at 6th International Conference on Web Engineering (ICWE '06), pp. 85–86, 2006.
32. R. Eshuis and R. Wieringa, "Verification support for workflow design with UML activity graphs," in Proceedings of the 22nd International Conference on Software Engineering (ICSE '02), pp. 166–176, 2002.

33. W. Hur, J.-Y. Jung, H. Kim, and S.-H. Kang, "Model-driven approach to workflow execution," in Proceedings of the 2nd International Conference on Business Process Management (BPM '04), vol. 3080 of Lecture Notes in Computer Science, pp. 261–273, 2004.
34. M. La Rosa, F. Gottschalk, M. Dumas, and W. M. P. van der Aalst, "Linking domain models and process models for reference model configuration," in Proceedings of the Business Process Management Workshop, vol. 4928 of Lecture Notes in Computer Science, pp. 417–430, 2008.
35. G. Wirtz, M. Weske, and H. Giese, "Extending UML with workflow modeling capabilities," in Proceedings of the 7th International Conference on Cooperative Information Systems (CoopIS '00), vol. 1901 of Lecture Notes in Computer Science, pp. 30–41, 2000.
36. M. Dumas and A. H. Hofstede, "UML activity diagrams as a workflow specification language," in Proceedings of the 4th International Conference on the Unified Modeling Language (UML '01), vol. 2185 of Lecture Notes in Computer Science, pp. 76–90, 2001.
37. P. Hruby, "Specification of workflow management systems with UML," in Proceedings of the Workshop on Implementation and Application of Object-Oriented Workflow Management Systems (OOPSLA '98), 1998.
38. M. Brambilla, A. Deutsch, L. Sui, and V. Vianu, "The role of visual tools in a web application design and verification framework: a visual notation for LTL formulae," in Proceedings of the Tool Presentation at 5th International Conference on Web Engineering (ICWE '05), vol. 3579 of Lecture Notes in Computer Science, pp. 557–568, 2005.
39. D. Domingos, A. Rito-Silva, and P. Veiga, "Workflow access control from a business perspective," in Proceedings of the 6th International Conference on Enterprise Information Systems (ICEIS '04), vol. 3, pp. 18–25, 2004.
40. W. M. P. van der Aalst and A. Kumar, "A reference model for team-enabled workflow management systems," Data and Knowledge Engineering, vol. 38, no. 3, pp. 335–363, 2001.
41. A. Knapp, N. Koch, G. Zhang, and H.-M. Hassler, "Modeling business processes in web applications with argoUWE," in Proceedings of the 7th International Conference on the Unified Modeling Language (UML '04), vol. 3273 of Lecture Notes in Computer Science, pp. 69–83, 2004.
42. T. Takemura and T. Tamai, "Rigorous business process modeling with OCL," in Proceedings of the OCL Workshop in Model Driven Engineering Languages and Systems (MoDELS '06), 2006.

Software Test Automation in Practice: Empirical Observations

Jussi Kasurinen, Ossi Taipale and Kari Smolander

ABSTRACT

The objective of this industry study is to shed light on the current situation and improvement needs in software test automation. To this end, 55 industry specialists from 31 organizational units were interviewed. In parallel with the survey, a qualitative study was conducted in 12 selected software development organizations. The results indicated that the software testing processes usually follow systematic methods to a large degree, and have only little immediate or critical requirements for resources. Based on the results, the testing processes have approximately three fourths of the resources they need, and have access to a limited, but usually sufficient, group of testing tools. As for the test automation, the situation is not as straightforward: based on our study, the applicability of test automation is still limited and its adaptation to testing contains practical difficulties in usability. In this study, we analyze and discuss these limitations and difficulties.

Introduction

Testing is perhaps the most expensive task of a software project. In one estimate, the testing phase took over 50% of the project resources [1]. Besides causing immediate costs, testing is also importantly related to costs related to poor quality, as malfunctioning programs and errors cause large additional expenses to software producers [1, 2]. In one estimate [2], software producers in United States lose annually 21.2 billion dollars because of inadequate testing and errors found by their customers. By adding the expenses caused by errors to software users, the estimate rises to 59.5 billion dollars, of which 22.2 billion could be saved by making investments on testing infrastructure [2]. Therefore improving the quality of software and effectiveness of the testing process can be seen as an effective way to reduce software costs in the long run, both for software developers and users.

One solution for improving the effectiveness of software testing has been applying automation to parts of the testing work. In this approach, testers can focus on critical software features or more complex cases, leaving repetitive tasks to the test automation system. This way it may be possible to use human resources more efficiently, which consequently may contribute to more comprehensive testing or savings in the testing process and overall development budget [3]. As personnel costs and time limitations are significant restraints of the testing processes [4, 5], it also seems like a sound investment to develop test automation to get larger coverage with same or even smaller number of testing personnel. Based on market estimates, software companies worldwide invested 931 million dollars in automated software testing tools in 1999, with an estimate of at least 2.6 billion dollars in 2004 [6]. Based on these figures, it seems that the application of test automation is perceived as an important factor of the test process development by the software industry.

The testing work can be divided into manual testing and automated testing. Automation is usually applied to running repetitive tasks such as unit testing or regression testing, where test cases are executed every time changes are made [7]. Typical tasks of test automation systems include development and execution of test scripts and verification of test results. In contrast to manual testing, automated testing is not suitable for tasks in which there is little repetition [8], such as explorative testing or late development verification tests. For these activities manual testing is more suitable, as building automation is an extensive task and feasible only if the case is repeated several times [7, 8]. However, the division between automated and manual testing is not as straightforward in practice as it seems; a large concern is also the testability of the software [9], because every piece of code can be made poorly enough to be impossible to test it reliably, therefore making it ineligible for automation.

Software engineering research has two key objectives: the reduction of costs and the improvement of the quality of products [10]. As software testing represents a significant part of quality costs, the successful introduction of test automation infrastructure has a possibility to combine these two objectives, and to overall improve the software testing processes. In a similar prospect, the improvements of the software testing processes are also at the focus point of the new software testing standard ISO 29119 [11]. The objective of the standard is to offer a company-level model for the test processes, offering control, enhancement and follow-up methods for testing organizations to develop and streamline the overall process.

In our prior research project [4, 5, 12–14], experts from industry and research institutes prioritized issues of software testing using the Delphi method [15]. The experts concluded that process improvement, test automation with testing tools, and the standardization of testing are the most prominent issues in concurrent cost reduction and quality improvement. Furthermore, the focused study on test automation [4] revealed several test automation enablers and disablers which are further elaborated in this study. Our objective is to observe software test automation in practice, and further discuss the applicability, usability and maintainability issues found in our prior research. The general software testing concepts are also observed from the viewpoint of the ISO 29119 model, analysing the test process factors that create the testing strategy in organizations. The approach to achieve these objectives is twofold. First, we wish to explore the software testing practices the organizations are applying and clarify the current status of test automation in the software industry. Secondly, our objective is to identify improvement needs and suggest improvements for the development of software testing and test automation in practice. By understanding these needs, we wish to give both researchers and industry practitioners an insight into tackling the most hindering issues while providing solutions and proposals for software testing and automation improvements.

The study is purely empirical and based on observations from practitioner interviews. The interviewees of this study were selected from companies producing software products and applications at an advanced technical level. The study included three rounds of interviews and a questionnaire, which was filled during the second interview round. We personally visited 31 companies and carried out 55 structured or semistructured interviews which were tape-recorded for further analysis. The sample selection aimed to represent different polar points of the software industry; the selection criteria were based on concepts such as operating environments, product and application characteristics (e.g., criticality of the products and applications, real time operation), operating domain and customer base.

The paper is structured as follows. First, in Section 2 we introduce comparable surveys and related research. Secondly, the research process and the qualitative and quantitative research methods are described in Section 3. Then the survey results are presented in Section 4 and the interview results are presented in Section 5. Finally, the results and observations and their validity are discussed in Section 6 and closing conclusions are discussed in Section 7.

Related Research

Besides our prior industry-wide research in testing [4, 5, 12–14], software testing practices and test process improvement have also been studied by others, like Ng et al. [16] in Australia. Their study applied the survey method to establish knowledge on such topics as testing methodologies, tools, metrics, standards, training and education. The study indicated that the most common barrier to developing testing was the lack of expertise in adopting new testing methods and the costs associated with testing tools. In their study, only 11 organizations reported that they met testing budget estimates, while 27 organizations spent 1.5 times the estimated cost in testing, and 10 organizations even reported a ratio of 2 or above. In a similar vein, Torkar and Mankefors [17] surveyed different types of communities and organizations. They found that 60% of the developers claimed that verification and validation were the first to be neglected in cases of resource shortages during a project, meaning that even if the testing is important part of the project, it usually is also the first part of the project where cutbacks and downscaling are applied.

As for the industry studies, a similar study approach has previously been used in other areas of software engineering. For example, Ferreira and Cohen [18] completed a technically similar study in South Africa, although their study focused on the application of agile development and stakeholder satisfaction. Similarly, Li et al. [19] conducted research on the COTS-based software development process in Norway, and Chen et al. [20] studied the application of open source components in software development in China. Overall, case studies covering entire industry sectors are not particularly uncommon [21, 22]. In the context of test automation, there are several studies and reports in test automation practices (such as [23–26]). However, there seems to be a lack of studies that investigate and compare the practice of software testing automation in different kinds of software development organizations.

In the process of testing software for errors, testing work can be roughly divided into manual and automated testing, which both have individual strengths and weaknesses. For example, Ramler and Wolfmaier [3] summarize the difference between manual and automated testing by suggesting that automation should be used to prevent further errors in working modules, while manual testing is better suited for finding new and unexpected errors. However, how and where the test automation should be used is not so straightforward issue, as the application of test automation seems to be a strongly diversified area of interest. The application of test automation has been studied for example in test case generation [27, 28], GUI testing [29, 30] and workflow simulators [31, 32] to name a few areas. Also according to Bertolino [33], test automation is a significant area of interest in current testing research, with a focus on improving the degree of automation by developing advanced techniques for generating the test inputs, or by finding support procedures such as error report generation to ease the supplemental workload. According to the same study, one of the dreams involving software testing is 100% automated testing. However, for example Bach's [23] study observes that this cannot be achieved, as all automation ultimately requires human intervention, if for nothing else, then at least to diagnose results and maintain automation cases.

The pressure to create resource savings are in many case the main argument for test automation. A simple and straightforward solution for building automation is to apply test automation just on the test cases and tasks that were previously done manually [8]. However, this approach is usually unfeasible. As Persson and Yilmaztürk [26] note, the establishment of automated testing is a costly, high risk project with several real possibilities for failure, commonly called as "pitfalls." One of the most common reasons why creating test automation fails, is that the software is not designed and implemented for testability and reusability, which leads to architecture and tools with low reusability and high maintenance costs. In reality, test automation sets several requisites on a project and has clear enablers and disablers for its suitability [4, 24]. In some reported cases [27, 34, 35], it was observed that the application of test automation with an ill-suited process model may be even harmful to the overall process in terms of productivity or cost-effectiveness.

Models for estimating testing automation costs, for example by Ramler and Wolfmaier [3], support decision-making in the tradeoff between automated and manual testing. Berner et al. [8] also estimate that most of the test cases in one project are run at least five times, and one fourth over 20 times. Therefore the test cases, which are done constantly like smoke tests, component tests and integration tests, seem like ideal place to build test automation. In any case, there seems to be a consensus that test automation is a plausible tool for enhancing quality, and consequently, reducing the software development costs in the long run if used correctly.

Our earlier research on the software test automation [4] has established that test automation is not as straightforward to implement as it may seem. There are several characteristics which enable or disable the applicability of test automation. In this study, our decision was to study a larger group of industry organizations and widen the perspective for further analysis. The objective is to observe, how the companies have implemented test automation and how they have responded to the issues and obstacles that affect its suitability in practice. Another objective is to analyze whether we can identify new kind of hindrances to the application of test automation and based on these findings, offer guidelines on what aspects should be taken into account when implementing test automation in practice.

Research Process

Research Population and Selection of the Sample

The population of the study consisted of organization units (OUs). The standard ISO/IEC 15504-1 [36] specifies an organizational unit (OU) as a part of an organization that is the subject of an assessment. An organizational unit deploys one or more processes that have a coherent process context and operates within a coherent set of business goals. An organizational unit is typically part of a larger organization, although in a small organization, the organizational unit may be the whole organization.

The reason to use an OU as the unit for observation was that we wanted to normalize the effect of the company size to get comparable data. The initial population and population criteria were decided based on the prior research on the subject. The sample for the first interview round consisted of 12 OUs, which were technically high level units, professionally producing software as their main process. This sample also formed the focus group of our study. Other selection criteria for the sample were based on the polar type selection [37] to cover different types of organizations, for example different businesses, different sizes of the company, and different kinds of operation. Our objective of using this approach was to gain a deep understanding of the cases and to identify, as broadly as possible, the factors and features that have an effect on software testing automation in practice.

For the second round and the survey, the sample was expanded by adding OUs to the study. Selecting the sample was demanding because comparability is not determined by the company or the organization but by comparable processes in the OUs. With the help of national and local authorities (the network of the

Finnish Funding Agency for Technology and Innovation) we collected a population of 85 companies. Only one OU from each company was accepted to avoid the bias of over-weighting large companies. Each OU surveyed was selected from a company according to the population criteria. For this round, the sample size was expanded to 31 OUs, which also included the OUs from the first round. The selection for expansion was based on probability sampling; the additional OUs were randomly entered into our database, and every other one was selected for the survey. In the third round, the same sample as in the first round was interviewed. Table 1 introduces the business domains, company sizes and operating areas of our focus OUs. The company size classification is taken from [38].

Table 1. Description of the interviewed focus OUs.

OU	Business	Company size[1]/Operation
Case A	MES[1] producer and electronics manufacturer	Small/National
Case B	Internet service developer and consultant	Small/National
Case C	Logistics software developer	Large/National
Case D	ICT consultant	Small/National
Case E	Safety and logistics system developer	Medium/National
Case F	Naval software system developer	Medium/International
Case G	Financial software developer	Large/National
Case H	MES[1] producer and logistics service systems provider	Medium/International
Case I	SME[2] business and agriculture ICT service provider	Small/National
Case J	Modeling software developer	Large/International
Case K	ICT developer and consultant	Large/International
Case L	Financial software developer	Large/International

[1]Manufacturing Execution System; [2]Small and Medium-sized Enterprise, definition [38].

Interview Rounds

The data collection consisted of three interview rounds. During the first interview round, the designers responsible for the overall software structure and/or module interfaces were interviewed. If the OU did not have separate designers, then the interviewed person was selected from the programmers based on their role in the process to match as closely as possible to the desired responsibilities. The interviewees were also selected so that they came from the same project, or from positions where the interviewees were working on the same product. The interviewees were not specifically told not to discuss the interview questions together, but this behavior was not encouraged either. In a case where an interviewee asked for the questions or interview themes beforehand, the person was allowed access to them in order to prepare for the meeting. The interviews in all three rounds lasted about an hour and had approximately 20 questions related to the test processes or test organizations. In two interviews, there was also more than one person present.

The decision to interview designers in the first round was based on the decision to gain a better understanding on the test automation practice and to see whether our hypothesis based on our prior studies [4, 5, 12–14] and supplementing literature review were still valid. During the first interview round, we interviewed 12 focus OUs, which were selected to represent different polar types in the software industry. The interviews contained semi-structured questions and were tape-recorded for qualitative analysis. The initial analysis of the first round also provided ingredients for the further elaboration of important concepts for the latter rounds. The interview rounds and the roles of the interviewees in the case OUs are described in Table 2.

Table 2. Interviewee roles and interview rounds.

Round type	Number of interviews	Interviewee role	Description
(1) Semistructured	12 focus OUs	Designer or Programmer	The interviewee is responsible for software design or has influence on how software is implemented
(2) Structured/ Semistructured	31 OUs quantitative, including 12 focus OUs qualitative	Project or testing manager	The interviewee is responsible for software development projects or test processes of software products
(3) Semistructured	12 focus OUs	Tester	The interviewee is a dedicated software tester or is responsible for testing the software product

The purpose of the second combined interview and survey round was to collect multiple choice survey data and answers to open questions which were based on the first round interviews. These interviews were also tape-recorded for the qualitative analysis of the focus OUs, although the main data collection method for this round was a structured survey. In this round, project or testing managers from 31 OUs, including the focus OUs, were interviewed. The objective was to collect quantitative data on the software testing process, and further collect material on different testing topics, such as software testing and development. The collected survey data could also be later used to investigate observations made from the interviews and vice versa, as suggested in [38]. Managers were selected for this round, as they tend to have more experience on software projects, and have a better understanding of organizational and corporation level concepts and the overall software process beyond project-level activities.

The interviewees of the third round were testers or, if the OU did not have separate testers, programmers who were responsible for the higher-level testing tasks. The interviews in these rounds were also semi-structured and concerned the work of the interviewees, problems in testing (e.g., increasing complexity of the systems), the use of software components, the influence of the business orientation, testing resources, tools, test automation, outsourcing, and customer influence for the test processes.

The themes in the interview rounds remained similar, but the questions evolved from general concepts to more detailed ones. Before proceeding to the next interview round, all interviews with the focus OUs were transcribed and analyzed because new understanding and ideas emerged during the data analysis. This new understanding was reflected in the next interview rounds. The themes and questions for each of the interview rounds can be found on the project website http://www2.it.lut.fi/project/MASTO/.

Grounded Analysis Method

The grounded analysis was used to provide further insight into the software organizations, their software process and testing policies. By interviewing people of different positions from the production organization, the analysis could gain additional information on testing- and test automation-related concepts like different testing phases, test strategies, testing tools and case selection methods. Later this information could be compared between organizations, allowing hypotheses on test automation applicability and the test processes themselves.

The grounded theory method contains three data analysis steps: open coding, axial coding and selective coding. The objective for open coding is to extract the categories from the data, whereas axial coding identifies the connections between the categories. In the third phase, selective coding, the core category is identified and described [39]. In practice, these steps overlap and merge because the theory development process proceeds iteratively. Additionally, Strauss and Corbin [40] state that sometimes the core category is one of the existing categories, and at other times no single category is broad enough to cover the central phenomenon.

The objective of open coding is to classify the data into categories and identify leads in the data, as shown in Table 3. The interview data is classified to categories based on the main issue, with observation or phenomenon related to it being the codified part. In general, the process of grouping concepts that seem to pertain to the same phenomena is called categorizing, and it is done to reduce the number of units to work with [40]. In our study, this was done using ATLAS.ti software [41]. The open coding process started with "seed categories" [42] that were formed from the research question and objectives, based on the literature study on software testing and our prior observations [4, 5, 12–14] on software and testing processes. Some seed categories, like "knowledge management," "service-orientation" or "approach for software development" were derived from our earlier studies, while categories like "strategy for testing," "outsourcing," "customer impact" or "software testing tools" were taken from known issues or literature review observations.

Table 3. Open coding of the interview data.

Interview transcript	Codes, Category: Code
"Well, I would hope for *stricter control or management for implementing our testing strategy*, as I am not sure if our *testing covers everything and is it sophisticated enough*. On the other hand, we do have *strictly limited resources, so it can be enhanced only to some degree*, we cannot test everything. And perhaps, recently we have had, in the newest versions, some regression testing, going through all features, seeing if nothing is broken, but *in several occasions this has been left unfinished because time has run out*. So there, on that issue we should focus."	Enhancement proposal: Developing testing strategy Strategy for testing: Ensuring case coverage Problem: Lack of resources Problem: Lack of time

The study followed the approach introduced by Seaman [43], which notes that the initial set of codes (seed categories) comes from the goals of the study, the research questions, and predefined variables of interest. In the open coding, we added new categories and merged existing categories to others if they seemed unfeasible or if we found a better generalization. Especially at the beginning of the analysis, the number of categories and codes quickly accumulated and the total number of codes after open coding amounted to 164 codes in 12 different categories. Besides the test process, software development in general and test automation, these categories also contained codified observations on such aspects as the business orientation, outsourcing, and product quality.

After collecting the individual observations to categories and codes, the categorized codes were linked together based on the relationships observed in the interviews. For example, the codes "Software process: Acquiring 3rd party modules," "Testing strategy: Testing 3rd party modules," and "Problem: Knowledge management with 3rd party modules" were clearly related and therefore we connected them together in axial coding. The objective of axial coding is to further develop categories, their properties and dimensions, and find causal, or any kinds of, connections between the categories and codes.

For some categories, the axial coding also makes it possible to define dimension for the phenomenon, for example "Personification-Codification" for "Knowledge management strategy," where every property could be defined as a point along the continuum defined by the two polar opposites. For the categories that are given dimension, the dimension represented the locations of the property or the attribute of a category [40]. Obviously for some categories, which were used to summarize different observations like enhancement proposals or process problems, defining dimensions was unfeasible. We considered using dimensions for some categories like "criticality of test automation in testing process" or "tool sophistication level for automation tools" in this study, but discarded them later as they yielded only little value to the study. This decision was based on the observation that values of both dimensions were outcomes of the applied test automation strategy, having no effect on the actual suitability or applicability of test automation to the organization's test process.

Our approach for analysis of the categories included Within-Case Analysis and Cross-Case-Analysis, as specified by Eisenhardt [37]. We used the tactic of selecting dimensions and properties with within-group similarities coupled with inter-group differences [37]. In this strategy, our team isolated one phenomenon that clearly divided the organizations to different groups, and searched for explaining differences and similarities from within these groups. Some of the applied features were, for example, the application of agile development methods, the application of test automation, the size [38] difference of originating companies and service orientation. As for one central result, the appropriateness of OU as a comparison unit was confirmed based on our size difference-related observations on the data; the within-group- and inter-group comparisons did yield results in which the company size or company policies did not have strong influence, whereas the local, within-unit policies did. In addition, the internal activities observed in OUs were similar regardless of the originating company size, meaning that in our study the OU comparison was indeed feasible approach.

We established and confirmed each chain of evidence in this interpretation method by discovering sufficient citations or finding conceptually similar OU activities from the case transcriptions. Finally, in the last phase of the analysis, in selective coding, our objective was to identify the core category [40]?a central phenomenon?and systematically relate it to other categories and generate the hypothesis and the theory. In this study, we consider test automation in practice as the core category, to which all other categories were related as explaining features of applicability or feasibility.

The general rule in grounded theory is to sample until theoretical saturation is reached. This means (1) no new or relevant data seem to emerge regarding a category, (2) the category development is dense, insofar as all of the paradigm elements are accounted for, along with variation and process, and (3) the relationships between categories are well established and validated [40]. In our study, the saturation was reached during the third round, where no new categories were created, merged, or removed from the coding. Similarly, the attribute values were also stable, that is, the already discovered phenomena began to repeat themselves in the collected data. As an additional way to ensure the validity of our study and avoid validity threats [44], four researchers took part in the data analysis. The bias caused by researchers was reduced by combining the different views of the researchers (observer triangulation) and a comparison with the phenomena observed in the quantitative data (methodological triangulation) [44, 45].

The Survey Instrument Development and Data Collection

The survey method described by Fink and Kosecoff [46] was used as the research method in the second round. An objective for a survey is to collect information from people about their feelings and beliefs. Surveys are most appropriate when information should come directly from the people [46]. Kitchenham et al. [47] divide comparable survey studies into exploratory studies from which only weak conclusions can be drawn, and confirmatory studies from which strong conclusions can be drawn. We consider this study as an exploratory, observational, and cross-sectional study that explores the phenomenon of software testing automation in practice and provides more understanding to both researchers and practitioners.

To obtain reliable measurements in the survey, a validated instrument was needed, but such an instrument was not available in the literature. However, Dybå [48] has developed an instrument for measuring the key factors of success in software process improvement. Our study was constructed based on the key factors of this instrument, and supplemented with components introduced in the standards ISO/IEC 29119 [11] and 25010 [49]. We had the possibility to use the current versions of the new standards because one of the authors is a member of the JTC1/SC7/WG26, which is developing the new software testing standard. Based on these experiences a measurement instrument derived from the ISO/IEC 29119 and 25010 standards was used.

The survey consisted of a questionnaire (available at http://www2.it.lut.fi/project/MASTO/) and a face-to-face interview. Selected open-ended questions were located at the end of the questionnaire to cover some aspects related to our qualitative study. The classification of the qualitative answers was planned in advance.

The questionnaire was planned to be answered during the interview to avoid missing answers because they make the data analysis complicated. All the interviews were tape-recorded, and for the focus organizations, further qualitatively analyzed with regard to the additional comments made during the interviews. Baruch [50] also states that the average response rate for self-assisted questionnaires is 55.6%, and when the survey involves top management or organizational representatives the response rate is 36.1%. In this case, a self-assisted, mailed questionnaire would have led to a small sample. For these reasons, it was rejected, and personal interviews were selected instead. The questionnaire was piloted with three OUs and four private persons.

If an OU had more than one respondent in the interview, they all filled the same questionnaire. Arranging the interviews, traveling and interviewing took two months of calendar time. Overall, we were able to accomplish 0.7 survey interviews per working day on an average. One researcher conducted 80% of the interviews, but because of the overlapping schedules also two other researchers participated in the interviews. Out of the contacted 42 OUs, 11 were rejected because they did not fit the population criteria in spite of the source information, or it was impossible to fit the interview into the interviewee's schedule. In a few individual cases, the reason for rejection was that the organization refused to give an interview. All in all, the response rate was, therefore, 74%.

Testing and Test Automation in Surveyed Organizations

General Information of the Organizational Units

The interviewed OUs were parts of large companies (55%) and small and medium-sized enterprises (45%). The OUs belonged to companies developing information systems (11 OUs), IT services (5 OUs), telecommunication (4 OUs), finance (4 OUs), automation systems (3 OUs), the metal industry (2 OUs), the public sector (1?OU), and logistics (1?OU). The application domains of the companies are presented in Figure 1. Software products represented 63% of the turnover, and services (e.g., consulting, subcontracting, and integration) 37%.

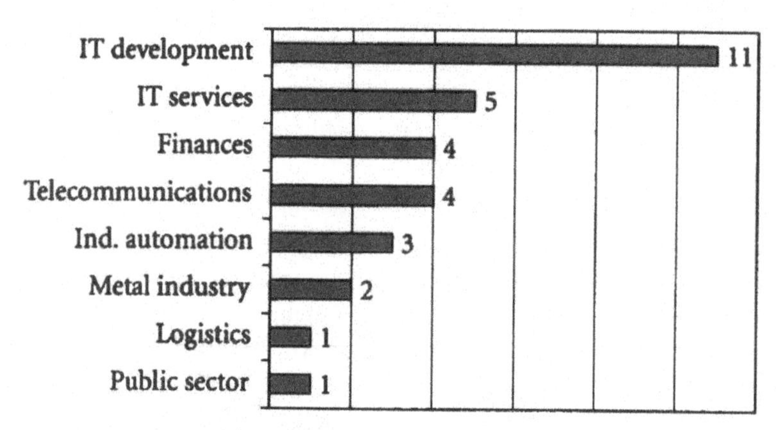

Figure 1. Application domains of the companies.

The maximum number of personnel in the companies to which the OUs belonged was 350, 000, the minimum was four, and the median was 315. The median of the software developers and testers in the OUs was 30 persons. OUs applied automated testing less than expected, the median of the automation in testing being 10%. Also, the interviewed OUs utilized agile methods less than expected: the median of the percentage of agile (reactive, iterative) versus plan driven methods in projects was 30%. The situation with human resources was better than what was expected, as the interviewees estimated that the amount of human resources in testing was 75%. When asking what percent of the development effort was spent on testing, the median of the answers was 25%. The cross-sectional situation of development and testing in the interviewed OUs is illustrated in Table 4.

Table 4. Interviewed OUs.

	Max.	Min.	Median
Number of employees in the company.	350 000	4	315
Number of SW developers and testers in the OU.	600	0[1]	30
Percentage of automation in testing.	90	0	10
Percentage of agile (reactive, iterative) versus plan driven methods in projects.	100	0	30
Percentage of existing testers versus resources need.	100	10	75
How many percent of the development effort is spent on testing?	70	0[2]	25

[1]0 means that all of the OUs developers and testers are acquired from 3rd parties.
[2]0 means that no project time is allocated especially for testing.

The amount of testing resources was measured by three figures; first the interviewee was asked to evaluate the percentage from total project effort allocated solely to testing. The survey average was 27%, the maximum being 70% and the minimum 0%, meaning that the organization relied solely on testing efforts carried out in parallel with development. The second figure was the amount of test resources compared to the organizational optimum. In this figure, if the company had two testers and required three, it would have translated as 66% of resources. Here the average was 70%; six organizations (19%) reported 100% resource availability. The third figure was the number of automated test cases compared to all of the test cases in all of the test phases the software goes through before its release. The average was 26%, varying between different types of organizations and project types. The results are presented in Figure 2, in which the qualitative study case OUs are also presented for comparison. The detailed descriptions for each case organization are available in the appendix.

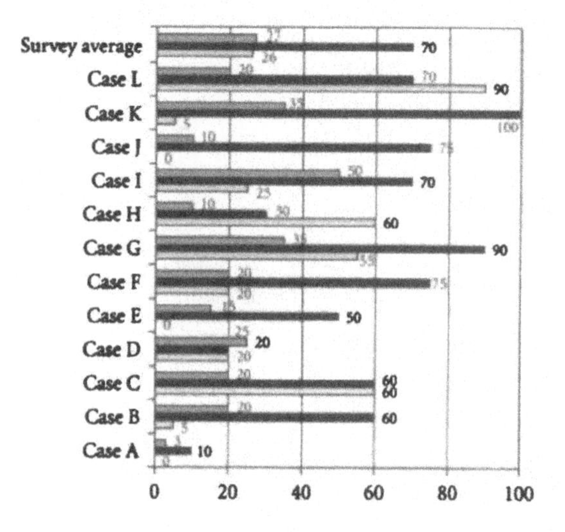

Figure 2. Amount of test resources and test automation in the focus organizations of the study and the survey average.

General Testing Items

The survey interviewed 31 organization managers from different types of software industry. The contributions of the interviewees were measured using a five-point Likert scale where 1 denoted "I fully disagree" and 5 denoted "I fully agree." The interviewees emphasized that quality is built in development (4.3) rather than in testing (2.9). Then the interviewees were asked to estimate their organizational testing practices according to the new testing standard ISO/IEC 29119 [11], which identifies four main levels for testing processes: the test policy, test strategy, test management and testing. The test policy is the company level guideline which defines the management, framework and general guidelines, the test strategy is an adaptive model for the preferred test process, test management is the control level for testing in a software project, and finally, testing is the process of conducting test cases. The results did not make a real difference between the lower levels of testing (test management level and test levels) and higher levels of testing (organizational test policy and organizational test strategy). All in all, the interviewees were rather satisfied with the current organization of testing. The resulted average levels from quantitative survey are presented in Figure 3.

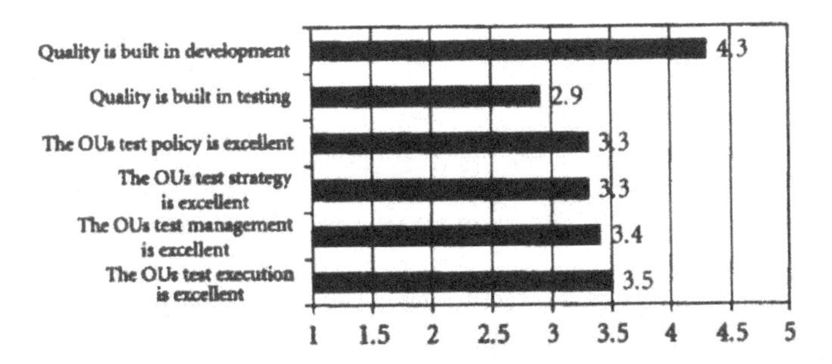

Figure 3. Levels of testing according to the ISO/IEC 29119 standard.

Besides the organization, the test processes and test phases were also surveyed. The five-point Likert scale with the same one to five–one being fully disagree and five fully agree–grading method was used to determine the correctness of different testing phases. Overall, the latter test phases—system, functional testing—were considered excellent or very good, whereas the low-level test phases such as unit testing and integration received several low-end scores. The organizations were satisfied or indifferent towards all test phases, meaning that there were no strong focus areas for test organization development. However, based on these results it seems plausible that one effective way to enhance testing would be to support low-level testing in unit and integration test phases. The results are depicted in Figure 4.

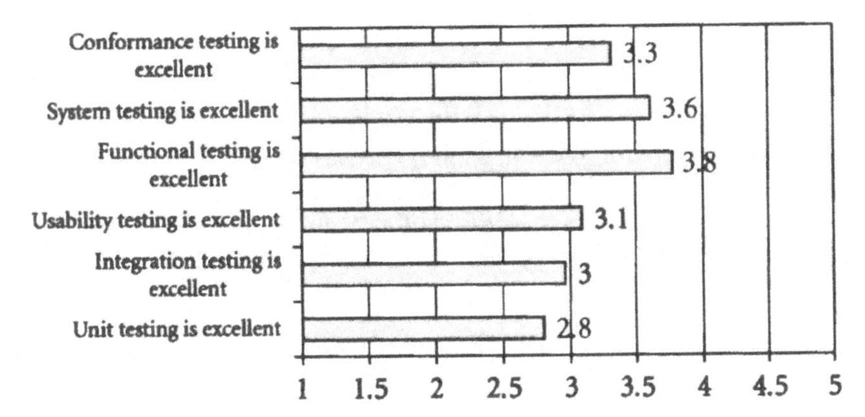

Figure 4. Testing phases in the software process.

Finally, the organizations surveyed were asked to rate their testing outcomes and objectives (Figure 5). The first three items discussed the test processes of a typical software project. There seems to be a strong variance in testing schedules and time allocation in the organizations. The outcomes 3.2 for schedule and 3.0 for time allocation do not give any information by themselves, and overall, the direction of answers varied greatly between "Fully disagree" and "Fully agree." However, the situation with test processes was somewhat better; the result 3.5 may also not be a strong indicator by itself, but the answers had only little variance, 20-OUs answering "somewhat agree" or "neutral." This indicates that even if the time is limited and the project schedule restricts testing, the testing generally goes through the normal, defined, procedures.

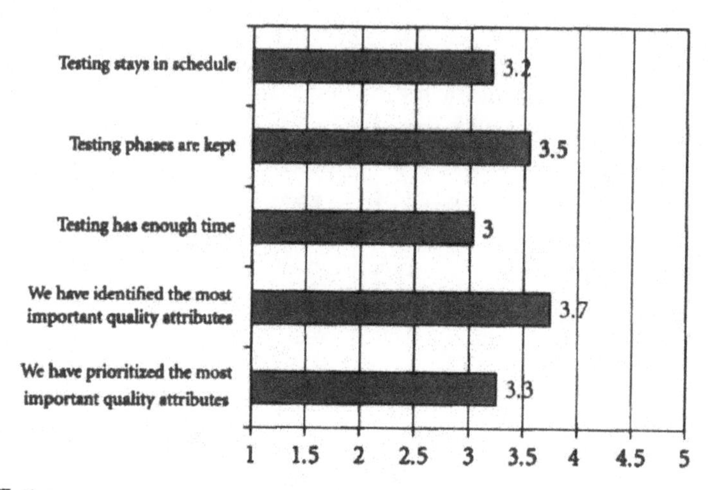

Figure 5. Testing process outcomes.

The fourth and fifth items were related to quality aspects, and gave insights into the clarity of testing objectives. The results of 3.7 for the identification of quality attributes indicate that organizations tend to have objectives for the test processes and apply quality criteria in development. However, the prioritization of their quality attributes is not as strong (3.3) as identification.

Testing Environment

The quality aspects were also reflected in the employment of systematic methods for the testing work. The majority (61%) of the OUs followed a systematic method or process in the software testing, 13% followed one partially, and 26% of the OUs did not apply any systematic method or process in testing. Process practices were derived from, for example, TPI (Test Process Improvement) [51] or the Rational Unified Process (RUP) [52]. Few Agile development process methods such as Scrum [53] or XP (eXtreme Programming) [54] were also mentioned.

A systematic method is used to steer the software project, but from the viewpoint of testing, the process also needs an infrastructure on which to operate. Therefore, the OUs were asked to report which kind of testing tools they apply to their typical software processes. The test management tools, tools which are used to control and manage test cases and allocate testing resources to cases, turned out to be the most popular category of tools; 15 OUs out of 31 reported the use of this type of tool. The second in popularity were manual unit testing tools (12 OUs), which were used to execute test cases and collect test results. Following them were tools to implement test automation, which were in use in 9 OUs, performance testing tools used in 8 OUs, bug reporting tools in 7 OUs and test design tools in 7 OUs. Test design tools were used to create and design new test cases. The group of other tools consisted of, for example, electronic measurement devices, test report generators, code analyzers, and project management tools. The popularity of the testing tools in different survey organizations is illustrated in Figure 6.

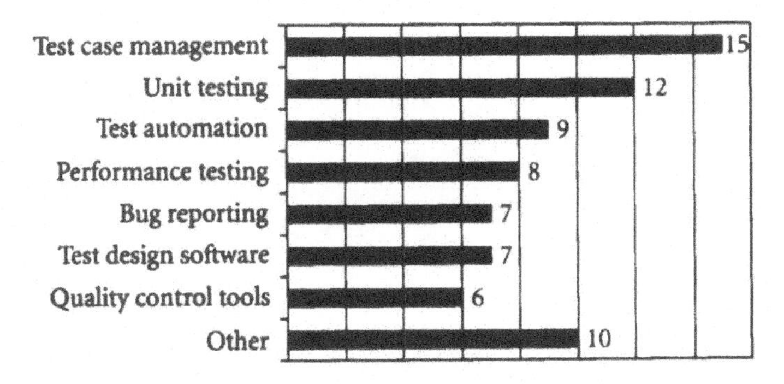

Figure 6. Popularity of the testing tools according to the survey.

The respondents were also asked to name and explain the three most efficient application areas of test automation tools. Both the format of the open-ended questions and the classification of the answers were based on the like best (LB) technique adopted from Fink and Kosecoff [46]. According to the LB technique, respondents were asked to list points they considered the most efficient. The primary selection was the area in which the test automation would be the most beneficial to the test organization, the secondary one is the second best area of application, and the third one is the third best area. The interviewees were also allowed to name only one or two areas if they were unable to decide on three application areas. The results revealed the relative importance of software testing tools and methods.

The results are presented in Figure 7. The answers were distributed rather evenly between different categories of tools or methods. The most popular category was unit testing tools or methods (10 interviewees). Next in line were regression testing (9), tools to support testability (9), test environment tools and methods (8), and functional testing (7). The group "others" (11) consisted of conformance testing tools, TTCN-3 (Testing and Test Control Notation version 3) tools, general test management tools such as document generators and methods of unit and integration testing. The most popular category, unit testing tools or methods, also received the most primary application area nominations. The most common secondary area of application was regression testing. Several categories ranked third, but concepts such as regression testing, and test environment-related aspects such as document generators were mentioned more than once. Also testability-related concepts—module interface, conformance testing—or functional testing—verification, validation tests—were considered feasible implementation areas for test automation.

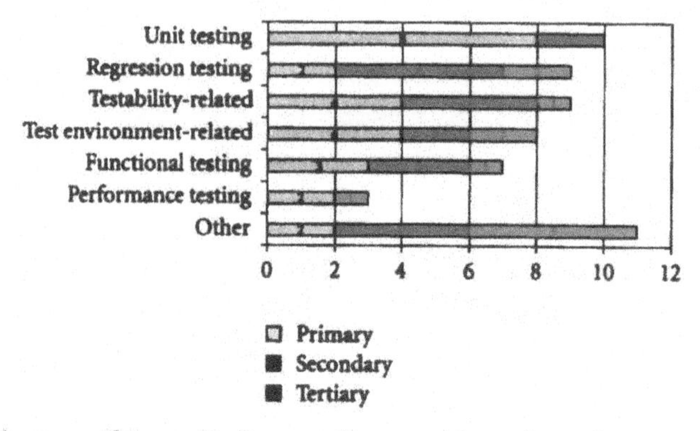

Figure 7. The three most efficient application areas of test automation tools according to the interviewees.

Summary of the Survey Findings

The survey suggests that interviewees were rather satisfied with their test policy, test strategy, test management, and testing, and did not have any immediate requirements for revising certain test phases, although low-level testing was slightly favoured in the development needs. All in all, 61% of the software companies followed some form of a systematic process or method in testing, with an additional 13% using some established procedures or measurements to follow the process efficiency. The systematic process was also reflected in the general approach to testing; even if the time was limited, the test process followed a certain path, applying the test phases regardless of the project limitations.

The main source of the software quality was considered to be in the development process. In the survey, the test organizations used test automation on an average on 26% of their test cases, which was considerably less than could be expected based on the literature. However, test automation tools were the third most common category of test-related tools, commonly intended to implement unit and regression testing. As for the test automation itself, the interviewees ranked unit testing tools as the most efficient tools of test automation, regression testing being the most common secondary area of application.

Test Automation Interviews and Qualitative Study

Besides the survey, the test automation concepts and applications were analyzed based on the interviews with the focus organizations. The grounded theory approach was applied to establish an understanding of the test automation concepts and areas of application for test automation in industrial software engineering. The qualitative approach was applied in three rounds, in which a developer, test manager and tester from 12 different case OUs were interviewed. Descriptions of the case OUs can be found in the appendix.

In theory-creating inductive research [55], the central idea is that researchers constantly compare theory and data iterating with a theory which closely fits the data. Based on the grounded theory codification, the categories identified were selected in the analysis based on their ability to differentiate the case organizations and their potential to explain the differences regarding the application of test automation in different contexts. We selected the categories so as to explore the types of automation applications and the compatibility of test automation services with the OUs testing organization. We conceptualized the most common test automation concepts based on the coding and further elaborated them to categories to either cater the essential features such as their role in the overall software process or their relation to test automation. We also concentrated on the OU differences in essential concepts such as automation tools, implementation issues or development strategies. This conceptualization resulted to the categories listed in Table 5.

Table 5. Test automation categories.

Category	Definition
Automation application	Areas of application for test automation in the software process
Role in software process	The observed roles of test automation in the company software process and the effect of this role
Test automation strategy	The observed method for selecting the test cases where automation is applied and the level of commitment to the application of test automation in the organizations
Automation development	The areas of active development in which the OU is introducing test automation
Automation tools	The general types of test automation tools applied
Automation issues	The items that hinder test automation development in the OU

The category "Automation application" describes the areas of software development, where test automation was applied successfully. This category describes the testing activities or phases which apply test automation processes. In the case where the test organization did not apply automation, or had so far only tested it for future applications, this category was left empty. The application areas were generally geared towards regression and stress testing, with few applications of functionality and smoke tests in use.

The category "Role in software process" is related to the objective for which test automation was applied in software development. The role in the software process describes the objective for the existence of the test automation infrastructure; it could, for example, be in quality control, where automation is used to secure module interfaces, or in quality assurance, where the operation of product functionalities is verified. The usual role for the test automation tools was in quality control and assurance, the level of application varying from third party-produced modules to primary quality assurance operations. On two occasions, the role of test automation was considered harmful to the overall testing outcomes, and on one occasion, the test automation was considered trivial, with no real return on investments compared to traditional manual testing.

The category "Test automation strategy" is the approach to how automated testing is applied in the typical software processes, that is, the way the automation was used as a part of the testing work, and how the test cases and overall test automation strategy were applied in the organization. The level of commitment to applying automation was the main dimension of this category, the lowest level being individual users with sporadic application in the software projects, and the highest being the application of automation to the normal, everyday testing infrastructure, where test automation was used seamlessly with other testing methods and had specifically assigned test cases and organizational support.

The category of "Automation development" is the general category for OU test automation development. This category summarizes the ongoing or recent efforts and resource allocations to the automation infrastructure. The type of new development, introduction strategies and current development towards test automation are summarized in this category. The most frequently chosen code was "general increase of application," where the organization had committed itself to test automation, but had no clear idea of how to develop the automation infrastructure. However, one OU had a development plan for creating a GUI testing environment, while two organizations had just recently scaled down the amount of automation as a result of a pilot project. Two organizations had only recently introduced test automation to their testing infrastructure.

The category of "Automation tools" describes the types of test automation tools that are in everyday use in the OU. These tools are divided based on their technological finesse, varying from self-created drivers and stubs to individual proof-of-concept tools with one specified task to test suites where several integrated components are used together for an effective test automation environment. If the organization had created the tools by themselves, or customized the acquired tools to the point of having new features and functionalities, the category was supplemented with a notification regarding in-house-development.

Finally, the category of "Automation issues" includes the main hindrances which are faced in test automation within the organization. Usually, the given issue was related to either the costs of test automation or the complexity of introducing automation to the software projects which have been initially developed without regards to support for automation. Some organizations also considered the efficiency of test automation to be the main issue, mostly contributing to the fact that two of them had just recently scaled down their automation infrastructure. A complete list of test automation categories and case organizations is given in Table 6.

Table 6. Test automation categories affecting the software process in case OUs.

OU	Category					
	Automation application	Role in software process	Test automation strategy	Automation development	Automation tools	Automation issues
Case A	GUI testing, regression testing	Functionality verification	Part of the normal test infrastructure	General increase of application	Individual tools, test suite, in-house development	Complexity of adapting automation to test processes
Case B	Performance, smoke testing	Quality control tool	Part of the normal test infrastructure	GUI testing, unit testing	Individual tools, in-house development	Costs of automation implementation
Case C	Functionality, regression testing, documentation automation	Quality control tool	Part of the normal test infrastructure	General increase of application	Test suite, in-house development	Cost of automation maintenance
Case D	Functionality testing	Quality control for secondary modules	Project-related cases	Upkeep for existing parts	Individual tools	Costs of automation implementation
Case E	System stress testing	Quality assurance tool	Part of the normal test infrastructure	General increase of application	Test suite	Costs of implementing new automation
Case F	Unit and module testing, documentation automation	QC, overall effect harmful	Individual users	Recently scaled down	Individual tools	Manual testing seen more efficient
Case G	Regression testing for use cases	Quality assurance tool	Part of the normal test infrastructure	General increase of application	Test suite	Cost of automation maintenance
Case H	Regression testing for module interfaces	Quality control for secondary modules	Part of the normal test infrastructure	General increase of application	Test suite, in-house development	Underestimation of the effect of automated testing on quality
Case I	Functionality testing	Quality control tool	Project-related cases	Application pilot in development	Proof-of-concept tools	Costs of automation implementation
Case J	Automation not in use	QA, no effect observed	Individual users	Application pilot in development	Proof-of-concept tools	No development incentive
Case K	Small scale system testing	QC, overall effect harmful	Individual users	Recently scaled down	Self-created tools; drivers and stubs	Manual testing seen more efficient
Case L	System stress testing	Verifies module compatibility	Project-related cases	Adapting automation to the testing strategy	Individual tools, in-house development	Complexity of adapting automation to test processes

We elaborated further these properties we observed from the case organizations to create hypotheses for the test automation applicability and availability. These resulting hypotheses were shaped according to advice given by Eisenhardt [37] for qualitative case studies. For example, we perceived the quality aspect as really important for the role of automation in software process. Similarly, the resource needs, especially costs, were much emphasized in the automation issues category. The purpose of the hypotheses below is to summarize and explain the features of test automation that resulted from the comparison of differences and similarities between the organizations.

Hypothesis 1 (Test Automation should be Considered more as a Quality Control Tool rather than a Frontline Testing Method)

The most common area of application observed was functionality verification, that is, regression testing and GUI event testing. As automation is time-consuming and expensive to create, these were the obvious ways to create test cases which had the minimal number of changes per development cycle. By applying this strategy, organizations could set test automation to confirm functional properties with suitable test cases, and acquire such benefits as support for change management and avoid unforeseen compatibility issues with module interfaces.

"Yes, regression testing, especially automated. It is not manually "hammered in" every time, but used so that the test sets are run, and if there is anything abnormal, it is then investigated." –Manager, Case G

"… had we not used it [automation tests], it would have been suicidal." –Designer, Case D

"It's [automated stress tests] good for showing bad code, how efficient it is and how well designed … stress it enough and we can see if it slows down or even breaks completely." –Tester, Case E

However, there seemed to be some contradicting considerations regarding the applicability of test automation. Cases F, J, and K had recently either scaled down their test automation architecture or considered it too expensive or inefficient when compared to manual testing. In some cases, automation was also considered too bothersome to configure for a short-term project, as the system would have required constant upkeep, which was an unnecessary addition to the project workload.

"We really have not been able to identify any major advancements from it [test automation]." –Tester, Case J

"It [test automation] just kept interfering." –Designer, Case K

Both these viewpoints indicated that test automation should not be considered a "frontline" test environment for finding errors, but rather a quality control tool to maintain functionalities. For unique cases or small projects, test automation is too expensive to develop and maintain, and it generally does not support single test cases or explorative testing. However, it seems to be practical in larger projects, where verifying module compatibility or offering legacy support is a major issue.

Hypothesis 2 (Maintenance and Development Costs are Common Test Automation Hindrances that Universally affect all test Organizations Regardless of their Business Domain or Company Size)

Even though the case organizations were selected to represent different types of organizations, the common theme was that the main obstacles in automation adoption were development expenses and upkeep costs. It seemed to make no difference whether the organization unit belonged to a small or large company, as in the OU levels they shared common obstacles. Even despite the maintenance and development hindrances, automation was considered a feasible tool in many organizations. For example, Cases I and L pursued the development of some kind of automation to enhance the testing process. Similarly, Cases E and H, which already had a significant number of test automation cases, were actively pursuing a larger role for automated testing.

"Well, it [automation] creates a sense of security and controllability, and one thing that is easily underestimated is its effect on performance and optimization. It requires regression tests to confirm that if something is changed, the whole thing does not break down afterwards." –Designer, Case H

In many cases, the major obstacle for adopting test automation was, in fact, the high requirements for process development resources.

"Shortage of time, resources ... we have the technical ability to use test automation, but we don't." –Tester, Case J

"Creating and adopting it, all that it takes to make usable automation ... I believe that we don't put any effort into it because it will end up being really expensive." –Designer, Case J

In Case J particularly, the OU saw no incentive in developing test automation as it was considered to offer only little value over manual testing, even if they otherwise had no immediate obstacles other than implementation costs. Also cases F and K reported similar opinions, as they both had scaled down the amount of automation after the initial pilot projects.

"It was a huge effort to manually confirm why the results were different, so we took it [automation] down." –Tester, Case F

"Well, we had gotten automation tools from our partner, but they were so slow we decided to go on with manual testing." –Tester, Case K

Hypothesis 3 (Test Automation is Applicable to most of the Software Processes, but Requires Considerable Effort from the Organization Unit)

The case organizations were selected to represent the polar types of software production operating in different business domains. Out of the focus OUs, there were four software development OUs, five IT service OUs, two OUs from the finance sector and one logistics OU. Of these OUs, only two did not have any test automation, and two others had decided to strategically abandon their test automation infrastructure. Still, the business domains for the remaining organizations which applied test automation were heterogeneously divided, meaning that the business domain is not a strong indicator of whether or not test automation should be applied.

It seems that test automation is applicable as a test tool in any software process, but the amount of resources required for useful automation compared to the overall development resources is what determines whether or not automation should be used. As automation is oriented towards quality control aspects, it may be unfeasible to implement in small development projects where quality control is manageable with manual confirmation. This is plausible, as the amount of required resources does not seem to vary based on aspects beyond the OU characteristics, such as available company resources or testing policies applied. The feasibility of test automation seems to be rather connected to the actual software process objectives, and fundamentally to the decision whether the quality control aspects gained from test automation supersede the manual effort required for similar results.

"... before anything is automated, we should calculate the maintenance effort and estimate whether we will really save time, instead of just automating for automation's sake." –Tester, Case G

"It always takes a huge amount of resources to implement." –Designer, Case A

"Yes, developing that kind of test automation system is almost as huge an effort as building the actual project." –Designer, Case I

Hypothesis 4 (The Available Repertoire of Testing Automation Tools is Limited, Forcing OUs to Develop the Tools themselves, which Subsequently Contributes to the Application and Maintenance Costs)

There were only a few case OUs that mentioned any commercial or publicly available test automation programs or suites. The most common approach to test automation tools was to first acquire some sort of tool for proof-of-concept piloting, then develop similar tools as in-house-production or extend the functionalities beyond the original tool with the OU's own resources. These resources for in-house-development and upkeep for self-made products are one of the components that contribute to the costs of applying and maintaining test automation.

"Yes, yes. That sort of [automation] tools have been used, and then there's a lot of work that we do ourselves. For example, this stress test tool ..." –Designer, Case E

"We have this 3rd party library for the automation. Well, actually, we have created our own architecture on top of it ..." –Designer, Case H

"Well, in [company name], we've-, we developed our own framework to, to try and get around some of these, picking which tests, which group of tests should be automated." –Designer, Case C

However, it should be noted that even if the automation tools were well-suited for the automation tasks, the maintenance still required significant resources if the software product to which it was connected was developing rapidly.

"Well, there is the problem [with automation tool] that sometimes the upkeep takes an incredibly large amount of time." –Tester, Case G

"Our system keeps constantly evolving, so you'd have to be constantly recording [maintaining tools] ..." –Tester, Case K

Discussion

An exploratory survey combined with interviews was used as the research method. The objective of this study was to shed light on the status of test automation and to identify improvement needs in and the practice of test automation. The survey revealed that the total effort spent on testing (median 25%) was less than expected. The median percentage (25%) of testing is smaller than the 50%–60% that is often mentioned in the literature [38, 39]. The comparable low percentage may indicate that that the resources needed for software testing are still underestimated even though testing efficiency has grown. The survey also indicated that companies used fewer resources on test automation than expected: on an average 26% of all of the test cases apply automation. However, there seems to be ambiguity as to which activities organizations consider test automation, and how automation should be applied in the test organizations. In the survey, several organizations reported that they have an extensive test automation infrastructure, but this did not reflect on the practical level, as in the interviews with testers particularly, the figures were considerably different. This indicates that the test automation does not have strong strategy in the organization, and has yet to reach maturity in several test organizations. Such concepts as quality assurance testing and stress testing seem to be particularly unambiguous application areas, as Cases E and L demonstrated. In Case E, the management did not consider stress testing an automation application, whereas testers did. Moreover, in Case L the large automation infrastructure did not reflect on the individual project level, meaning that the automation strategy may strongly vary between different projects and products even within one organization unit.

The qualitative study which was based on interviews indicated that some organizations, in fact, actively avoid using test automation, as it is considered to be expensive and to offer only little value for the investment. However, test automation seems to be generally applicable to the software process, but for small projects the investment is obviously oversized. One additional aspect that increases the investment are tools, which unlike in other areas of software testing, tend to be developed in-house or are heavily modified to suit specific automation needs. This development went beyond the localization process which every new software tool requires, extending even to the development of new features and operating frameworks. In this context it also seems plausible that test automation can be created for several different test activities. Regression testing, GUI testing or unit testing, activities which in some form exist in most development projects, all make it possible to create successful automation by creating suitable tools for the task, as in each phase can be found elements that have sufficient stability or unchangeability. Therefore it seems that the decision on applying automation is not only connected to the enablers and disablers of test automation [4], but rather on tradeoff of required effort and acquired benefits; In small projects or with low amount of reuse the effort becomes too much for such investment as applying automation to be feasible.

The investment size and requirements of the effort can also be observed on two other occasions. First, test automation should not be considered as an active testing tool for finding errors, but as a tool to guarantee the functionality of already existing systems. This observation is in line with those of Ramler and Wolfmaier [3], who discuss the necessity of a large number of repetitive tasks for the automation to supersede manual testing in cost-effectiveness, and of Berner et al. [8], who notify that the automation requires a sound application plan and well-documented, simulatable and testable objects. For both of these requirements, quality control at module interfaces and quality assurance on system operability are ideal, and as it seems, they are the most commonly used application areas for test automation. In fact, Kaner [56] states that 60%–80% of the errors found with test automation are found in the development phase for the test cases, further supporting the quality control aspect over error discovery.

Other phenomena that increase the investment are the limited availability and applicability of automation tools. On several occasions, the development of the automation tools was an additional task for the automation-building organization that required the organization to allocate their limited resources to the test automation tool implementation. From this viewpoint it is easy to understand why some case organizations thought that manual testing is sufficient and even more efficient when measured in resource allocation per test case. Another approach which could explain the observed resistance to applying or using test automation was also discussed in detail by Berner et al. [8], who stated that organizations tend to have inappropriate strategies and overly ambitious objectives for test automation development, leading to results that do not live up to their expectations, causing the introduction of automation to fail. Based on the observations regarding the development plans beyond piloting, it can also be argued that the lack of objectives and strategy also affect the successful introduction processes. Similar observations of "automation pitfalls" were also discussed by Persson and Yilmaztürk [26] and Mosley and Posey [57].

Overall, it seems that the main disadvantages of testing automation are the costs, which include implementation costs, maintenance costs, and training costs. Implementation costs included direct investment costs, time, and human resources. The correlation between these test automation costs and the effectiveness of the infrastructure are discussed by Fewster [24]. If the maintenance of testing automation is ignored, updating an entire automated test suite can cost as much, or even more than the cost of performing all the tests manually, making automation a bad investment for the organization. We observed this phenomenon in two case organizations. There is also a connection between implementation costs and maintenance costs [24]. If the testing automation system is designed with the minimization of maintenance costs in mind, the implementation costs increase, and vice versa. We noticed the phenomenon of costs preventing test automation development in six cases. The implementation of test automation seems to be possible to accomplish with two different approaches: by promoting either maintainability or easy implementation. If the selected focus is on maintainability, test automation is expensive, but if the approach promotes easy implementation, the process of adopting testing automation has a larger possibility for failure. This may well be due to the higher expectations and assumption that the automation could yield results faster when promoting implementation over maintainability, often leading to one of the automation pitfalls [26] or at least a low percentage of reusable automation components with high maintenance costs.

Conclusions

The objective of this study was to observe and identify factors that affect the state of testing, with automation as the central aspect, in different types of organizations. Our study included a survey in 31 organizations and a qualitative study in 12 focus organizations. We interviewed employees from different organizational positions in each of the cases.

This study included follow-up research on prior observations [4, 5, 12–14] on testing process difficulties and enhancement proposals, and on our observations on industrial test automation [4]. In this study we further elaborated on the test automation phenomena with a larger sample of polar type OUs, and more focused approach on acquiring knowledge on test process-related subjects. The survey revealed that test organizations use test automation only in 26% of their test cases, which was considerably less than could be expected based on the literature. However, test automation tools were the third most common category of test-related tools, commonly intended to implement unit and regression testing. The results indicate that adopting test automation in software organization is a demanding effort. The lack of existing software repertoire, unclear objectives for overall development and demands for resource allocation both for design and upkeep create a large threshold to overcome.

Test automation was most commonly used for quality control and quality assurance. In fact, test automation was observed to be best suited to such tasks, where the purpose was to secure working features, such as check module interfaces for backwards compatibility. However, the high implementation and maintenance requirements were considered the most important issues hindering test automation development, limiting the application of test automation in most OUs. Furthermore, the limited availability of test automation tools and the level of commitment required to develop a suitable automation infrastructure caused additional expenses. Due to the high maintenance requirements and low return on investments in small-scale application, some organizations had actually discarded their automation systems or decided not to implement test automation. The lack of a common strategy for applying automation was also evident in many interviewed OUs. Automation applications varied even within the organization, as was observable in the differences when comparing results from different stakeholders. In addition, the development strategies were vague and lacked actual objectives. These observations can also indicate communication gaps [58] between stakeholders of the overall testing strategy, especially between developers and testers.

The data also suggested that the OUs that had successfully implemented test automation infrastructure to cover the entire organization seemed to have difficulties in creating a continuance plan for their test automation development. After the adoption phases were over, there was an ambiguity about how to continue, even if the organization had decided to further develop their test automation infrastructure. The overall objectives were usually clear and obvious—cost savings and better test coverage—but in practice there were only few actual development ideas and novel concepts. In the case organizations this was observed in the vagueness of the development plans: only one of the five OUs which used automation as a part of their normal test processes had development plans beyond the general will to increase the application.

The survey established that 61% of the software companies followed some form of a systematic process or method in testing, with an additional 13% using some established procedures or measurements to follow the process efficiency. The main source of software quality was considered to reside in the development process, with testing having much smaller impact in the product outcome. In retrospect of the test levels introduced in the ISO/IEC29119 standard, there seems to be no one particular level of the testing which should be the research and development interest for best result enhancements. However, the results from the self-assessment of the test phases indicate that low-level testing could have more potential for testing process development.

Based on these notions, the research and development should focus on uniform test process enhancements, such as applying a new testing approach and creating an organization-wide strategy for test automation. Another focus area should be the development of better tools to support test organizations and test processes in the low-level test phases such as unit or integration testing. As for automation, one tool project could be the development of a customizable test environment with a common core and with an objective to introduce less resource-intensive, transferable and customizable test cases for regression and module testing.

Appendix

Case Descriptions

Case A (Manufacturing Execution System (MES) Producer and Electronics Manufacturer)

Case A produces software as a service (SaaS) for their product. The company is a small-sized, nationally operating company that has mainly industrial customers. Their software process is a plan-driven cyclic process, where the testing is embedded to the development itself, having only little amount of dedicated resources. This organization unit applied test automation as a user interface and regression testing tool, using it for product quality control. Test automation was seen as a part of the normal test strategy, universally used in all software projects. The development plan for automation was to generally increase the application, although the complexity of the software and module architecture was considered major obstacle on the automation process.

Case B (Internet Service Developer and Consultant)

Case B organization offers two types of services; development of Internet service portals for the customers like communities and public sector, and consultation in the Internet service business domain. The origination company is small and operates on a national level. Their main resource on the test automation is in the performance testing as a quality control tool, although addition of GUI test automation has also been proposed. The automated tests are part of the normal test process, and the overall development plan was to increase the automation levels especially to the GUI test cases. However, this development has been hindered by the cost of designing and developing test automation architecture.

Case C (Logistics Software Developer)

Case C organization focuses on creating software and services for their origin company and its customers. This organization unit is a part of a large-sized, nationally operating company with large, highly distributed network and several clients. The test automation is widely used in several testing phases like functionality testing, regression testing and document generation automation. These investments are used for quality control to ensure the software usability and correctness. Although the OU is still aiming for larger test automation infrastructure, the large amount of related systems and constant changes within the inter-module communications is causing difficulties in development and maintenance of the new automation cases.

Case D (ICT Consultant)

Case D organization is a small, regional software consultant company, whose customers mainly compose of small business companies and the public sector. Their organization does some software development projects, in which the company develops services and ICT products for their customers. The test automation comes mainly trough this channel, as the test automation is mainly used as a conformation test tool for the third party modules. This also restricts the amount of test automation to the projects, in which these modules are used. The company currently does not have development plans for the test automation as it is considered unfeasible investment for the OU this size, but they do invest on the upkeep of the existing tools as they have usage as a quality control tool for the acquired outsider modules.

Case E (Safety and Logistics System Developer)

Case E organization is a software system developer for safety and logistics systems. Their products have high amount of safety critical features and have several interfaces on which to communicate with. The test automation is used as a major quality assurance component, as the service stress tests are automated to a large degree. Therefore the test automation is also a central part of the testing strategy, and each project has defined set of automation cases. The organization is aiming to increase the amount of test automation and simultaneously develop new test cases and automation applications for the testing process. The main obstacle for this development has so far been the costs of creating new automation tools and extending the existing automation application areas.

Case F (Naval Software System Developer)

The Case F organization unit is responsible for developing and testing naval service software systems. Their product is based on a common core, and has considerable requirements for compatibility with the legacy systems. This OU has tried test automation on several cases with application areas such as unit- and module testing, but has recently scaled down test automation for only support aspects such as the documentation automation. This decision was based on the resource requirements for developing and especially maintaining the automation system, and because the manual testing was in this context considered much more efficient as there were too much ambiguity in the automation-based test results.

Case G (Financial Software Developer)

Case G is a part of a large financial organization, which operates nationally but has several internationally connected services due to their business domain. Their software projects are always aimed as a service portal for their own products, and have to pass considerable verification and validation tests before being introduced to the public. Because of this, the case organization has sizable test department when compared to other case companies in this study, and follows rigorous test process plan in all of their projects. The test automation is used in the regression tests as a quality assurance tool for user interfaces and interface events, and therefore embedded to the testing strategy as a normal testing environment. The development plans for the test automation is aimed to generally increase the amount of test cases, but even the existing test automation infrastructure is considered expensive to upkeep and maintain.

Case H (Manufacturing Execution System (MES) Producer and Logistics Service System Provider)

Case H organization is a medium-sized company, whose software development is a component for the company product. Case organization products are used in logistics service systems, usually working as a part of automated processes. The case organization applies automated testing as a module interface testing tool, applying it as a quality control tool in the test strategy. The test automation infrastructure relies on the in-house-developed testing suite, which enables organization to use the test automation to run daily tests to validate module conformance. Their approach on the test automation has been seen as a positive enabler, and the general trend is towards increasing automation cases. The main test automation disability is considered to be that the quality control aspect is not visible when working correctly and therefore the effect of test automation may be underestimated in the wider organization.

Case I (Small and Medium-Sized Enterprise (SME) Business and Agriculture ICT-Service Provider)

The Case I organization is a small, nationally operating software company which operates on multiple business domain. Their customer base is heterogeneous, varying from finances to the agriculture and government services. The company is currently not utilizing test automation in their test process, but they have development plans for designing quality control automation. For this development they have had some individual proof-of-concept tools, but currently the overall testing resources limit the application process.

Case J (Modeling Software Developer)

Case J organization develops software products for civil engineering and architectural design. Their software process is largely plan-driven with rigorous verification and validation processes in the latter parts of an individual project. Even though the case organization itself has not implemented test automation, on the corporate level there are some pilot projects where regression tests have been automated. These proof-of-concept-tools have been introduced to the case OU and there are intentions to apply them in the future, but there has so far been no incentive for adoption of the automation tools, delaying the application process.

Case K (ICT Developer and Consultant)

Case K organization is a large, international software company which offers software products for several business domains and government services. Case organization has previously piloted test automation, but decided against adopting the system as it was considered too expensive and resource-intensive to maintain when compared to the manual testing. However, some of these tools still exist, used by individual developers along with test drivers and interface studs in unit and regression testing.

Case L (Financial Software Developer)

Case L organization is a large software provider for their corporate customer which operates on the finance sector. Their current approach on software process is plan-driven, although some automation features has been tested on a few secondary processes. The case organization does not apply test automation as is, although some module stress test cases have been automated as pilot tests. The development plan for test automation is to generally implement test automation as a part of their testing strategy, although amount of variability and interaction in the module interfaces is considered difficult to implement in test automation cases.

Acknowledgement

This study is a part of the ESPA project (http://www.soberit.hut.fi/espa/), funded by the Finnish Funding Agency for Technology and Innovation (project number 40125/08) and by the participating companies listed on the project web site.

References

1. E. Kit, Software Testing in the Real World: Improving the Process, Addison-Wesley, Reading, Mass, USA, 1995.
2. G. Tassey, "The economic impacts of inadequate infrastructure for software testing," RTI Project 7007.011, U.S. National Institute of Standards and Technology, Gaithersburg, Md, USA, 2002.
3. R. Ramler and K. Wolfmaier, "Observations and lessons learned from automated testing," in Proceedings of the International Workshop on Automation of Software Testing (AST '06), pp. 85–91, Shanghai, China, May 2006.
4. K. Karhu, T. Repo, O. Taipale, and K. Smolander, "Empirical observations on software testing automation," in Proceedings of the 2nd International Conference

on Software Testing, Verification, and Validation (ICST '09), pp. 201–209, Denver, Colo, USA, April 2009.

5. O. Taipale and K. Smolander, "Improving software testing by observing causes, effects, and associations from practice," in Proceedings of the International Symposium on Empirical Software Engineering (ISESE '06), Rio de Janeiro, Brazil, September 2006.
6. B. Shea, "Sofware testing gets new respect," InformationWeek, July 2000.
7. E. Dustin, J. Rashka, and J. Paul, Automated Software Testing: Introduction, Management, and Performance, Addison-Wesley, Boston, Mass, USA, 1999.
8. S. Berner, R. Weber, and R. K. Keller, "Observations and lessons learned from automated testing," in Proceedings of the 27th International Conference on Software Engineering (ICSE '05), pp. 571–579, St. Louis, Mo, USA, May 2005.
9. J. A. Whittaker, "What is software testing? And why is it so hard?" IEEE Software, vol. 17, no. 1, pp. 70–79, 2000.
10. L. J. Osterweil, "Software processes are software too, revisited: an invited talk on the most influential paper of ICSE 9," in Proceedings of the 19th IEEE International Conference on Software Engineering, pp. 540–548, Boston, Mass, USA, May 1997.
11. ISO/IEC and ISO/IEC 29119-2, "Software Testing Standard?Activity Descriptions for Test Process Diagram," 2008.
12. O. Taipale, K. Smolander, and H. Kälviäinen, "Cost reduction and quality improvement in software testing," in Proceedings of the 14th International Software Quality Management Conference (SQM '06), Southampton, UK, April 2006.
13. O. Taipale, K. Smolander, and H. Kälviäinen, "Factors affecting software testing time schedule," in Proceedings of the Australian Software Engineering Conference (ASWEC '06), pp. 283–291, Sydney, Australia, April 2006.
14. O. Taipale, K. Smolander, and H. Kälviäinen, "A survey on software testing," in Proceedings of the 6th International SPICE Conference on Software Process Improvement and Capability dEtermination (SPICE '06), Luxembourg, May 2006.
15. N. C. Dalkey, The Delphi Method: An Experimental Study of Group Opinion, RAND, Santa Monica, Calif, USA, 1969.
16. S. P. Ng, T. Murnane, K. Reed, D. Grant, and T. Y. Chen, "A preliminary survey on software testing practices in Australia," in Proceedings of the Australian

Software Engineering Conference (ASWEC '04), pp. 116–125, Melbourne, Australia, April 2004.

17. R. Torkar and S. Mankefors, "A survey on testing and reuse," in Proceedings of IEEE International Conference on Software—Science, Technology and Engineering (SwSTE '03), Herzlia, Israel, November 2003.

18. C. Ferreira and J. Cohen, "Agile systems development and stakeholder satisfaction: a South African empirical study," in Proceedings of the Annual Research Conference of the South African Institute of Computer Scientists and Information Technologists (SAICSIT '08), pp. 48–55, Wilderness, South Africa, October 2008.

19. J. Li, F. O. Bjørnson, R. Conradi, and V. B. Kampenes, "An empirical study of variations in COTS-based software development processes in the Norwegian IT industry," Empirical Software Engineering, vol. 11, no. 3, pp. 433–461, 2006.

20. W. Chen, J. Li, J. Ma, R. Conradi, J. Ji, and C. Liu, "An empirical study on software development with open source components in the Chinese software industry," Software Process Improvement and Practice, vol. 13, no. 1, pp. 89–100, 2008.

21. R. Dossani and N. Denny, "The Internet's role in offshored services: a case study of India," ACM Transactions on Internet Technology, vol. 7, no. 3, 2007.

22. K. Y. Wong, "An exploratory study on knowledge management adoption in the Malaysian industry," International Journal of Business Information Systems, vol. 3, no. 3, pp. 272–283, 2008.

23. J. Bach, "Test automation snake oil," in Proceedings of the 14th International Conference and Exposition on Testing Computer Software (TCS '99), Washington, DC, USA, June 1999.

24. M. Fewster, Common Mistakes in Test Automation, Grove Consultants, 2001.

25. A. Hartman, M. Katara, and A. Paradkar, "Domain specific approaches to software test automation," in Proceedings of the 6th Joint Meeting of the European Software Engineering Conference and the ACM SIGSOFT Symposium on the Foundations of Software Engineering (ESEC/FSE '07), pp. 621–622, Dubrovnik, Croatia, September 2007.

26. C. Persson and N. Yilmaztürk, "Establishment of automated regression testing at ABB: industrial experience report on 'avoiding the pitfalls'," in Proceedings of the 19th International Conference on Automated Software Engineering (ASE '04), pp. 112–121, Linz, Austria, September 2004.

27. M. Auguston, J. B. Michael, and M.-T. Shing, "Test automation and safety assessment in rapid systems prototyping," in Proceedings of the 16th IEEE International Workshop on Rapid System Prototyping (RSP '05), pp. 188–194, Montreal, Canada, June 2005.
28. A. Cavarra, J. Davies, T. Jeron, L. Mournier, A. Hartman, and S. Olvovsky, "Using UML for automatic test generation," in Proceedings of the International Symposium on Software Testing and Analysis (ISSTA '02), Roma, Italy, July 2002.
29. M. Vieira, J. Leduc, R. Subramanyan, and J. Kazmeier, "Automation of GUI testing using a model-driven approach," in Proceedings of the International Workshop on Automation of Software Testing, pp. 9–14, Shanghai, China, May 2006.
30. Z. Xiaochun, Z. Bo, L. Juefeng, and G. Qiu, "A test automation solution on gui functional test," in Proceedings of the 6th IEEE International Conference on Industrial Informatics (INDIN '08), pp. 1413–1418, Daejeon, Korea, July 2008.
31. D. Kreuer, "Applying test automation to type acceptance testing of telecom networks: a case study with customer participation," in Proceedings of the 14th IEEE International Conference on Automated Software Engineering, pp. 216–223, Cocoa Beach, Fla, USA, October 1999.
32. W. D. Yu and G. Patil, "A workflow-based test automation framework for web based systems," in Proceedings of the 12th IEEE Symposium on Computers and Communications (ISCC '07), pp. 333–339, Aveiro, Portugal, July 2007.
33. A. Bertolino, "Software testing research: achievements, challenges, dreams," in Proceedings of the Future of Software Engineering (FoSE '07), pp. 85–103, Minneapolis, Minn, USA, May 2007.
34. M. Blackburn, R. Busser, and A. Nauman, "Why model-based test automation is different and what you should know to get started," in Proceedings of the International Conference on Practical Software Quality, Braunschweig, Germany, September 2004.
35. P. Santos-Neto, R. Resende, and C. Pádua, "Requirements for information systems model-based testing," in Proceedings of the ACM Symposium on Applied Computing, pp. 1409–1415, Seoul, Korea, March 2007.
36. ISO/IEC and ISO/IEC 15504-1, "Information Technology?Process Assessment?Part 1: Concepts and Vocabulary," 2002.
37. K. M. Eisenhardt, "Building theories from case study research," The Academy of Management Review, vol. 14, no. 4, pp. 532–550, 1989.

38. EU and European Commission, "The new SME definition: user guide and model declaration," 2003.
39. G. Paré and J. J. Elam, "Using case study research to build theories of IT implementation," in Proceedings of the IFIP TC8 WG 8.2 International Conference on Information Systems and Qualitative Research, pp. 542–568, Chapman & Hall, Philadelphia, Pa, USA, May-June 1997.
40. A. Strauss and J. Corbin, Basics of Qualitative Research: Grounded Theory Procedures and Techniques, SAGE, Newbury Park, Calif, USA, 1990.
41. ATLAS.ti, The Knowledge Workbench, Scientific Software Development, 2005.
42. M. B. Miles and A. M. Huberman, Qualitative Data Analysis, SAGE, Thousand Oaks, Calif, USA, 1994.
43. C. B. Seaman, "Qualitative methods in empirical studies of software engineering," IEEE Transactions on Software Engineering, vol. 25, no. 4, pp. 557–572, 1999.
44. C. Robson, Real World Research, Blackwell, Oxford, UK, 2nd edition, 2002.
45. N. K. Denzin, The Research Act: A Theoretical Introduction to Sociological Methods, McGraw-Hill, New York, NY, USA, 1978.
46. A. Fink and J. Kosecoff, How to Conduct Surveys: A Step-by-Step Guide, SAGE, Beverly Hills, Calif, USA, 1985.
47. B. A. Kitchenham, S. L. Pfleeger, L. M. Pickard, et al., "Preliminary guidelines for empirical research in software engineering," IEEE Transactions on Software Engineering, vol. 28, no. 8, pp. 721–734, 2002.
48. T. Dybå, "An instrument for measuring the key factors of success in software process improvement," Empirical Software Engineering, vol. 5, no. 4, pp. 357–390, 2000.
49. ISO/IEC and ISO/IEC 25010-2, "Software Engineering?Software product Quality Requirements and Evaluation (SQuaRE) Quality Model," 2008.
50. Y. Baruch, "Response rate in academic studies—a comparative analysis," Human Relations, vol. 52, no. 4, pp. 421–438, 1999.
51. T. Koomen and M. Pol, Test Process Improvement: A Practical Step-by-Step Guide to Structured Testing, Addison-Wesley, Reading, Mass, USA, 1999.
52. P. Kruchten, The Rational Unified Process: An Introduction, Addison-Wesley, Reading, Mass, USA, 2nd edition, 1998.
53. K. Schwaber and M. Beedle, Agile Software Development with Scrum, Prentice-Hall, Upper Saddle River, NJ, USA, 2001.

54. K. Beck, Extreme Programming Explained: Embrace Change, Addison-Wesley, Reading, Mass, USA, 2000.
55. B. Glaser and A. L. Strauss, The Discovery of Grounded Theory: Strategies for Qualitative Research, Aldine, Chicago, Ill, USA, 1967.
56. C. Kaner, "Improving the maintainability of automated test suites," Software QA, vol. 4, no. 4, 1997.
57. D. J. Mosley and B. A. Posey, Just Enough Software Test Automation, Prentice-Hall, Upper Saddle River, NJ, USA, 2002.
58. D. Foray, Economics of Knowledge, MIT Press, Cambridge, Mass, USA, 2004.

A Strategy for Automatic Quality Signing and Verification Processes for Hardware and Software Testing

Mohammed I. Younis and Kamal Z. Zamli

ABSTRACT

We propose a novel strategy to optimize the test suite required for testing both hardware and software in a production line. Here, the strategy is based on two processes: Quality Signing Process and Quality Verification Process, respectively. Unlike earlier work, the proposed strategy is based on integration of black box and white box techniques in order to derive an optimum test suite during the Quality Signing Process. In this case, the generated optimal test suite significantly improves the Quality Verification Process. Considering both processes, the novelty of the proposed strategy is the fact that the optimization and reduction of test suite is performed by selecting only mutant killing test cases from cumulating t-way test cases. As such, the proposed strategy can potentially enhance the quality of product with minimal cost in terms of overall resource usage and time execution. As a case study, this paper describes the step-by-step application of the strategy for testing a 4-bit Magnitude Comparator Integrated Circuits in a production line. Comparatively, our result demonstrates that the proposed strategy outperforms the traditional block partitioning strategy with the mutant score of 100% to 90%, respectively, with the same number of test cases.

Introduction

In order to ensure acceptable quality and reliability of any embedded engineering products, many inputs parameters as well as software/hardware configurations need to be tested against for conformance. If the input combinations are large, exhaustive testing is next to impossible due to combinatorial explosion problem.

As illustration, consider the following small-scale product, a 4-bit Magnitude Comparator IC. Here, the Magnitude Comparator IC consists of 8 bits for inputs and 3 bits for outputs. It is clear that each IC requires 256 test cases for exhaustive testing. Assuming that each test case takes one second to run and be observed, the testing time for each IC is 256 seconds. If there is a need to test one million chips, the testing process will take more than 8 years using a single line of test.

Now, let us assume that we received an order of delivery for one million qualified (i.e., tested) chips within two weeks. As an option, we can do parallel testing. However, parallel testing can be expensive due to the need for 212 testing lines. Now, what if there are simultaneous multiple orders? Here, as the product demand grows in numbers, parallel testing can also become impossible. Systematic random testing could also be another option. In random testing, test cases are chosen randomly from some input distribution (such as a uniform distribution) without exploiting information from the specification or previously chosen test cases. More recent results have favored partition testing over random testing in many practical cases. In all cases, random testing is found to be less effective than the investigated partition testing methods [1].

A systematic solution to this problem is based on Combinatorial Interaction Testing (CIT) strategy. The CIT approach can systematically reduce the number of test cases by selecting a subset from exhaustive testing combination based on the strength of parameter interaction coverage (t) [2]. To illustrate the CIT approach, consider the web-based system example (see Table 1) [3].

Table 1. Web-based system example.

Parameter 1	Parameter 2	Parameter 3	Parameter 4
Netscape	Windows XP	LAN	Sis
IE	Windows VISTA	PPP	Intel
Firefox	Windows 2008	ISDN	VIA

Considering full strength interaction $t = 4$ (i.e., interaction of all parameters) for testing yields exhaustive combinations of 34 = 81 possibilities. Relaxing the interaction strength to $t = 3$ yields 27 test cases, a saving of nearly 67 percent. Here, all the 3-way interaction elements are all covered by at least one test. If the interaction is relaxed further to $t = 2$, then the number of combination possibilities is reduced even further to merely 9 test cases, a saving of over 90 percent.

In the last decade, CIT strategies were focused on 2-way (pairwise) testing. More recently, several strategies (e.g., Jenny [4], TVG [5], IPOG [6], IPOD [7], IPOF [8], DDA [9], and GMIPOG [10]) that can generate test suite for high degree interaction ($2 \le t \le 6$).

Being predominantly black box, CIT strategy is often criticized for not being efficiently effective for highly interacting parameter coverage. Here, the selected test cases sometimes give poor coverage due to the wrong selection of parameter strength. In order to address this issue, we propose to integrate the CIT strategy with that of fault injection strategy. With such integration, we hope to effectively measure the effectiveness of the test suite with the selection of any particular parameter strength. Here, the optimal test case can be selected as the candidate of the test suite only if it can help detect the occurrence of the injected fault. In this manner, the desired test suite is the most optimum for evaluating the System Under Test (SUT).

The rest of this paper is organized as follows. Section 2 presents related work on the state of the art of the applications of t-way testing and fault injection tools. Section 3 presents the proposed minimization strategy. Section 4 gives a step-by-step example as prove of concept involving the 4-bit Magnitude Comparator. Section 5 demonstrates the comparison with our proposed strategy and the traditional block partitioning strategy. Finally, Section 6 describes our conclusion and suggestion for future work.

Related Work

Mandl was the first researcher who used pairwise coverage in the software industry. In his work, Mandl adopts orthogonal Latin square for testing an Ada compiler [11]. Berling and Runeson use interaction testing to identify real and false targets in target identification system [12]. Lazic and Velasevic employed interaction testing on modeling and simulation for automated target-tracking radar system [13]. White has also applied the technique to test graphical user interfaces (GUIs) [14]. Other applications of interaction testing include regression testing through the graphical user interface [15] and fault localization [16, 17]. While earlier work has indicated that pairwise testing (i.e., based on 2-way interaction of variables) can be effective to detect most faults in a typical software system, a counter argument suggests such conclusion infeasible to generalize to all software system faults. For example, a test set that covers all possible pairs of variable values can typically detect 50% to 75% of the faults in a program [18–20]. In other works it is found that 100% of faults are detectable by a relatively low degree of interaction, typically 4-way combinations [21–23].

More recently, a study by The National Institute of Standards and Technology (NIST) for error-detection rates in four application domains included medical devices, a Web browser, an HTTP server, and a NASA-distributed database reported that 95% of the actual faults on the test software involve 4-way interaction [24, 25]. In fact, according to the recommendation from NIST, almost all of the faults detected with 6-way interaction. Thus, as this example illustrates, system faults caused by variable interactions may also span more than two parameters, up to 6-way interaction for moderate systems.

All the aforementioned related work in CIT applications highlighted the potential of adopting the CIT strategies for both software/hardware testing. While the CIT strategies can significantly partition the exhaustive test space into manageable manner, additional reduction can still be possible particularly by systematically examining the effectiveness of each test case in the test suite, that is, by exploiting fault injection techniques.

The use of fault injection techniques for software and hardware testing is not new. Tang and Chen [26], Boroday [27], and Chandra et al. [28] study circuit testing in hardware environment, proposing test coverage that includes each 2^t of the input settings for each subset of t inputs. Seroussi and Bshouty [29] give a comprehensive treatment for circuit testing. Dumer [30] examines the related question of isolating memory faults and uses binary covering arrays. Finally, Ghosh and Kelly give a survey to include a number of studies and tools that have been reported in the area of failure mode identification [31]. These studies help in the long-term improvement of the software development process as the recurrence of the same failures can be prevented. Failure modes can be specific to a system or be applicable to systems in general. They can be used in testing for fault tolerance, as realistic faults are needed to perform effective fault injection testing. Additionally, Ghosh and Kelly also describe a technique that injects faults in Java software by manipulating the bytecode level for third party software components used by the developers.

Proposed Strategy

The proposed strategy consists for two processes, namely, Test Quality Signing (TQS) process and Test Verification process (TV). Briefly, the TQS process deals with optimizing the selection of test suite for fault injection as well as performs the actual injection whilst the TV process analyzes for conformance (see Figure 1).

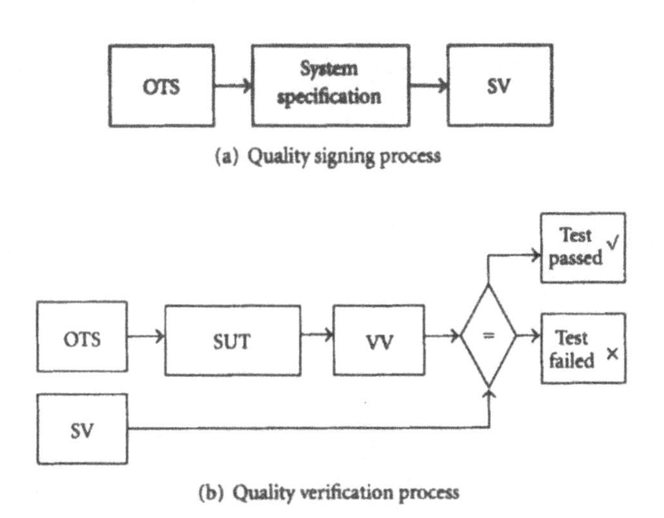

Figure 1. The quality signing and verification processes.

As implied earlier, the TQS process aims to derive an effective and optimum test suite and works as follows:

1. Start with an empty Optimized Test Suite (OTS), and empty Signing Vector (SV).
2. Select the desired software class (for software testing). Alternatively, build an equivalent software class for the Circuit Under Test (CUT) (for hardware testing).
3. Store these faults in fault list (FL).
4. Inject the class with all possible faults.
5. Let N be maximum number of parameters.
6. Initialize CIT strategy with strength of coverage (t) equal one (i.e., t = 1).

7. Let CIT strategy partition the exhaustive test space. The portioning involves generating one test case at a time for t coverage. If t coverage criteria are satisfied, then t = t + 1.
8. CIT strategy generates one Test Case (TC).
9. Execute TC.
10. If TC detects any fault in FL, remove the detected fault(s) from FL, and add TC and its specification output(s) to OTS and SV, respectively.
11. If FL is not empty or t<=N, go to 7.
12. The desired optimized test suite and its corresponding output(s) are stored in OTS and SV, respectively.

The TV process involves the verification of fault free for each unit. TV process for a single unit works as follows:

(1) for i=1..Size(OTS) each TC in OTS do:

  (a) Subject the SUT to TC[i], store the output in Verification Vector VV[i].

  (b) If VV[i] = SV [i], continue. Else, go to 3.

(2) Report that the cut has been passing in the test. Go to 4.

(3) Report that the cut has failed the test.

(4) The verification process ends.

As noted in the second step of the TQS process, the rationale for taking equivalent software class for the CUT is to ensure that the cost and control of the fault injection be more practical and manageable as opposed to performing it directly to a real hardware circuit. Furthermore, the derivation of OTS is faster in software than in hardware. Despite using equivalent class for the CUT, this verification process should work for both software and hardware systems. In fact, it should be noted that the proposed strategy could also be applicable in the context of N-version programming (e.g., the assessment of student programs for the same assignment) and not just hardware production lines. The concept of N-version programming was introduced by Chen and Avizienis with the central conjecture that the "independence of programming efforts will greatly reduce the probability of identical software faults occurring in two or more versions of the program" [32, 33].

Case Study

As proof of concept, we have adopted GMIPOG [10] as our CIT strategy implementation, and MuJava version 3 (described in [34, 35]) as our fault injection strategy implementation.

Briefly, GMIPOG is a combinatorial test generator based on specified inputs and parameter interaction. Running on a Grid environment, GMIPOG adopts both the horizontal and vertical extension mechanism (i.e., similar to that of IPOG [6]) in order to derive the required test suite for a given interaction strength. While there are many useful combinatorial test generators in the literature (e.g., Jenny [3], TConfig [4], TVG [5], IPOG [6], IPOD [7], IPOF [8], DDA [9]), the rationale for choosing GMIPOG is the fact that it supports high degree of interaction and can be run in cumulative mode (i.e., support one-test-at-a-time approach with the capability to vary t automatically until the exhaustive testing is reached).

Complementary to GMIPOG, MuJava is a fault injection tool that permits mutated Java code (i.e., based on some defined operators) to be injected into the running Java program. Here, the reason for choosing MuJava stemmed from the fact that it is a public domain Java tool freely accessible for download in the internet [35].

Using both tools (i.e., GMIPOG and MuJava), a case study problem involving a 4-bit Magnitude Comparator IC will be discussed here in order to evaluate the proposed strategy. A 4-bit Magnitude Comparator consists of 8 inputs (two four bits inputs, namely, a0…a3, and b0…b3. where a0 and b0 are the most significant bits), 4 xnor gates (or equivalent to 4xor with 4 not gates), five not gates, five and gates, three or gates, and three outputs. The actual circuit realization of the Magnitude Comparator is given in Figure 2. Here, it should be noted that this version of the circuit is a variant realization (implementation) of the Magnitude Comparator found in [36]. The equivalent class of the Magnitude Comparator is given in Figure 3 (using the Java-programming language).

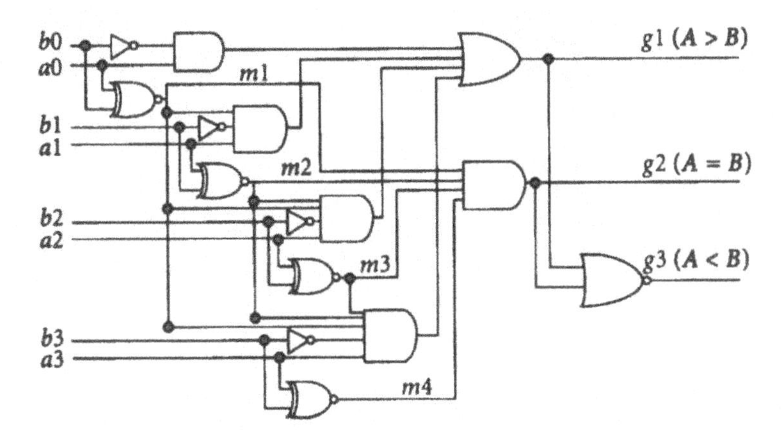

Figure 2. Schematic diagram for the 4-bit magnitude comparator.

```
public class Comparator {
//Comparator takes two four bits numbers (A&B), where A = a0a1a2a3
//B=b0b1b2b3. Here, a0 and b0 are the most significant bits.
//The function returns an output string that s
//g1, g2, and g3 represent the logical outputs of A > B, A = B, and A < B respectively.
//the code symbols (!, ^, |, and &)
//represent the logical operator for Not, Xor, Or, and And respectively.
        public static String compare
            (boolean a0, boolean a1, boolean a2, boolean a3,
            boolean b0, boolean b1, boolean b2, boolean b3) {
            boolean  g1,g2,g3;
            boolean  m1,m2,m3,m4;
              String s = null;
              m1 =!(a0 ^ b0);
              m2 =!(a1 ^ b1);
              m3 =!(a2 ^ b2);
              m4 =!(a3 ^ b3);
              g1 = (a0 &!b0)| (m1&a1 &!b1) |(m1&m2&a2 &!b2)| (m1&m2 &m3&a3 &!b3);
              g2 = (m1&m2 &m3&m4);
              g3 =!(g1|g2);
              s = g1 +"" +g2 +"" +g3; // just to return output strings for MuJava compatibility
              return s;
    }
}
```

Figure 3. Equivalent class Java program for the 4-bit magnitude comparator.

Here, it is important to ensure that the software implementation obeys the hardware implementation strictly. By doing so, we can undertake the fault injection and produce the OTS in the software domain without affecting the logical of relation and parameter interactions of the hardware implementation.

Now, we apply the TQS process; as illustrated in Section 3. Here, there are 80 faults injected in the system. To assist our work, we use GMIPOG [10] to produce the TC in a cumulative mode. Following the steps in TQS process, Table 2 demonstrates the derivation of OTS. Here, it should be noted that the first 36 test cases can remove all the faults. Furthermore, only the first 12 test cases when t = 4 are needed to catch that last two live mutants. The efficiency of integration GMIPOG with MuJava can be observed (by taken only the effective TC) in the last column in Table 2.

Table 2. Derivation of OTS for the 4-bit Magnitude Comparator.

t =	Cumulative Test Size	Live Mutant	Killed Mutant	% Mutant Score	Effective test size
1	2	15	65	81.25	2
2	9	5	75	93.75	6
3	24	2	78	97.50	8
4	36	0	80	100.00	9

Table 3 gives the desired OTS and SV, where T and F represent true and false, respectively. In this case, TQS process reduces the test size to nine test cases only, which significantly improves the TV process.

Table 3. OTS and SV for the 4-bit Magnitude Comparator.

#TC	OTS TC (a0...a3, b0...b3)	SV Outputs ($A > B, A = B, A < B$)	Accumulative faults detected/80
1	FFFFFFFF	F T F	53
2	TTTTTTTT	F T F	65
3	FTTTTTTT	F F T	68
4	TTFTFTFT	T F F	71
5	TTFFTFTT	T F F	72
6	TTTFTTFF	T F F	75
7	TTFTTTTF	F F T	77
8	FFTTTTTF	F F T	78
9	TFTTTFTF	T F F	80

To illustrate how the verification process is done (see Figure 2), assume that the second output (i.e., A=B) is out-of-order (i.e., malfunction). Suppose that A=B output is always on (i.e., short circuit to "VCC"). This fault cannot be detected as either TC1 or TC2 (according to Table 2). Nevertheless, when TC3, the output vector ("VV") of faulty IC, is FTT, and the SV is FFT, the TV process can straightforwardly detects that the IC is malfunctioning (i.e., cut fails).

To consider the effectiveness of the proposed strategy in the production line, we return to our illustrative example given in Section 1. Here, the reduction of exhaustive test from 256 test cases to merely nine test cases is significantly important. In this case, the TV process requires only 9 seconds instead of 256 seconds for considering all tests. Now, using one testing line and adopting our strategy for two weeks can test (14X24X60X60/9 = 134400) chips. Hence, to deliver one millions tested ICs' during these two weeks, our strategy requires eight parallel testing lines instead of 212 testing lines (if the test depends on exhaustive testing strategy). Now, if we consider the saving efforts factor as the size of exhaustive test suite minus optimized test suite to the size of exhaustive test suite, we would obtain the saving efforts factor of 256-9/256=96.48%.

Comparison

In this section, we demonstrate the possible test reduction using block partitioning approach [1, 37] for comparison purposes. Here, the partitions could be two 4-bit numbers, with block values =0, 0<x<15, =15 and 9 test cases would give all combination coverage. In this case, we have chosen x=7 as a representative value. Additionally, we have also run a series of 9 tests where x is chosen at random between 0 and 15. The results of the generated test cases and their corresponding cumulative faults detected are tabulated in Tables 4 and 5, respectively.

Table 4. Cumulative faults detected when x = 7.

#TC	TC (a0...a3, b0...b3)	Cumulative faults detected /80
1	FFFFFFFF	53
2	FFFFFTTT	54
3	FFFFTTTT	54
4	FTTTFFFF	59
5	FTTTFTTT	67
6	FTTTTTTT	70
7	TTTTFFFF	71
8	TTTTFTTT	71
9	TTTTTTTT	72

Table 5. Cumulative faults detected when x is randomly selective.

#TC	TC (a0...a3, b0...b3)	Cumulative faults detected /80
1	FFFFFFFF	53
2	FFFFFTTF	55
3	FFFFTTTT	55
4	TFTTFFFF	59
5	TFFTFTTT	61
6	TFTFTTTT	61
7	TTTTFFFF	61
8	TTTTTFFF	64
9	TTTTTTTT	72

Referring to Tables 4 and 5, we observe that block partitioning techniques have achieved the mutant score of 90%. For comparative purposes, it should be noted that our proposed strategy achieved a mutant score of 100% with the same number of test cases.

Conclusion

In this paper, we present a novel strategy for automatic quality signing and verification technique for both hardware and software testing. Our case study in hardware production line demonstrated that the proposed strategy could improve the saving efforts factor significantly. In fact, we also demonstrate that our proposed strategy outperforms the traditional block partitioning strategy in terms of achieving better mutant score with the same number of test cases. As such, we can also potentially predict benefits in terms of the time and cost saving if the strategy is applied as part of software testing endeavor.

Despite giving a good result (i.e., as demonstrated in earlier sections), we foresee a number of difficulties as far as adopting mutation testing is concerned. In general, mutation testing does not scale well. Applying mutation testing in large programs can result in very large numbers of mutations making it difficult to find a good test suite to kill all the mutants. We are addressing this issue as part of our future work by dealing with variable strength interaction testing.

Finally, we also plan to investigate the application of our proposed strategy for computer-aided software application and hardware design tool.

Acknowledgements

The authors acknowledge the help of Jeff Offutt, Jeff Lei, Raghu Kacker, Rick Kuhn, Myra B. Cohen, and Sudipto Ghosh for providing them with useful comments and the background materials. This research is partially funded by the USM: Post Graduate Research Grant—T-Way Test Data Generation Strategy Utilizing Multicore System, USM GRID—The Development and Integration of Grid Services & Applications, and the fundamental research grants—"Investigating Heuristic Algorithm to Address Combinatorial Explosion Problem" from the Ministry of Higher Education (MOHE). The first author, Mohammed I. Younis, is the USM fellowship recipient.

References

1. M. Grindal, J. Offutt, and S. F. Andler, "Combination testing strategies: a survey," Tech. Rep. ISETR-04-05, GMU, July 2004.
2. M. I. Younis, K. Z. Zamli, and N. A. M. Isa, "Algebraic strategy to generate pairwise test set for prime number parameters and variables," in Proceedings of the International Symposium on Information Technology (ITSim '08), vol. 4, pp. 1662–1666, IEEE Press, Kuala Lumpur, Malaysia, August 2008.
3. M. I. Younis, K. Z. Zamli, and N. A. M. Isa, "IRPS: an efficient test data generation strategy for pairwise testing," in Proceedings of the 12th International Conference on Knowledge-Based and Intelligent Information & Engineering Systems (KES '08), vol. 5177 of Lecture Notes in Computer Science, pp. 493–500, 2008.

4. Jenny tool, June 2009, http://www.burtleburtle.net/bob/math/.
5. TVG tool, June 2009, http://sourceforge.net/projects/tvg/.
6. Y. Lei, R. Kacker, D. R. Kuhn, V. Okun, and J. Lawrence, "IPOG: a general strategy for T-way software testing," in Proceedings of the International Symposium and Workshop on Engineering of Computer Based Systems, pp. 549–556, Tucson, Ariz, USA, March 2007.
7. Y. Lei, R. Kacker, D. R. Kuhn, V. Okun, and J. Lawrence, "IPOG-IPOG-D: efficient test generation for multi-way combinatorial testing," Software Testing Verification and Reliability, vol. 18, no. 3, pp. 125–148, 2008.
8. M. Forbes, J. Lawrence, Y. Lei, R. N. Kacker, and D. R. Kuhn, "Refining the in-parameter-order strategy for constructing covering arrays," Journal of Research of the National Institute of Standards and Technology, vol. 113, no. 5, pp. 287–297, 2008.
9. R. C. Bryce and C. J. Colbourn, "A density-based greedy algorithm for higher strength covering arrays," Software Testing Verification and Reliability, vol. 19, no. 1, pp. 37–53, 2009.
10. M. I. Younis, K. Z. Zamli, and N. A. M. Isa, "A strategy for grid based T-Way test data generation," in Proceedings the 1st IEEE International Conference on Distributed Frameworks and Application (DFmA '08), pp. 73–78, Penang, Malaysia, October 2008.
11. R. Mandl, "Orthogonal latin squares: an application of experiment design to compiler testing," Communications of the ACM, vol. 28, no. 10, pp. 1054–1058, 1985.
12. T. Berling and P. Runeson, "Efficient evaluation of multifactor dependent system performance using fractional factorial design," IEEE Transactions on Software Engineering, vol. 29, no. 9, pp. 769–781, 2003.
13. L. Lazic and D. Velasevic, "Applying simulation and design of experiments to the embedded software testing process," Software Testing Verification and Reliability, vol. 14, no. 4, pp. 257–282, 2004.
14. L. White and H. Almezen, "Generating test cases for GUI responsibilities using complete interaction sequences," in Proceedings of the International Symposium on Software Reliability Engineering (ISSRE '00), pp. 110–121, IEEE Computer Society, San Jose, Calif, USA, 2000.
15. A. M. Memon and M. L. Soffa, "Regression testing of GUIs," in Proceedings of the 9th Joint European Software Engineering Conference (ESEC) and the 11th SIGSOFT Symposium on the Foundations of Software Engineering (FSE-11), pp. 118–127, ACM, September 2003.

16. C. Yilmaz, M. B. Cohen, and A. A. Porter, "Covering arrays for efficient fault characterization in complex configuration spaces," IEEE Transactions on Software Engineering, vol. 32, no. 1, pp. 20–34, 2006.

17. M. S. Reorda, Z. Peng, and M. Violanate, Eds., System-Level Test and Validation of Hardware/Software Systems, Advanced Microelectronics Series, Springer, London, UK, 2005.

18. R. Brownlie, J. Prowse, and M. S. Phadke, "Robust testing of AT&T PMX/StarMail using OATS," AT&T Technical Journal, vol. 71, no. 3, pp. 41–47, 1992.

19. S. R. Dalal, A. Jain, N. Karunanithi, et al., "Model-based testing in practice," in Proceedings of the International Conference on Software Engineering, pp. 285–294, 1999.

20. K.-C. Tai and Y. Lei, "A test generation strategy for pairwise testing," IEEE Transactions on Software Engineering, vol. 28, no. 1, pp. 109–111, 2002.

21. D. R. Wallace and D. R. Kuhn, "Failure modes in medical device software: an analysis of 15 years of recall data," International Journal of Reliability, Quality, and Safety Engineering, vol. 8, no. 4, pp. 351–371, 2001.

22. D. R. Kuhn and M. J. Reilly, "An investigation of the applicability of design of experiments to software testing," in Proceedings of the 27th NASA/IEEE Software Engineering Workshop, pp. 91–95, IEEE Computer Society, December 2002.

23. D. R. Kuhn, D. R. Wallace, and A. M. Gallo Jr., "Software fault interactions and implications for software testing," IEEE Transactions on Software Engineering, vol. 30, no. 6, pp. 418–421, 2004.

24. D. R. Kuhn and V. Okun, "Pseudo-exhaustive testing for software," in Proceedings of the 30th Annual IEEE/NASA Software Engineering Workshop (SEW '06), pp. 153–158, April 2006.

25. R. Kuhn, Y. Lei, and R. Kacker, "Practical combinatorial testing: beyond pairwise," IT Professional, vol. 10, no. 3, pp. 19–23, 2008.

26. D. T. Tang and C. L. Chen, "Iterative exhaustive pattern generation for logic testing," IBM Journal of Research and Development, vol. 28, no. 2, pp. 212–219, 1984.

27. S. Y. Boroday, "Determining essential arguments of Boolean functions," in Proceedings of the International Conference on Industrial Mathematics (ICIM '98), pp. 59–61, Taganrog, Russia, 1998.

28. A. K. Chandra, L. T. Kou, G. Markowsky, and S. Zaks, "On sets of Boolean n-vectors with all k-projections surjective," Acta Informatica, vol. 20, no. 1, pp. 103–111, 1983.

29. G. Seroussi and N. H. Bshouty, "Vector sets for exhaustive testing of logic circuits," IEEE Transactions on Information Theory, vol. 34, no. 3, pp. 513–522, 1988.

30. I. I. Dumer, "Asymptotically optimal codes correcting memory defects of fixed multiplicity," Problemy Peredachi Informatskii, vol. 25, pp. 3–20, 1989.

31. S. Ghosh and J. L. Kelly, "Bytecode fault injection for Java software," Journal of Systems and Software, vol. 81, no. 11, pp. 2034–2043, 2008.

32. A. A. Avizienis, The Methodology of N-Version Programming, Software Fault Tolerance, John Wiley & Sons, New York, NY, USA, 1995.

33. L. Chen and A. Avizienis, "N-version programming: a fault-tolerance approach to reliability of software operation," in Proceedings of the 18th IEEE International Symposium on Fault-Tolerant Computing, pp. 3–9, 1995.

34. Y.-S. Ma, J. Offutt, and Y. R. Kwon, "MuJava: an automated class mutation system," Software Testing Verification and Reliability, vol. 15, no. 2, pp. 97–133, 2005.

35. MuJava Version 3, June 2009, http://cs.gmu.edu/~offutt/mujava/.

36. M. M. Mano, Digital Design, Prentice Hall, Upper Saddle River, NJ, USA, 3rd edition, 2002.

37. L. Copeland, A Practitioner's Guide to Software Test Design, STQE Publishing, Norwood, Mass, USA, 2004.

A Tester-Assisted Methodology for Test Redundancy Detection

Negar Koochakzadeh and Vahid Garousi

ABSTRACT

Test redundancy detection reduces test maintenance costs and also ensures the integrity of test suites. One of the most widely used approaches for this purpose is based on coverage information. In a recent work, we have shown that although this information can be useful in detecting redundant tests, it may suffer from large number of false-positive errors, that is, a test case being identified as redundant while it is really not. In this paper, we propose a semiautomated methodology to derive a reduced test suite from a given test suite, while keeping the fault detection effectiveness unchanged. To evaluate the methodology, we apply the mutation analysis technique to measure the fault detection effectiveness of the reduced test suite of a real Java project. The results confirm that the proposed manual interactive inspection process leads to a reduced test suite with the same fault detection ability as the original test suite.

Introduction

In today's large-scale software systems, test (suite) maintenance is an inseparable part of software maintenance. As a software system evolves, its test suites need to be updated (maintained) to verify new or modified functionality of the software. That may cause test code to erode [1, 2]; it may become complex and unmanageable [3] and increase the cost of test maintenance. Decayed parts of test suite that cause test maintenance problems are referred to as test smells [4].

Redundancy (among test cases) is a discussed but a seldom-studied test smell. A redundant test case is one, which if removed, will not affect the fault detection effectiveness of the test suite. Another type of test redundancy discussed in the literature (e.g., [5, 6]) is test code duplication. This type of redundancy is similar to conventional source code duplication and is of syntactic nature. We refer to the above two types of redundancy as semantic and syntactic test redundancy smells, respectively. In this work, we focus on the semantic redundancy smell which is known to be more challenging to detect in general than the syntactic one [5].

Redundant test cases can have serious consequences on test maintenance. By modifying a software unit in the maintenance phase, testers need to investigate the test suite to find all relevant test cases which test that feature and update them correctly with the unit. Finding all of the related test cases increases the cost of maintenance. From the other hand, if test maintenance (updating) is not conducted carefully, the integrity of the entire test suite will be under question. For example, we can end up in a situation in which two test cases test the same features of a unit, if one of them is updated correctly with the unit and not the other one, one test may fail while the other may pass, making the test results ambiguous and conflicting.

The motivation for test redundancy detection is straightforward. By detecting and dealing with redundant test case (e.g., carefully removing them), we reduce test maintenance cost and the risk of loosing integrity in our test suite, while fault detection capability of our test suite remains constant.

One of the most widely used approaches in the literature (e.g., [6–11]) for test redundancy detection, also referred to as test minimization, is based on coverage information. The rationale followed is that, if several test cases in a test suite execute the same program elements, the test suite can then be reduced to a smaller suite that guarantees equivalent test coverage ratio [6].

However, test redundancy detection based on coverage information does not guarantee to keep fault detection capability of a given test suite. Evaluation results from our previous work [12] showed that although coverage information can be very useful in test redundancy detection, detecting redundancy only based on this information may lead to a test suite which is weaker in detecting faults than the original one.

Considering fault detection capability of a test case for the purpose of redundancy detection is thus very important. To achieve this purpose, we propose a collaborative process between testers and a proposed redundancy detection engine to guide the tester to use valuable coverage information in a proper and useful way.

The output of the process is a reduced test suite. We claim that if testers play their role carefully in this process, fault detection effectiveness of this reduced test set would be equal to the original set.

High amount of human effort should be spent on inspecting a test suite manually. However, the proposed process in this paper tries to use the coverage information in a constructive fashion to reduce the required tester efforts. More automation can be added to this process later to save more cost and thus the proposed process should be considered as the first step to reduce required human effort for test redundancy detection.

To evaluate our methodology, we apply the mutation technique in a case study in which common types of faults are injected. Then original and reduced test set are then executed to detect faulty versions of the systems. The results show similar capability of fault detection for those two test sets.

The remainder of this paper is structured as follows. We review the related works in Section 2. Our recent previous work [12] which evaluated the precision of test redundancy detection based on coverage information is summarized in Section 3. The need for knowledge collaboration between human testers and the proposed redundancy detection engine is discussed in Section 4. To leverage and share knowledge between the automated engine and human tester, we propose a collaborative process for redundancy detection in Section 5. In Section 6, we show the results of our case study and evaluate the results using the mutation technique. Efficiency, precision, and a summary of the proposed process are discussed in Section 7. Finally, we conclude the paper in Section 8 and discuss the future works.

Related Works

We first review the related works on test minimization and test redundancy detection. We then provide a brief overview of the literature on semiautomated processes that collaborate with software engineers to complete tasks in software engineering and specifically in software testing.

There are numerous techniques that address test suite minimization by considering different types of test coverage criteria (e.g., [6–11]). In all of those works, to achieve the maximum possible test reduction, the smallest test set which covers the same part of the system was created [7]. The problem of finding the smallest test set has been shown to be NP-complete [13]. Therefore, in order to find an approximation to the minimum cardinality test set, heuristics are usually used in the literature (e.g., [7, 9]).

A few works have applied data flow coverage criteria (e.g., [7, 10]) while a few others have applied control flow criteria (e.g., [6, 9, 11]).

In [7], in addition to the experiment which was performed for all-definition-use coverage criterion on a relatively simple program (LOC is unknown), the authors mentioned that all the possible coverage criteria should be considered in order to detect redundant test cases more precisely. The authors were able to reduce 40% of the size of the test suite under study based on coverage information.

Coverage criteria used in [10] were predicate-use, computation-use, definition-use, and all-uses. The authors applied their approach on 10 Unix programs (with average LOC of 354) and 91% of the original test suites were reduced in total.

The control flow coverage criteria used in [6, 9, 11] are Branch [6], statement [9], and MC/DC [11]. In [9], mutation analysis was used to assess and evaluate the fault detection effectiveness of the reduced test suites. The ratios of reduction reported in these works were 50%, 34%, and 10%, respectively. The Systems Under Tests (SUTs) used in [6, 9] were small scale (avg. LOC of 29 and 231, resp.), while [11] used a medium size space program as its SUT with 9,564 LOC.

The need to evaluate test redundancy detection by assessing fault detection effectiveness was mentioned in [6, 11]. In those works, faults were manually injected into the SUTs to generate mutants. Then the mutation scores of original and reduced test sets were compared. Reference [6] concludes that test minimization based on coverage information can reduce the ability of fault detection, while [11] showed opposite conclusions.

In [6], faults were seeded to the SUTs manually by modifying mostly a single line of code (first order mutation), while in a few other cases, the authors modified between two and five lines of code (k-order mutation). As mentioned in [6], ten people (mostly without knowledge of each other's work) had tried to introduce faults that were as realistic as possible, based on their experience with real programs.

In [11], the manually injected faults (18 of them) were obtained from the error-log maintained during its testing and integration phase. Eight faults were in the "logic omitted or incorrect" category, seven faults belong to the type of "computational problems," and the remaining three faults had "data handling problems" [11].

In our previous work [12], an experiment was performed with 4 real Java programs to evaluate coverage-based test redundancy detection. The objects of study were JMeter, FitNesse, Lurgee and Allelogram with LOC of 69,424, 22,673, 7,050, and 3,296, respectively. Valuable lessons learned from our previous experiment revealed that coverage information cannot be the only source of knowledge to precisely detect test redundancy. Lessons are summarized in Section 3 of this paper.

To the best of the authors' knowledge, there has been no existing work to improve the shortcomings (imprecision) of coverage-based redundancy detection. In this paper, we are proposing a semiautomated process for this purpose.

Semiautomated decision supports systems leverage human-computer interaction which put together the knowledge of human users and intelligent systems to support decision-making tasks. Hybrid knowledge is very effective in such situations where the computational intelligence provides a set of qualified and diversified solutions and human experts are involved interactively in the decision-making process for final decision [14].

A logical theory of human-computer interaction has been suggested by Milner [15]. Besides, the ways in which open systems' behavior can be expressed by the composition of collaborative components is explained by Arbab [16]. There are various semiautomated systems designed for software engineering such as user-centered software design [17].

There have also been semiautomated systems used specifically in software testing. For instance, test case generation tools require tester's assistance in providing test oracles [18]. Another example of collaborative tool for testing is manual testing frameworks [19]. In these tools, testers perform test cases manually while system records them for later uses. The process proposed in this paper is a semiautomated framework with the purpose of finding test redundancy in software maintenance phase.

Coverage-Based Redundancy Detection can be Imprecise

In our previous work [12], we performed an experiment to evaluate test redundancy detection based only on coverage information. We formulated two experimental metrics for coverage-based measurement of test redundancy in the context of JUnit test suites. We then evaluated the approach by measuring the redundancy of four real Java projects (FitNesse, Lurgee, Allelogram, and JMeter). The automated test redundancy measures were compared with manual redundancy decisions derived from inspection performed by a human software tester.

In this paper, we use the term test artifact for different granularity levels supported in JUnit (Figure 1). Three levels of package, class, and methods are grouping mechanism for test cases that have been introduced in JUnit.

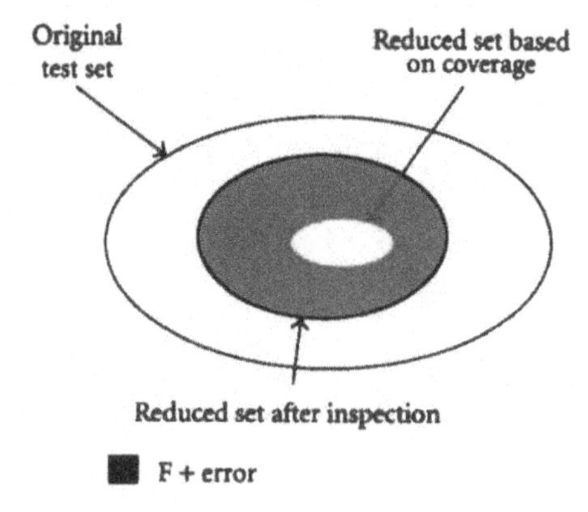

Figure 1. Test granularity in JUnit.

The results from that study [12] showed that measuring test redundancy based only on coverage information is vulnerable to imprecision given the current implementation of JUnit unit test framework and also coverage tools. The following discussion explains the root causes.

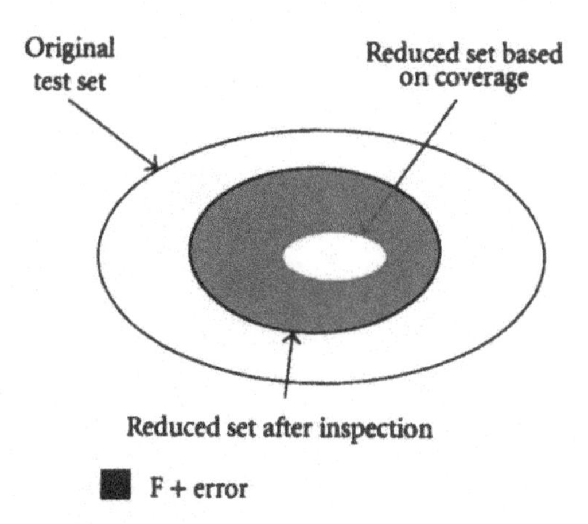

Figure 2. False-Positive Error in Test Redundancy Detection based on Coverage Information.

In the SUTs we analyzed in [12], about 50% of test artifacts, manually recognized as nonredundant, had been detected as redundant tests by our coverage-based redundancy metrics. In a Venn diagram notation, Figure 2 compares a hypothetical original test set with two reduced sets showing high number of false-positive errors. Three main reasons discovered in [12] to justify the errors are discussed next.

(1) Test redundancy detection based on coverage information in all previous works have been done by only considering limited number of coverage criteria. This fact that two test cases may cover the same part of SUT according to one coverage criterion but not the other one causes impreciseness in test redundancy detection only by considering one coverage criterion.

(2) In JUnit, each test case contains four phases: setup, exercise, verify, and teardown [4]. In the setup phase the required state of the SUT for the purpose of a particular test case is setup. In the exercise phase, the SUT is exercised. In the teardown phase the SUT state is rolled back into the state before running the test. In these three phases SUT is covered while in the verification phase only a comparison between expected and actual outputs is performed and SUT is not covered. Therefore, there might be some test cases with the same covered part of SUT with various verifications. In this case, coverage information may lead to detecting a nonredundant test as redundant.

(3) Coverage information is calculated only based on the SUT instrumented for coverage measurement. External resources (e.g., libraries) are not usually instrumented. There are cases in which two test methods cover different libraries. In such cases, the coverage information of the SUT alone is not enough to measure redundancies.

Another reason of impreciseness in redundancy detection based on coverage information mentioned in [12] was some limitations in coverage tools implementation. For example, the coverage tool that we used in [12] was CodeCover [20]. The early version of this tool (version 1.0.0.0) was unable to instrument return and throw statements due to a technical limitation. Hence, the earlier version of the tool excluded covering of such statements from coverage information. This type of missing values can lead to detecting a nonredundant test as redundant. However, this limitation has now been resolved in the newest version of CodeCover (version 1.0.0.1 released on April 2009) and we have updated our redundancy detection framework by using the latest version of this tool. Since in [12] this problem was a root cause of false positive error, here we just report this as a possible reason of impreciseness in redundancy detection, while in this paper we do not have this issue.

Algorithm 1 shows the source code of two test methods from Allelogram test suite as an example of incorrect redundancy detection by only applying coverage information. In this example, test method testAlleleOrderDoesntMatter covers a subset of covered items by the test method testOffset both in setup and exercise phases. The setup phase includes calling Genotype (new double) constructor. The exercise phase contains calling getAdjestedAlleleValues(int) method by passing the created Genotype object, which both are called in the second test method as well. However, the assertion goal in the first test is completely different from the assertion goal in the second one. In the first test method, the goal is comparing the output value of getAdjestedAlleleValues method for two Genotype objects, while in second one, one of the goals is checking the size of output list from the getAdjestedAlleleValues method. Therefore, although according to coverage information the first test method is redundant, in reality it is nonredundant.

```
public void testAlleleOrderDoesntMatter () {
   Genotype g1 = new Genotype(new double [ ] {0,1});
   Genotype g2 = new Genotype(new double [ ] [13]);
   assertTrue (g1.getAdjustedAlleleValues (2).
           equals(g2.getAdjustedAlleleValues (2)));
}
public void testOffset (){
   Genotype g = new Genotype(new double [ ]{0,1});
   g.offsetBy (0.5);
   List<Double> adjusted =
   g.getAdjustedAlleleValues (2);
   assertEquals (2, adjusted.size ());
   assertEquals (0.5, adjusted.get (0));
   assertEquals (1.5, adjusted.get (1));
   g.clearOffset();
   adjusted = g.getAdjustedAlleleValues (2);
   assertEquals (0.0, adjusted.get (0));
   assertEquals (1.0, adjusted.get (1));
}
```

Algorithm 1: Source code of two test methods in the Allelogram test suite.

The Need for Knowledge Collaboration with Testers

Reduced test set based on coverage information contains those test artifacts that cover at least one coverable item not covered by any other test artifact. Therefore these test artifacts contribute to achieving more coverage and according to the concept of test coverage, they may increase the fault detection capability of the test suites.

Based on the above discussion, it is worthwhile to use coverage information for test redundancy detection to reduce the number of test artifacts that might be redundant.

On the other side, high ratio of false-positive errors shows that the coverage-based results alone are not reliable and we may inaccurately detect many nonredundant test artifacts as redundant ones.

The above advantages and disadvantages of coverage-based redundancy detection have motivated us to improve the test redundancy detection process by leveraging knowledge from human testers. The three main root causes of imprecision discussed in Section 3 should be considered in such a tester-assisted approach.

First, the more coverage criteria are applied, the more precise test redundancy will be detected. However, all of the existing test coverage tools support a limited number of coverage criteria. White-box criteria are more usually supported, while there are only a few tools supporting black-box criteria (e.g., JFeature [21]). In addition, usually there are no precise formal specifications for some units in some systems. Thus, automated measurement of black-box coverage is impossible in those cases. Also, there is a lack of coverage tools which automatically measure both white-box and black-box coverage criteria at the same time. Combing the coverage results from various coverage tools might be a solution. However, lack of formal specification for many real projects makes it very challenging for us testers to consider automated measurement of black-box coverage for the purpose of redundancy detection in this work. For projects with full formal specifications, if test minimization is performed precisely with respect to all available coverage criteria, loss of fault detection ability can be minimized or eliminated altogether. However, since formal specifications are not available for many real projects, we propose to involve human testers in the process of test redundancy detection.

For this purpose, testers can use their knowledge to write formal specification for the SUT and use them in black-box coverage tools, or apply black-box coverage manually. For instance, if test t1 covers a subset of covered items by t2, and the main goal of t1 is to check whether there is an exception thrown by the SUT while t2 has a different goal, t1 is not redundant. In other words, the inputs of two above tests are from different equivalence classes (i.e., a black-box coverage criterion should be applied).

Second, the verification phase of JUnit test methods should be analyzed separately. As explained in Section 3, this phase is independent of coverage information, and is thus a precision threat to redundancy detection. Assertion statements in JUnit tests should be compared to find if they cause redundancy or not. In some cases, the actual and expected values in assert statements have complicated data flow. In such cases, comparing assertions in verification phase would require sophisticated source code analysis (e.g., data flow analysis). For example, the actual outcomes of the two assertEquals statements (located in two test methods) in Figure 3 are the same: adjusted.get(). However, determining whether their expected outcomes (a and 1.5) have the same value or not would require data flow analysis in this example. Automating such an analysis is possible, but is challenging while in this step we use human tester for this purpose by leaving its automation as a future work.

```
...
double a = getDefaultAdjusted(0);
...
assertEquals(a, adjusted.get(0));
...
```

```
...
assertEquals(1.5, adjusted.get(0));
...
```

Figure 3. The challenge of comparing assertions: excerpts from the test suite of Allelogram.

Third, in addition to the SUT, all the external libraries used should be considered. However, as the source code of those libraries is not probably available, we need to instrument the class files in Java systems or to monitor coverage through the JVM. As per our investigations, automating this instrumentation and calculating coverage information for the external libraries and combining them with coverage information of the source code of the SUT is challenging and is thus considered as a future work. At this step, we propose the human tester to analyze the test code to find out how an external library affects test results and consider that in comparing test artifacts.

As explained previously, although it is possible to increase the degree of automation to cover the shortcoming of redundancy detection only based on limited number of coverage criteria, there is one main reason that does not allow full automation for this process, which is the lack of precise and formal specification for real world project. In other words, in the process of test redundancy detection the existence of human testers is necessary to confirm the real redundancy of those test artifacts detected as redundant by the system. The human tester has to conduct a manual inspection with guidelines proposed in this work and has to consider the three root causes to prevent false positive errors.

Using the three above guidelines helps testers to collaborate more effectively in the proposed redundancy detection process by analyzing test codes. Testers who have developed test artifacts are the best source of knowledge to decide about test redundancy by considering the above three lessons. However, other test experts can also use our methodology to find the redundancy of a test suite through manual inspection. For instance, in the experiment of this work, the input test suite was created by the developers of an open source project while the first author has performed the process of test redundancy detection.

A Collaborative Process for Redundancy Detection

To systematically achieve test redundancy detection with lower false-positive error, we propose a collaborative process between an automated redundancy detection system and human testers. The system will help the tester to inspect test artifacts with the least required amount of effort to find the actually redundant tests by using the benefits from coverage information while the fault detection capability of the reduced test suite is not reduced.

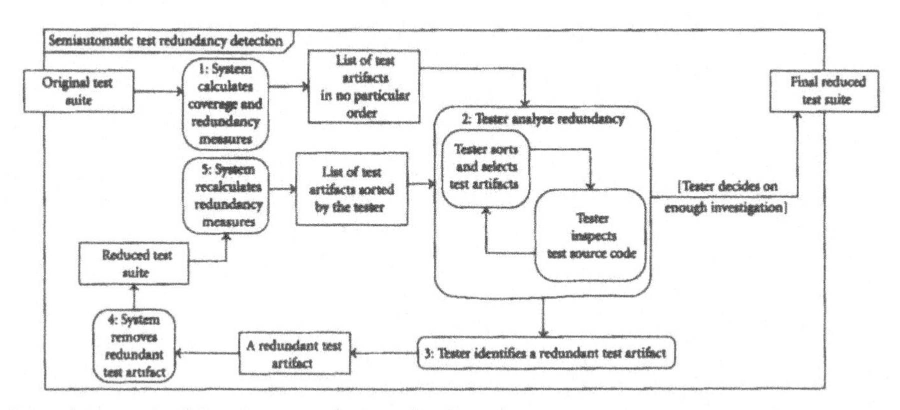

Figure 4. Proposed collaborative process for test redundancy detection.

Figure 4 illustrates the activity diagram of the proposed interactive redundancy detection process. The input of this process is the original test suite of a SUT. Since human knowledge is involved, the precision of the inspection conducted by the human tester is paramount. If the tester follows the process and the three above guidelines carefully, the output would be a reduced test with the same fault detection effectiveness as the original one.

As the first step in this process, redundancy detection system uses a coverage tool to calculate coverage information, which is used later to calculate two redundancy metrics (discussed next).

Two redundancy metrics were proposed in [12]: Pair Redundancy and Suite Redundancy. The Pair Redundancy is defined between two test artifacts and is the ratio of covered items in SUT by the first test artifact with respect to the second one. In Suite Redundancy, this ratio is considered for one test artifact with respect to all other tests in the test suite.

Equations (1) and (2) define the Pair and Suite Redundancy metrics, respectively. In both of these equations, CoveredItemsi(tj) is the set of code items (e.g., statement and branch) covered by test artifact tj, according to a given coverage criterion i (e.g., statement coverage). CoverageCriteria in these two equations is the set of available coverage criteria used during the redundancy detection process.

Based on the design rationale of the above metrics, their values are always a real number in the range of [0…1]. This enables us to measure redundancy in a quantitative domain (i.e., partial redundancy is supported too).

However, the results from [12] show that this type of partial redundancy is not precise and may mislead the tester in detecting the redundancy of the test. For instance, suppose that two JUnit test methods have similar setups with different exercises. If for example 90% of the test coverage is in the common setup the pair redundancy metrics would indicate that they are 90% redundant with respect to each other. However different exercises in these tests separate their goals and thus they should not be considered as redundant with respect to each other while 90% redundancy can mislead the tester about their redundancy.

Equation (1) shows Redundancy of test artifact (tj) with respect to another one (tk):

$$PR\left(t_j, t_k\right) = \left(\sum_{i \in CoverageCriteria} \left|CoveredItems_i\left(t_j\right) \cap CoveredItems_i\left(t_k\right)\right|\right) / \left(\sum_{i \in CoverageCriteria} \left|CoveredItems_i\left(t_j\right)\right|\right) \tag{1}$$

equation (2) shows Redundancy of one test artifact (t_j) with respect to all others:

$$SR\left(t_j\right) = \left(\sum_{i \in CoverageCriteria} \left|CoveredItems_i\left(t_j\right) \cap CoveredItems_i\left(TS - t_j\right)\right|\right) / \left(\sum_{i \in CoverageCriteria} \left|CoveredItems_i\left(t_j\right)\right|\right) \tag{2}$$

However, partial redundancy concept can be useful in some cases to warn testers to refactor test code. To find these cases, in [12], we have offered to separate phases in a test case. As this approach is considered as a future work, in this work we do not consider partial redundancy concept. A test artifact can be redundant or nonredundant. The suite redundancy metric is used as a binary measure to separate test artifacts into these two groups: redundant, and nonredundant. If SR value of a test artifact = 1, that test is considered as redundant otherwise it is nonredundant.

In some cases, a test artifact does not cover any type of items (according to the considered coverage criteria). In [12], we have found that these cases may occur for various reasons, for example, (1) a test case may only cover items outside the SUT (e.g., an external library), (2) a test case may verify (assert) a condition without exercising anything from the SUT, or (3) a test method may be completely empty (developed by mistake). In these cases, the nominator and the denominator of both above metrics (PR and SR) will be zero (thus causing the 0-divide-by-0 problem). We assign the value of NaN (Not a Number) to the SR metric for these cases leaving them to be manually inspected to determine the reason.

After calculating coverage and redundancy metrics, the system prepares a list of test artifacts in no particular order. All the information about coverage ratios, number of covered items and redundancy metrics (both SR for each test and PR for each test pair) is available for exploration by the tester.

Step 2 in the process is the tester's turn. He/she should inspect the tests which are identified as a redundant test by the SR value (=1) to find out whether they are really redundant or not. This manual redundancy analysis should be performed for each test artifact separately. Therefore tester needs to choose a test from a set of candidate redundant tests.

The sequence in which test artifacts are inspected may affect the final precision of the process. Test sequencing often becomes important for an application that has internal state. Dependency between test artifacts may cause the erratic test smell in which one or more tests behave erratically (the test result depends on the result of other tests) [4]. However, in this work we do not consider this smell (our case study does not have this problem and thus we did not have any constraints for sequencing the test artifacts).

Our experience with manual redundancy detection in our case study (discussed in next section) helps us to find that the locality principle of test artifacts is an important factor that should be considered in test sequencing. In other words, for instance, test methods inside one test class have more likelihood of redundancy with respect to each other and should be inspected simultaneously.

There can be different strategies for ordering test artifacts and picking one to inspect at a time. One strategy can be defined according to number of covered items by each test artifact. As discussed next ascending and descending orders of number of coverage items each may have their own benefits.

A test expert may prefer to first choose a test artifact with higher redundancy probability. In this case, we hypothesize that the ascending order based on number of covered items is more suitable. The rationale behind this hypothesis is that the likelihood of covering fewer code items (e.g., statement, branch) by more than one test artifact is more than covering more items by the same test artifacts. Relationship between numbers of covered items by a test artifact with probability of redundancy of that test needs to be analyzed in an experiment. However, this is not the main goal of this paper and we leave it as a future work.

Descending order can have its own benefits. A test expert may believe that having test cases with more covered items would lead to the eager test smell (i.e., a test with too many assertions [22]). In this case, he/she would prefer to first analyze a test that covers more items in the SUT.

Finding a customized order of two above extreme cases by considering their benefits and costs is not discussed in this paper. Also other factors more than redundancy and coverage information may be useful in finding a proper test order.

Another strategy for sorting the existing test cases would be according to their execution time. If one of the objectives of reducing test suite is reducing the execution time, by this strategy test cases which need more time to be executed have more priority of redundancy candidates. However, we believe that in unit testing level execution time of test cases is not as important as other smells like being eager.

After picking appropriate test artifact, tester can use PR values of that test with respect to other tests. This information guides tester to inspect source code of that test case and compare it with source code of those tests with higher PR values. Without this information, manual inspection would take much more time from testers since he/she may not have any idea how to find another test to compare the source code together.

As discussed in Section 4, the main reason of need for human knowledge is to cover shortcomings of coverage-based redundancy detection. Therefore testers should be thoroughly familiar with these shortcomings and attempt at covering them.

After redundancy analysis, the test is identified as redundant or not. If it was detected as redundant by tester (Step 3), system removes it from original test set (Step 4). In this step, the whole collaborative process between system and tester should be repeated. Removing one test from test suite changes the value of CoveredItemsi(TS-tj) in (2). Therefore system should recalculate Suite Redundancy metric for all of the available tests (Step 5). In Section 6 we show how removing a redundant test detected by tester and recalculating the redundancy information can help the tester not to be misled by initial redundancy information and reduce the required effort of the tester.

Stopping condition of this process depends on tester's discretion. To find this stopping point, tester needs to compare the cost of process with savings in test maintenance costs resulting from test redundancy detection. Process cost at any point of the process can be measured by the time and effort that testers have spent in the process.

Test maintenance tasks have two types of costs which should be estimated: (1) costs incurred by updating (synchronizing) test code and SUT code, and (2) costs due to fixing integrity problems in test suite (e.g., one of two test cases testing the same SUT feature fails, while the other passes). Having redundant tests can lead testers to updating more than a test for each modification. Secondly, as a result of having redundant tests, the test suites would suffer from integrity issues, since the tester might have missed to update all the relevant tests.

To estimate the above two cost factors, one might perform change impact analysis on the SUT, and subsequently effort-prediction analysis (using techniques such as [23]) on SUT versus test code changes.

To decide about stopping point of the process, a tester would need to measure the process costs spent so far and to also estimate the maintenance costs containing both the above-discussed cost factors. By comparing them, he/she may decide to either stop or to continue the proposed process.

In the outset of this work, we have not systematically analyzed the above cost factors. As discussed before, we suggest testers to inspect all the tests with the value SR=1 as many as possible. However, according to high number of false-positive errors, other tests in this category (with SR=1) which were not inspected, should be considered as nonredundant. If the SR metric of a test artifact is less than 1, it means that there are some items in the SUT which are covered only by this test artifact. Thus, they should also be considered as nonredundant.

To automate the proposed process for test redundancy detection, we have modified the CodeCover coverage tool [20] to be able to measure our redundancy metrics. We refer to our extended tool as TeReDetect (Test Redundancy Detection tool). The tool shows a list of test artifacts containing coverage and redundancy information of each of them, it lets the tester to sort test artifacts according to his/her strategy (as explained before) and to introduce a real detected redundant test to the system for further metrics recalculation. After detecting a redundant test method, system automatically recalculates the redundancy metrics and updates the tester with new redundancy information for the next inspection iteration. A snapshot of the TeReDetect tool, during the process being applied to Allelogram, is shown in Figure 5.TeReDetect is an open source project (it has been extended to the SVN repository of CodeCover http://codecover.svn.sourceforge.net/svnroot/codecover). TeReDetect is not a standalone plug-in, rather it has been embedded inside the CodeCover plug-in. For instance, ManualRedundancyView.java is one of the extended classes for our tool which is available from http://codecover.svn.sourceforge.net/svnroot/codecover/trunk/code/eclipse/src/org/codecover/eclipse/views/.

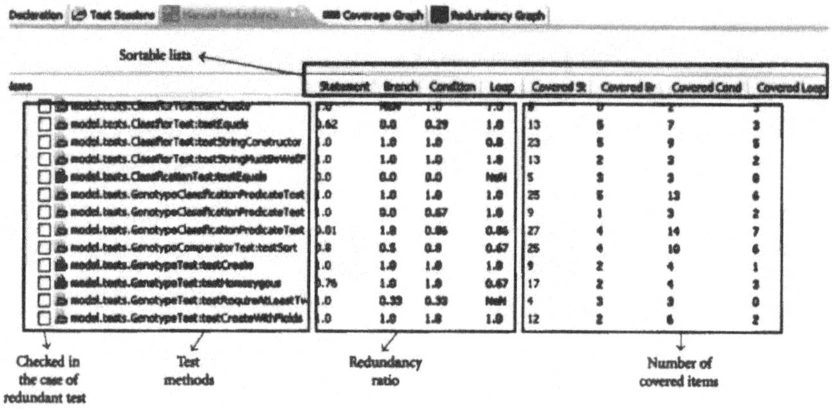

Figure 5. Snapshot of the TeReDetect tool.

Case Study

Performing the Proposed Process

We used Allelogram [24], an open-source SUT developed in Java, as the object of our case study. Allelogram is a program for processing genomes and is used by biological scientists [24]. Table 1 shows the size measures of this system.

Table 1. The size measures of Allelogram code.

SLOC	Number of packages	Number of classes	Number of methods
3,296	7	57	323

TABLE 2: The size measures of Allelogram test suite.

Test suite SLOC	Number of test packages	Number of test classes	Number of test methods
2,358	6	21	82

The unit test suite of Allelogram is also available through its project website [24] and is developed in JUnit. Table 2 lists the size metrics of its test suite. As the lowest implemented test level in JUnit is test method, we applied our redundancy detection process on the test method level in this SUT.

Table 2. The size measures of Allelogram test suite.

Test suite SLOC	Number of test packages	Number of test classes	Number of test methods
2,358	6	21	82

As the first step of proposed redundancy detection process, coverage metrics are measured. For this purpose, we used the CodeCover tool [20] in our experiment. This tool is an open-source coverage tool written in Java supporting the following four coverage criteria: statement, branch, condition (MC/DC), and loop. The loop coverage criterion, as supported by CodeCover, requires that each loop is executed 0 times, once, and more than once.

Table 3 shows the coverage metrics for our SUT. The first row in this table is the coverage ratios of the whole Allelogram system which are relatively low. We also looked at the code coverage of different packages in this system. Our analysis showed that the Graphical User Interface (GUI) package of this SUT is not tested (covered) at all by its test suite. This is most probably since JUnit is supposed to be used for unit testing and not GUI or functional testing. By excluding the GUI package from coverage measurement, we recalculated the coverage values shown in the second row of Table 3. These values show that the non-GUI parts of the system were tested quite thoroughly.

Table 3. Coverage information (%).

	Coverage (%)			
	Statement	Branch	Condition	Loop
Entire Allelogram	23.3	34.7	35.9	22.2
Without GUI components	68.0	72.9	71.4	43.0

The next step in the process is the calculation of suite-level redundancy for each test method and pairwise redundancy for each pair of test methods in the test suite of our SUT.

To automate the measurement of redundancy of each test method using the two metrics defined in Section 5 ((1) and (2)), we have modified CodeCover to calculate the metrics and export them into a text file, once it executes a test suite.

Table 4 reports the percentage of fully redundant test methods (those with SR = 1) according to each coverage criterion and also by considering all of the criteria together.

Table 4. The percentage of fully redundant test methods.

Coverage criteria	Percentage of fully redundant test methods
Statement	77%
Branch	84%
Condition	83%
Loop	87%
All	69%

As we expected, according to Table 4, ratio of full redundancy detected by considering each coverage criteria separately is higher than the case when all of them are considered. This confirms the fact that the more coverage criteria used in redundancy detection, the less false positive error can be achieved. In other words, All coverage criterion detects those tests as nonredundant that improve the coverage ratio values of at least one of the coverage criteria. As All criterion is more precise than the others, in the rest of our case study we consider the suite redundancy based on All criterion.

According to the suite redundancy result by considering all four coverage criteria (Table 4), 31% (100-69) of the tests in test suites of Allelogram are nonredundant. To confirm the nonredundancy of those methods, we randomly sampled a set of test methods in this group and inspected them. We found few cases that seem as redundant tests which are in fact true-negative errors as reported in [12]. However, according to our inspection and code analysis, such test methods cover at least one coverable item not covered by any other test method. For instance, a test method named testOneBin in Allelogram covers a loop only once while some other test methods cover that loop more than one time. Therefore, loop redundancy of this method is slightly less than 1 (0.91) and thus detected as nonredundant by our redundancy metrics. For the same test method, the other types of redundancy considering only statement, branch, and condition coverage are 1. In fact, the above test cases contribute to loop coverage and we thus mark it as nonredundant since it covers a loop in a way (only once) not covered by other test methods.

Having a candidate set of redundant test methods (redundant tests based on All criterion: 69%), tester needs to decide about their order to inspect their source code. In this study, the first author (a graduate student of software testing) manually inspected the test methods. Recall the heuristics discussed in Section 5 about the sorting strategy of test method in the proposed process: test methods with fewer numbers of covered items have higher likelihood of being redundant. We thus decided to order the tests in the ascending order of the number of covered items (e.g., statement). In this case, we hoped to find redundant test methods sooner which may lead to a reduction in the search space (discussed next).

As the next step, manual inspection of a test was performed by comparing the source code of the test with other tests having high pair redundancy with the current one. The main focus of this step should be detecting redundancy by covering the shortcomings of coverage-based redundancy detection discussed in Section 5.

Redundancy of one test affects the redundancy of others. For instance, if test method A is redundant because it covers the same functionality covered by test method B (while there are no other tests to cover this functionality), test method B cannot be redundant at the same time. Therefore, while both of them are candidates for being redundant tests according to coverage information, but only one of them should be considered redundant finally. We refer to such effects as inter-test-method-redundancy effects

By only using redundancy information from the beginning step of the process, tester would need to keep track of all the tests previously detected as redundant during the process and apply the inter-test-method-redundancy effects by him/her self. However, recalculating the coverage information, after each redundancy detection, can reduce the search space (as explained next). Therefore, detecting redundant tests one by one and subsequently recalculating redundancy metrics increase precision and efficiency of the tester.

In this case study, we manually inspected the whole test suite of Allelogram. Figure 6 illustrates the whole process results by showing the size of five different test sets manipulated during the process. Those five test sets are discussed next.

We divide test methods into two categories: redundancy known and redundancy unknown. The test artifacts in the redundancy-unknown set are pending inspection to determine whether they are redundant or not (Set 1). Redundancy-known set contains redundant (Set 2) and nonredundant test sets whose decisions have been finalized. Furthermore, the set of nonredundant tests inside redundancy-known category contains three different sets: those identified through inspection (Set 3), those identified without inspection (Set 4), and the ones that were identified by system as nonredundant after nonredundancy has been detected through inspection (Set 5).

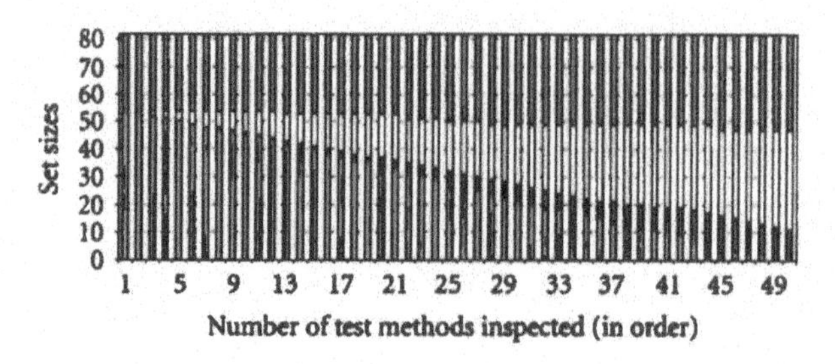

- Set 4-|nonredundant tests identified without inspection|
- Set 5-|nonredundant tests identified with unnecessary inspection|
- Set 3-|nonredundant tests identified with inspection|
- Set 2-|redundant tests|
- Set 1-|remaining tests pending inspection|

Figure 6. Labeling the test cases through the redundancy detection process.

At the beginning of the process, by calculating redundancy metrics based on coverage information, test methods are divided into two sets of Nonredundant Tests without Inspection and Remaining Tests Pending Inspection sets. As the figure shows, 28 test methods were recognized as nonredundant, while 54 (82-28) test methods needed to be inspected.

After each test method inspection, redundancy of that test is identified. This test method then leaves the Remaining Tests Pending Inspection set and Nonredundant test joins Nonredundant Tests with Inspection set while each redundant test joins Redundant Tests set. In the second case, redundancy metrics are recalculated.

In this case study, as shown in Figure 5, 11 test methods are recognized as redundant (test methods numbered in the x-axis as 7, 12, 19, 21, 24, 27, 36, 38, 40, 41, and 44). In these cases, new iterations of the process were performed by recalculating the redundancy metrics. In 5 cases (test methods numbered 12, 21, 24, 27, and 44), the recalculating led to search space reduction (5 test methods left the Remaining Tests Pending Inspection set and joined the Nonredundant Tests without Inspection set). In 2 of them (test methods 21 and 44), recalculating caused 2 test methods to leave Nonredundant Tests with Inspection set and join Nonredundant Tests with Unnecessary Inspection set.

At the beginning of the process, the size of the Remaining Tests Pending Inspection set was 54 (our initial search space). However, through the process, recalculating reduced the number of test methods that needed to be inspected to 49. In this case study, we ordered test methods in the ascending order of number of their covered items.

The final result of the process is a reduced test set containing 71 test methods instead of 82 (the original test suite of Allelogram). Stopping point of this process is considered by inspecting all the redundant candidate test methods (with SR=1) and no cost estimation is applied for this purpose.

Evaluating the Proposed Process

To evaluate the preciseness of the proposed process, we considered the main purpose of test redundancy detection as discussed by many researchers. Test minimization should be performed in a way that the fault detection effectiveness of the test suite is preserved. Therefore, the process is successful if it does not reduce the fault detection capability.

One way to evaluate the above success factor of our test minimization approach is to inject probable faults in the SUT. Mutation is a technique that is widely used for this purpose ([25, 26]). The researches in [27, 28] show that the use of mutation operators is yielding trustworthy results and generated mutants can be used to predict the detection effectiveness of real faults.

In this work, we used the mutation analysis technique for the evaluation of the fault detection effectiveness of the reduced test suites generated by our technique. However, after completing this research project, we found out that, as another approach, we could also use the mutation analysis technique to detect test redundancy in a different alternative approach as follows. If the mutation scores of a given test suite with and without a particular test case are the same, then that test case is considered redundant. In other words, that test case does not kill (distinguish) any additional mutant. We plan to compare the above test redundancy detection approach with the one we conducted in this paper in a future work.

To inject simple faults into our case study, we used the MuClipse [29] tool which is a reincarnation of the MuJava [30] tool in the form of an Eclipse plug-in. Two main types of mutation operators are supported by MuClipse: method level (traditional) and class level (object oriented) [30].

To inject faults according to the traditional mutation operators, MuClipse replaces, inserts or deletes the primitive operators in the program. 15 different types of traditional mutation operators are available in MuClipse [29]. One example of this operators is the Arithmetic Operator Replacement (AOR) [31].

The strategy in object-oriented mutation operators is to handle all the possible syntactic changes for OO features by deleting, inserting, or changing the target syntactic element. 28 different types of OO mutation operators are available in MuClipse [29]. One example is Hiding variable deletion (IHD) which deletes a variable in a subclass that has the same name and type as a variable in the parent class [32].

All the available above mutation operators were used in this experiment. During this step, we found that MuClipse generates some mutants which failed to compile. These types of mutants are referred to as stillborn mutants which are syntactically incorrect and are killed by the compiler [29]. The total number of mutants for Allelogram that were not stillborn was 229.

To evaluate the fault detection effectiveness of the reduced test set by our proposed process compared to original test set, we calculated their mutation scores. We used MuClipse to execute all the created mutants with the two test sets (original and reduced). Table 5 shows the mutation score of three test sets: original test set, reduced test set only based on coverage information, and reduced test set through collaboration process with a tester.

Table 5. Mutation score of three test suites for Allelogram.

Test set	Cardinality	Mutation score
Original	82	51%
Reduced (coverage based)	28	20%
Reduced (collaborative process)	71	51%

The result shows that every mutant that is killed by original test set is killed by the reduced set (derived by the collaborative process) as well. In other words, the effectiveness of these two test sets is equal while the reduced set (solely based on coverage information) has 11 (82-71) less tests than the first one. That test suite thus has lower fault detection effectiveness.

Mutation score decreasing from 51% in original test set to 20% in the reduced set only based on coverage information confirms our discussion in Section 3 about impreciseness of test redundancy detection based only on coverage information.

Discussion

Effectiveness and Precision

Let us recall the main purpose of reducing the number of test cases in a test suite (Section 1): decreasing the cost of software maintenance. Thus, if the proposed methodology turns to be very time consuming, then it will not be worthwhile to be applied.

Although the best way to increase the efficiency of the process is to automate all required tasks, at this step we suppose that it is not practical to automate all of them. Thus, as we discuss next, human knowledge is currently needed in this process.

To perform manual inspection on test suite with the purpose of finding redundancy, testers need to spend time and effort on each test source code and compare them together. To decrease the amount of required effort, we have devised the proposed approach in a way to reduce the number of tests needed to be inspected (by using the suite redundancy metric). Our process also suggests useful information such as pair redundancy metric to help testers find other proper tests to compare with the test under inspection.

We believe that by using the above information, the efficiency of test redundancy detection has been improved. This improvement was seen on our case study while we first spent on average more than 15 minutes for each test method of Allelogram test suite before having our process. But inspecting them using the proposed process took on average less than 5 minutes per test method (the reason of time reduction is that in the later we knew other proper test methods to compare them with the current test). Since only one human subject (tester) performed the above two approaches, different parts of the Allelogram test suite were analyzed in each approach to avoid bias (due to learning and gaining familiarity) on time measurement.

However the above results are based on our preliminary experiment and it is thus inadequate to provide a general picture about the efficiency of the process. For a more systematic analysis in that direction, both time and effort should be measured more precisely with more than one subject on more than one object. Such an experiment is considered as a future work.

In addition to the efficiency of the process, precision of redundancy detection was also evaluated in our work. As explained in Section 6.2, this evaluation has been done in our case study by applying mutation technique. The result of analysis on one SUT confirmed the high precision of the process.

However, human's error is inevitable in collaborative processes which can affect the precision of the whole process. To decrease this type of error, the tester needs to be familiar with the written tests. Therefore, we suggest having the original test suite developers involved in the redundancy detection process if possible or that they be at least available for the possible questions during the process. In other words, a precise teamwork communication is required to detect correct test redundancy.

Cost/Benefit Analysis

According to above discussions, our redundancy detection technique has the following benefits:

(i) Reducing the size of test suite by keeping the fault detection effectiveness of that.

(ii) Preventing possible future integrity issues in the test suite.

(iii) Reducing test maintenance costs.

Different types of required costs in this process are summarized as follows:

(i) TeReDetect installation costs.

(ii) System execution time during the process (steps 1, 4, and 5 in Figure 4).

(iii) Redundancy analysis by human testers (steps 2 and 3 in Figure 4).

The first and second cost items are not considerable while the main part of the cost is about the third one which contains human efforts.

Table 6 shows an informal comparison of above costs and benefits in three approaches of full automation, full manual, and semiautomated process proposed in this paper. In the second and third approaches that human has a role, it is inevitable that the preciseness of human affects the benefits of the results.

Table 6. Cost/benefit comparison.

	Cost	Benefit
Full automation	Low	Imprecise reduced set
Full manual	High	Precise reduced set
Semiautomated	Mid	Precise reduced set

Scalability

In large-scale systems with many LOC and test cases, it is not usually feasible to look at and analyze the test cases for the entire system. However, as mentioned before, in TeReDetect it is possible to select a subset of test suite and also a subset of SUT. This functionality of TeReDetect increases the scalability of this tool to a great extent by making it possible to divide the process of redundancy detection into separate parts and assign each part to a tester. However a precise teamwork communication is required to make the whole process successful.

Flexible stopping point of the proposed process is another reason for its scalability. According to the tester's discretion, the process of redundancy detection may stop after analyzing the subset of test cases or continue for all existing tests. For instance, in huge systems, by considering the cost of redundancy detection, project manager may decide to analyze only the critical part of the system.

Threats to Validity

External Validity

Two issues limit the generalization of our results. The first one is the subject representativeness of our case study. In this paper the process has been done by the first author (a graduate student). More than one subject should be experimented in this process to be able to compare their results to each other. Also, this subject knew the exact objective of the study which is a threat to the result. The second issue is the object program representativeness. We have performed the process and evaluate the result on one SUT (Allelogram). More objects should be used in experiments to improve the result. Also our SUT is a random project chosen from the open source community. Other industrial programs with different characteristics may have different test redundancy behavior.

Internal Validity

The result about efficiency and precision of the proposed process might be from some other factors which we had no control or had not measured. For instance, the bias and knowledge of the tester while trying to find redundancy can be such a factor.

Conclusion and Future Works

Measuring and removing test redundancy can prevent the integrity issues of test suites and decrease the cost of test maintenance. Previous works on test set minimization believed that coverage information is useful resource to detect redundancy.

To evaluate the above idea we performed an experiment in [12]. The result shows that coverage information is not enough knowledge for detecting redundancy according to fault detection effectiveness. However, this information is a very useful starting point for further manual inspection by human testers.

Root-cause analysis of above observation in [12] has helped us to improve the precision of redundancy detection by covering the shortcomings in the process proposed in this paper.

We proposed a collaborative process between human testers and redundancy system based on coverage information. We also performed an experiment with that process on a real java project. This in turn led us to find out that the sharing the knowledge between the human user and the system can be useful for the purpose of test redundancy detection. We conclude that test redundancy detection can be performed more effectively when it is done in an interactive process.

The result of the case study performed in this paper shows that fault detection effectiveness of the reduced set is the same as the original test set while the cost of test maintenance for reduced one is less than the other (since the size of the first set is less than the second one).

The efficiency of this process in terms of time and effort is improved comparing to the case of manual inspection for finding test redundancy without this proposed process.

In this paper, the efficiency factor was discussed qualitatively. Therefore measuring precise time and efforts spent in this process is considered as a future experiment.

Finding the stopping point of the process needs maintenance and effort cost estimation which is not studied thoroughly in this work and is also considered as a future work.

As explained in Section 5, the order of the tests inspected in the proposed process can play an important role in the test reduction result. In this work we suggested a few strategies with their benefits to order the test while this needs to be studied more precisely. Also, test sequential constraints such as the case of dependent test cases are not discussed in this work.

Visualization of coverage and redundancy information can also improve the efficiency of this process extensively. We are now in the process of developing such a visualization technique to further help human testers in test redundancy detect processes.

In addition to above, some tasks which are now done manually in this proposed process could be automated in future works. One example is the automated detection of redundancy in the verification phase of JUnit test methods which will most probably require the development of sophisticated code analysis tools to compare the verification phase of two test methods.

Acknowledgements

The authors were supported by the Discovery Grant no. 341511-07 from the Natural Sciences and Engineering Research Council of Canada (NSERC).

V. Garousi was further supported by the Alberta Ingenuity New Faculty Award no. 200600673.

References

1. S. G. Eick, T. L. Graves, A. F. Karr, U. S. Marron, and A. Mockus, "Does code decay? Assessing the evidence from change management data," IEEE Transactions on Software Engineering, vol. 27, no. 1, pp. 1–12, 2001.
2. D. L. Parnas, "Software aging," in Proceedings of the International Conference on Software Engineering (ICSE '94), pp. 279–287, Sorrento, Italy, May 1994.
3. B. V. Rompaey, B. D. Bois, and S. Demeyer, "Improving test code reviews with metrics: a pilot study," Tech. Rep., Lab on Reverse Engineering, University of Antwerp, Antwerp, Belgium, 2006.
4. G. Meszaros, xUnit Test Patterns, Refactoring Test Code, Addison-Wesley, Reading, Mass, USA, 2007.
5. A. Deursen, L. Moonen, A. Bergh, and G. Kok, "Refactoring test code," in Proceedings of the 2nd International Conference on Extreme Programming and Flexible Processes in Software Engineering (XP '01), Sardinia, Italy, May 2001.
6. G. Rothermel, M. J. Harrold, J. Ostrin, and C. Hong, "An empirical study of the effects of minimization on the fault detection capabilities of test suites," in Proceedings of the Conference on Software Maintenance (ICSM '98), pp. 34–43, Bethesda, Md, USA, November 1998.
7. M. J. Harrold, R. Gupta, and M. L. Soffa, "Methodology for controlling the size of a test suite," ACM Transactions on Software Engineering and Methodology, vol. 2, no. 3, pp. 270–285, 1993.
8. J. A. Jones and M. J. Harrold, "Test-suite reduction and prioritization for modified condition/decision coverage," IEEE Transactions on Software Engineering, vol. 29, no. 3, pp. 195–209, 2003.
9. A. J. Offutt, J. Pan, and J. M. Voas, "Procedures for reducing the size of coverage-based test sets," in Proceedings of the 11th International Conference on Testing Computer Software (ICTCS '95), pp. 111–123, Washington, DC, USA, June 1995.
10. W. E. Wong, J. R. Morgan, S. London, and A. P. Mathur, "Effect of test set minimization on fault detection effectiveness," Software—Practice & Experience, vol. 28, no. 4, pp. 347–369, 1998.

11. W. E. Wong, J. R. Horgan, A. P. Mathur, and Pasquini, "Test set size minimization and fault detection effectiveness: a case study in a space application," in Proceedings of the IEEE Computer Society's International Computer Software and Applications Conference (COMPSAC '97), pp. 522–528, Washington, DC, USA, August 1997.
12. N. Koochakzadeh, V. Garousi, and F. Maurer, "Test redundancy measurement based on coverage information: evaluations and lessons learned," in Proceedings of the 2nd International Conference on Software Testing, Verification, and Validation (ICST '09), pp. 220–229, Denver, Colo, USA, April 2009.
13. M. R. Garey and D. S. Johnson, Computers and Intractability; A Guide to the Theory of NP-Completeness, W. H. Freeman, San Francisco, Calif, USA, 1990.
14. A. Ngo-The and G. Ruhe, "A systematic approach for solving the wicked problem of software release planning," Soft Computing, vol. 12, no. 1, pp. 95–108, 2008.
15. R. Milner, "Turing, computing, and communication," in Interactive Computation: The New Paradigm, pp. 1–8, Springer, Berlin, Germany, 2006.
16. F. Arbab, "Computing and Interaction," in Interactive Computation: The New Paradigm, pp. 9–24, Springer, Berlin, Germany, 2006.
17. M. Takaai, H. Takeda, and T. Nishida, "A designer support environment for cooperative design," Systems and Computers in Japan, vol. 30, no. 8, pp. 32–39, 1999.
18. Parasoft Corporation, "Parasoft Jtest," October 2009, http://www.parasoft.com/jsp/products/home.jsp?product=Jtest.
19. IBM Rational Corporation, "Rational manual tester," January 2009, http://www-01.ibm.com/software/awdtools/tester/manual/.
20. T. Scheller, "CodeCover," 2007, http://codecover.org/.
21. Nitinpatil, "JFeature," June 2009, https://jfeature.dev.java.net/.
22. B. V. Rompaey, B. D. Bois, S. Demeyer, and M. Rieger, "On the detection of test smells: a metrics-based approach for general fixture and eager test," IEEE Transactions on Software Engineering, vol. 33, no. 12, pp. 800–816, 2007.
23. L. C. Briand and J. Wüst, "Modeling development effort in object-oriented systems using design properties," IEEE Transactions on Software Engineering, vol. 27, no. 11, pp. 963–986, 2001.
24. C. Manaster, "Allelogram," August 2008, http://code.google.com/p/allelogram/.

25. R. A. DeMillo, R. J. Lipton, and F. G. Sayward, "Hints on test data selection: help for the practicing programmer," IEEE Computer, vol. 11, no. 4, pp. 34–41, 1978.

26. R. G. Hamlet, "Testing programs with the aid of a compiler," IEEE Transactions on Software Engineering, vol. 3, no. 4, pp. 279–290, 1977.

27. J. H. Andrews, L. C. Briand, Y. Labiche, and A. S. Namin, "Using mutation analysis for assessing and comparing testing coverage criteria," IEEE Transactions on Software Engineering, vol. 32, no. 8, pp. 608–624, 2006.

28. J. H. Andrews, L. C. Briand, and Y. Labiche, "Is mutation an appropriate tool for testing experiments?" in Proceedings of the 27th International Conference on Software Engineering (ICSE '05), pp. 402–411, 2005.

29. B. Smith and L. Williams, "MuClipse," December 2008, http://muclipse.sourceforge.net/.

30. J. Offutt, Y. S. Ma, and Y. R. Kwon, "MuJava," December 2008, http://cs.gmu.edu/~offutt/mujava/.

31. Y. S. Ma and J. Offutt, "Description of method-level mutation operators for java," December 2005, http://cs.gmu.edu/~offutt/mujava/mutopsMethod.pdf.

32. Y. S. Ma and J. Offutt, "Description of class mutation mutation operators for java," December 2005, http://cs.gmu.edu/~offutt/mujava/mutopsClass.pdf.

Automatic Generation of Web Applications from Visual High-Level Functional Web Components

Quan Liang Chen and Takao Shimomura

ABSTRACT

This paper presents high-level functional Web components such as frames, framesets, and pivot tables, which conventional development environments for Web applications have not yet supported. Frameset Web components provide several editing facilities such as adding, deleting, changing, and nesting of framesets to make it easier to develop Web applications that use frame facilities. Pivot table Web components sum up various kinds of data in two dimensions. They reduce the amount of code to be written by developers greatly. The paper also describes the system that implements these high-level functional components as visual Web components. This system assists designers in the development of Web applications based on the page-transition framework that models a Web application as a set of Web page transitions, and by using visual Web components, makes it easier to write processes to be executed when a Web page transfers to another.

Introduction

To develop a Web application, we need to write a lot of codes to perform processes such as display of Web pages, receipt of requests, execution of actions, session operations, database accesses, and business logic. On the other hand, before we develop a Web application, we have some images of the application (i.e., what it will be like or what it will look like) in our mind. The objective of our research is to make it possible to develop Web applications with less code by making full use of such image. Conventional development environments for Web applications only provide fundamental Web components such as text fields, buttons, checkboxes, anchors, and tables. They have not yet supported high-level functional Web components. This paper presents the automatic generation of Web applications that makes use of customizable visual high-level functional Web components such as frames, framesets, and pivot tables, which are close to our image and make it possible to develop Web applications with less code.

Web applications that need to show a lot of information on one Web page at a time often use inline frames, exchange displayed elements using their tabs, or enable users to see a hidden part of the page using the scroll bars of a Web browser. On the other hand, if Web applications use an HTML frame facility, they can display a lot of information on one Web page at a time and show it to users [1]. The users can easily grasp the outline of the provided information, and at the same time, they can see the contents of the frame of interest in detail by extending the frame to the whole page if they need to. Therefore, some Web applications that need to display a lot of information at a time such as computer-assisted instruction systems, Web-based chat systems, and the Help windows of various kinds of Web applications often use the HTML frame facility [2, 3].

The pivot tables sum up various kinds of data in two dimensions. They reduce the amount of code to be written by developers greatly. The paper describes how to implement high-level functional components as visual Web components and it also presents the system, which is an example of their implementation. This system assists designers in the development of Web applications based on the page-transition framework. This framework models a Web application as a set of Web page transitions, and by using visual Web components, makes it easier to write processes to be executed when a Web pag etransfers to another Web page.

Image-Oriented Design

Design Example of a Web Application

In the system, we design a Web application by copying our image of each Web page into a Web page window. Like home page building tools [4], we choose Web components such as hyperlinks, HTML tables, text fields, text areas, submit buttons, framesets, and pivot tables from menus (or buttons), and paste them into a Web page window that corresponds to each Web page. In this section, we consider a simple Web application favoriteCake that obtains the information of favorite cakes by means of questionnaires. This application consists of three Web pages, entryCake, chooseCake, and loveCake.

(1) Customers first enter their names in the entryCake page.

(2) They choose their favorite cake from a menu in the chooseCake page.

(3) The loveCake page shows the cumulative result and the history of their answers.

Figure 1(b) illustrates the design of this application, which consists of entryCake page, chooseCake page, loveCake page, cake DB table, and favoriteCake DB table. Figure 1(a) shows an example of its execution.

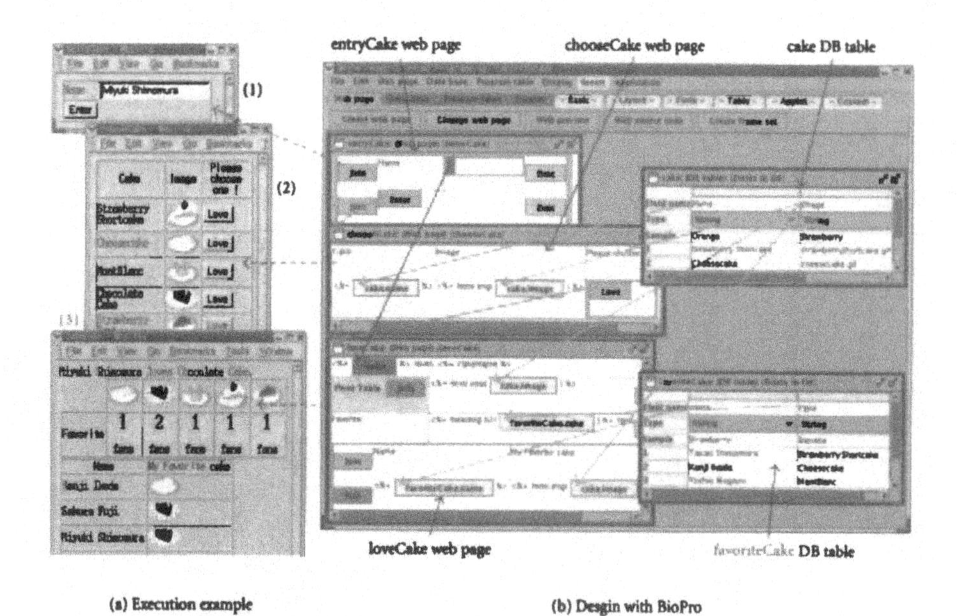

Figure 1. Questionnaire program favoriteCake.

We store the names and the images of a variety of cakes in a database table (hereafter, described as a DB table) cake, which is shown in the top right corner of Figure 1(b). The DB table favoriteCake, which is shown in the bottom right corner of Figure 1(b), stores the history of customers' answers. It stores the customer's name in the name field, and the customer's favorite cake in the cake field. For each kind of cake, the loveCake page displays the number of the customers who like it by using a pivot table (described in detail in Section 4).

We drag and drop each field of these DB tables into the Web pages. By this, the system inserts field references such as cake.name, cake.image, favoriteCake.name, and favoriteCake.cake at the dropped positions. At the execution of the application, these field references display the values of the corresponding fields in various forms such as text, images, buttons, and checkboxes. In addition to DB tables that are stored persistently, the system introduces the Program tables that are only used during the execution of the program. Developers can use these Program tables as visual components [5]. By dragging some fields of a Program table into a Web page, their values can be displayed when the application is executed. The Program table can be joined to other Program tables and DB tables. For example, a Program table can be used to store the contents of a shopping cart in online shopping applications, which is only used in a session of the application.

MVC Architecture

In the system, we develop Web applications based on the Model-View-Controller (MVC) architecture [6] as shown in Figure 2. The MVC architecture isolates business logic from user interface, resulting in an application where it is easier to modify either the visual appearance of the application or the underlying business rules without affecting the other. The model represents the data of the application and the business rules used to manipulate the data; the view corresponds to elements of the user interface; and the controller manages details involving the communication of user actions to the model.

We first design DB tables and Program tables. If database tables have already been created by some DBMS tools, we have only to recall them in the DB table windows of the design workspace. Next, we drag some of the fields of these tables into Web pages to display. Finally, we write business processes in the Web source windows. The system automatically generates methods necessary for accessing the Program table data. In summary, we design the data access layer (model) of the application using DB tables and Program tables, design its presentation layer (view) using Web page windows, and write its business-logic layer (controller) using Web source windows.

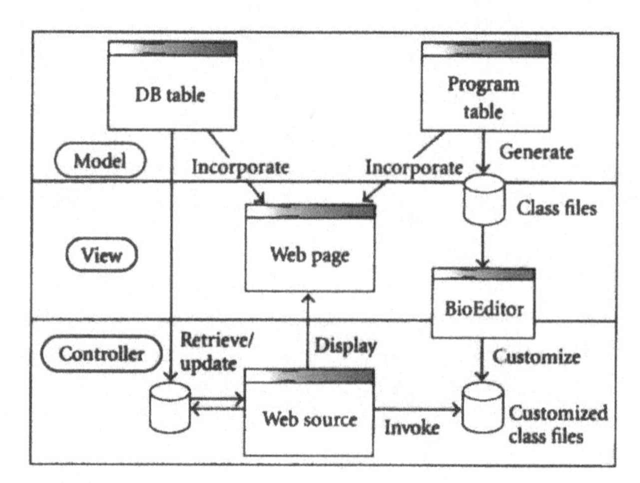

Figure 2. MVC architecture.

In any phase of the development, developers can verify their applications by displaying Web page previews, Web page transitions, Web frame references, and various kinds of field references. As shown in Figure 1(b), the system displays references between components as arrows whose colors indicate the types of the references.

Frame and Frameset Components

Requirements for Frames and Framesets

Web pages can contain framesets, and framesets contain frames. Moreover, the Web page that is displayed in a frame can also contain framesets. Because this relationship ranges among multiple pages, it is difficult to grasp with the conventional development environments. To make it easier to develop Web applications that use frame facilities, we take into account the following requirements.

1. We can nest frames in a frameset more than once, and we can create a frame pointing to another Web page that contains framesets.
2. We can easily understand the relationship between the frames and the Web pages that are displayed in the corresponding frames.
3. We can easily change the hierarchical structure of framesets (i.e., which frameset should contain which frames and framesets), easily add and delete frames, and easily change the Web pages that will be displayed inside frames by using the mouse dragging (see Figure 3(c)).

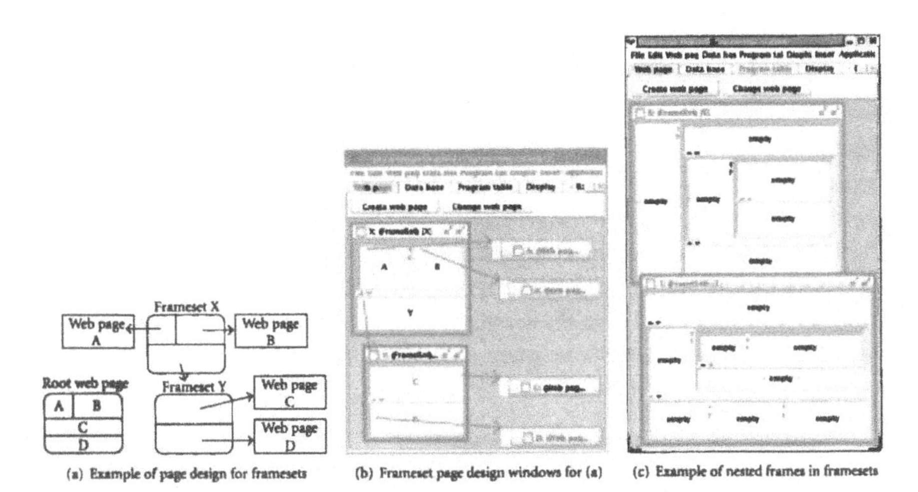

(a) Example of page design for framesets (b) Frameset page design windows for (a) (c) Example of nested frames in framesets

Figure 3. Frameset page design windows created by the BioPro system.

Introduction of Frameset Page Design Windows

We introduce Frameset page design windows as root Web pages that represent framesets, and link each frame included in those windows to an ordinary Web page design window. For example, in Figure 3(a), Frameset page design window Y has two frames, and its upper frame is linked to Web page design window C, and its lower frame is linked to Web page design window D. We make it possible to link each frame included in a Frameset page design window to not only an ordinary Web page design window, but also another Frameset page design window. This enables frames to be nested in a frameset more than once.

Figure 3(b) shows two Frameset page design windows and four Web page design windows created by a programmer using the BioPro system, as illustrated in Figure 3(a). Each frame of the Frameset page design window is linked with a line to the Web page design window that is displayed in it.

As shown in Figure 3(a), the Frameset page design window X has two frames, where its upper frame is divided into two other frames, left and right frames, each of which is linked to Web page design windows A and B, respectively. Its lower frame is linked to another Frameset page design window Y. When we execute the Web application that contains these pages, the Frameset page design window X is displayed as a root Web page that includes four Web pages A, B, C, and D as shown in Figure 3(a). In a Frameset page design window, we can easily divide, delete, and exchange frames by clicking or drag-and-dropping the mouse. Figure 3(c) shows an example of nested frames in Frameset page design windows.

Visual Programming for Frames and Framesets

As an example of a Web application that uses frames, let us consider a simple Web-based chat system. As shown in Figure 4(b), this Web page consists of two frames. The upper frame has a text field to enter a name, and buttons to enter and exit a chat room. Below them, it has an area to display the contents of chatting and its status. The lower frame has a text field to enter a chat message, and a button to send the message. To update the contents of chatting that vary any minute, the upper frame keeps being refreshed at a certain period of time. Because this Web page is divided into these two frames, the lower frame is not affected by refreshing the upper frame even when a user is entering a message.

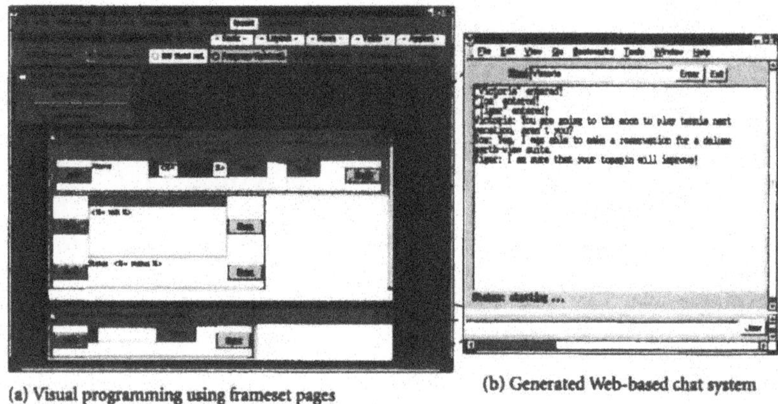

(a) Visual programming using frameset pages

(b) Generated Web-based chat system

Figure 4. Visual programming using frameset pages.

Figure 4(a) shows an example of visual programming of this Web-based chat system. The window (A) in the top left corner of Figure 4(a) is a Frameset page, which corresponds to the Frameset page Y in Figure 3(a). The upper frame is linked to chatFrame Web page to display chatFrame Web page. The lower frame is linked to utterFrame Web page to display utterFrame Web page. Windows (B) and (C) of Figure 4(a) show the Web pages of chatFrame and utterFrame, respectively.

Pivot Table Components

Visual Programming for Pivot Tables

This section introduces a pivot table that sums up various kinds of data in two dimensions and describes how we use this table as a visual Web component. The pivot table consists of three cells, row, column, and data. Into each cell of the pivot table, a field of either a DB table or a Program table is dragged. At the execution time of the application, the row and column cells of the pivot table are dynamically expanded to display the values of the corresponding field of the DB table or the Program table. The data cells of the expanded pivot table display the corresponding field values for each row and each column in a specified form.

Figure 5 illustrates how the pivot table is expanded at the execution time of the application. The way of expanding the pivot table changes depending on whether or not a field of a DB table (or a Program table) is assigned to each cell of the pivot table. Let {dij} be a set of field values in the data cell that corresponds to each row and column. The value of the set $\{d_{ij}\}$ is displayed in the data cell in a variety of forms such as Standard (the sum of the element values), Counter (the number of the elements, that is, $\#\{d_{ij}\}$), and Image (the image the first element's value refers to). In addition, developers can customize a way of displaying the data cell by creating a class and its methods that define how the data cell should be displayed. For example, as shown in Figure 6(a), we can display the name and the picture of the person whose name is the value of d_{ij}. If we further customize it to specify the data cell's format, background color, and component to be displayed as shown in Figure 6(b), checkboxes used in the meeting room reservation system will be changed to buttons as shown in Figure 6(c). Customization of visual functional Web components will be described in more detail in Section 5.

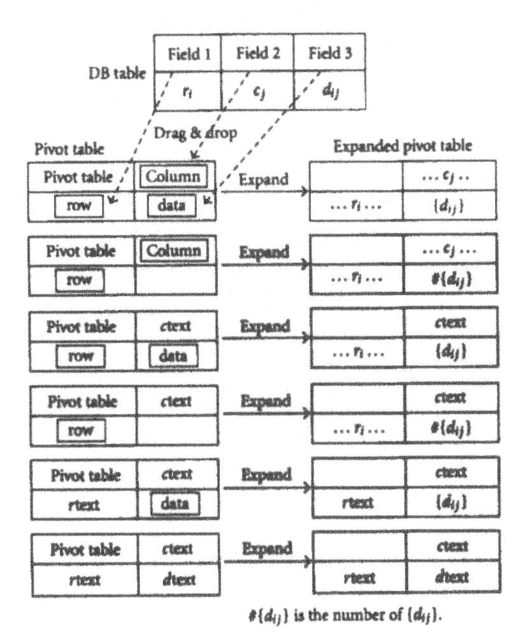

Figure 5. Pivot table expansion.

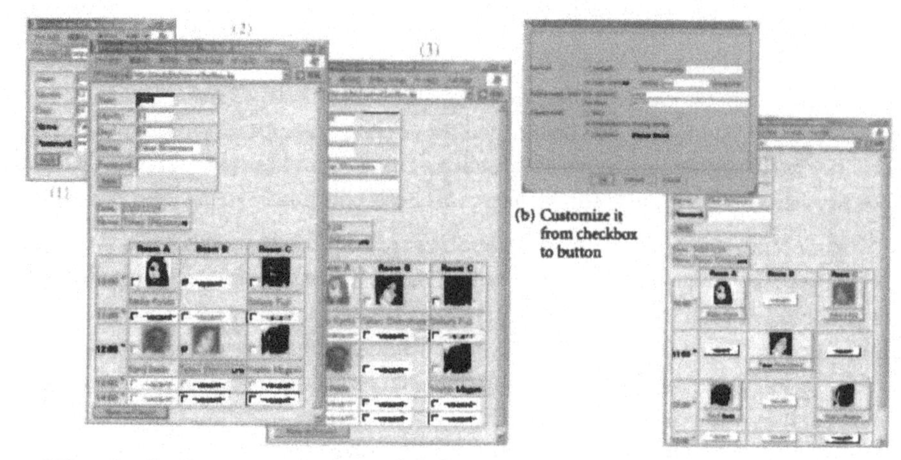

(a) Excecution of meeting-room reservation system with checkboxes

(c) The same system with buttons

Figure 6. Execution of meeting room reservation system.

A Reservation System for Meeting Rooms

Figure 7 shows an example of design for the meeting room reservation system that uses a pivot table, and Figure 6 shows an example of execution of this application. This meeting room reservation system works as follows:

1. Customers enter a date for making a reservation, and their names and passwords.
2. The system displays reservations for that date using the pictures of the people who have reservations.
3. The customers check a vacant room off to make a reservation or check their own pictures to cancel the reservation.

We record reservations for meeting rooms in a database. The DB table meeting (in the top right corner of Figure 7) records meeting rooms (room), periods of time (hour), and people who have reservations (name). This DB table meeting is a virtual table only used for designing the Web page, and the real database table is dynamically determined at the execution time of the application. The pivot table is arranged in the bottom left corner of Figure 7. We drag and drop the room field of the DB table meeting into the column cell of the pivot table to insert a field reference meeting.room. Similarly, we insert a field reference meeting.hour to the row cell of the pivot table, and a field reference meeting.name to the data cell. As shown in Figure 6(a), when this Web application is executed, this pivot table will be dynamically expanded so that each row will display a period of time; each column will display the name of a meeting room; and each data cell will display the name and the picture of the person who has a reservation for the corresponding row and column.

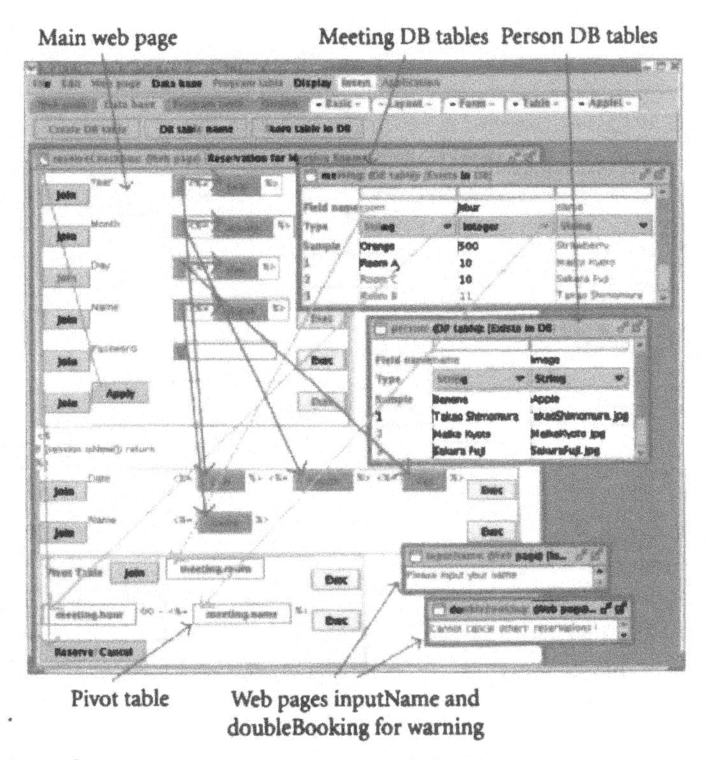

Figure 7. Design of meeting room reservation system.

In this example, we have displayed the data cells as checkboxes. Instead, we can also display them in a variety of forms such as text and buttons. If we display the data cells as submit buttons, we will not need the Reserve/Cancel button, and customers can make a reservation immediately by clicking on a vacant button, and can cancel the reservation immediately by clicking on their own pictures.

Customization of Visual Functional Web Components

Definition of Web Components

In the proposed system, we can create a new Web component and add it to the system. When we design a Web application, we can use these created Web components as visual components in the same way as we use other components such as pivot tables, DB tables, and Program tables. To create a new component, we define a class that extends WebComp class. We have only to define several methods to override those defined in the super WebComp class. Table 1 shows some methods the super WebComp class provides. We define a component name, write code to create a visual component (a JComponent object in Java), and specify how to change the component's properties. For the verification of relationships between components, we write the code that obtains the relationships of the new component with other components. For code generation, we define HTML/JSP code to display this component in a Web page, and so on.

Table 1. Definition of customizable visual Web components.

	Methods	Description
Creation	String getMenuName( )	Define a component name
	JComponent createWebComp( )	Create a visual component
	void changeProperties(JComponent comp)	Display a dialog to change the properties of a component comp
Verification	void addRefers(ArrayList refers)	Check and add a relationship of this component with other components
Code generation	void addParameterNames(HashMap pageToParams)	Add the names of parameters this page sends
	void addInitializeParameterCode(HashMap pageToCode)	Define code that will be executed to analyze received parameters if necessary
	void addJSP(JSP jsp)	Define HTML/JSP code to display a component in a Web page

For example, we here create a new Web component "Autograph." When we choose this component, a dialog will open to enter an autograph. Then, this component will be inserted into a Web page design window. When this Web application is executed, the entered autograph will be shown in italic. Figure 8 shows an Autograph class for this component. This class extends a WebCompAdapter class, which extends the WebComp class a part of whose methods have been shown in Table 1. The Autograph class defines getMenuName() method to specify its component name "Autograph," which will be displayed as the name of a menu item that corresponds to this component. CreateWebComp() method creates a JLabel object to display this visual component in a Web page design window. ChangeProperties() method invokes specifyProperties() method, which displays a dialog to enter this component's properties. AddJSP() method defines how to display this component in a Web page when this Web application is executed. This addJSP() method defines a <h2 style="font-style:italic"> tag to display the specified autograph in italic.

```
public class Autograph extends WebCompAdapter {
public String getMenuName( ) { return "Autograph"; }
public String autograph = "Quan Liang Chen";
public JComponent createWebComp(WebPage
containerWebPage, CellTextPane containerTextPane) {
JLabel label = new JLabel( );
label.setBackground(new Color(200, 255, 200));
label.setBorder(new LineBorder(Color.green));
label.setFont(new Font("SansSerif", Font.ITALIC, 10));
if (!specifyProperties(label)) return null;
return label;
}
public void changeProperties(JComponent comp,
int x, int y) {
specifyProperties(comp);
}
public void addJSP(JSP jsp) throws Exception {
String html = "<h2 style=\"font-style:italic\">" +
autograph + "</h2>; \n";
jsp.code += html;
}
private boolean specifyProperties(JComponent comp) {
String title = "Specify autograph";
String[ ]names = new String[ ]{ "Autograph", };
String[ ]values = { autograph, };
String[ ]defaultValues = { "Who am I?", };
values = new InputTextDialog(title, names, values,
defaultValues).getInputValues( );
if (values == null) return false;
autograph = values[0];
((JLabel) comp).setText(autograph);
return true;
}
}
```

Figure 8. Definition of Web component "Autograph."

Customization of Pivot Tables

A pivot table is one of visual functional Web components. High-level functional Web components are easy to use, and they can easily produce even complicated display of Web pages. However, the higher level they are on, the less flexible they will be. Therefore, it is important to provide a mechanism to customize those components. To customize the display of data cells in a pivot table, we can specify the name of a method for customization as shown in Figure 6(b). Figure 9 illustrates an example of such a method, pivot(). This method makes it possible to display a person's name and picture in the data cells of the pivot table as shown in Figures 6(a) and 6(c). Pivot() method has a parameter dataList of type ArrayList, which contains a list of data corresponding to this data cell. In this meeting room reservation system, it contains only one value, which is the name of the person who reserves a meeting room that corresponds to this data cell. This pivot() method retrieves the person's picture from DB table "person," and returns HTML code that consists of an <img> tag, a <hr> tag, and the person's name.

```
private static final String dbTableName = "person";
private static final String nameFieldName = "name";
private static final String imageFieldName = "image";
private static final String vacant = "-vacant-";
public String pivot(ArrayList dataList) throws Exception {
 String html = null;
 int size = dataList.size( );
 if (size == 0) {
 html = vacant;
 } else if (size >= 1) {
 String name = (String) dataList.get(0);
 java.sql.Statement stm = con.createStatement( );
 ResultSet result = stm.executeQuery("select" +
 imageFieldName + "from" + dbTableName +
 "where" + nameFieldName + "=" + "'" + name + "'");
 if (result.next( )) {
 String image = result.getString(1);
 html = "<img src=\"" + Std.resourceDirName( ) +
 image + "\" width=\"50\" >;<;hr>" + name;
 } else {
 html = name;
 }
 }
 return html;
}
```

Figure 9. Definition of a method pivot() to customize the data cell display of a pivot table

Implementation of Visual Functional Web Components

Implementation of Frameset Hierarchy

To develop Frameset page design windows, we define FrameSet class and Frame class. As shown in Figure 10, both of FrameSet class and Frame class extend abstract class FrameOrSet. In FrameOrSet class, variable parent refers to its parent frameset or frame. GetComponent() method returns JComponent that displays its frameset or frame. A frameset is displayed as JPanel that contains a JSplitPane, and a frame is displayed as a button in a Frameset page design window. GetHtml() method returns the HTML code that represents its frameset or frame. In FrameSet class, frameset array points to its child frameset or frame. In Frame class, variable srcPage points to a Web page or a Frameset page. Drop() method enables a user to drag and drop a frame to another frame to exchange them.

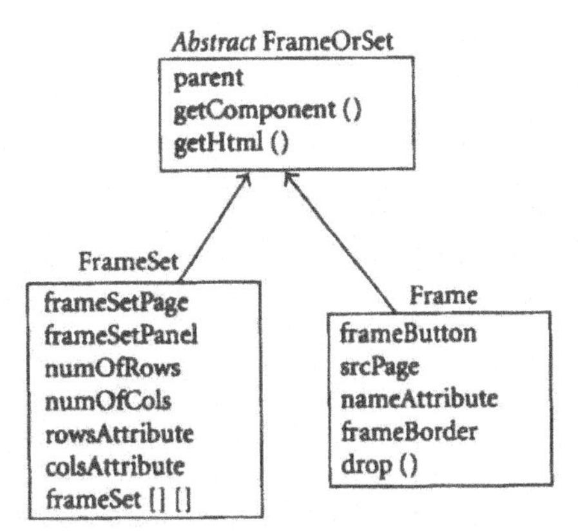

Figure 10. FrameSet and Frame that extend abstract FrameOrSet.

Figure 11 illustrates the frameset hierarchy of the two Frameset pages shown in Figure 3(b), where Frame and FrameSet instances also have a pointer that refers to their parent frame or frameset because they extend the abstract FrameOrSet class shown in Figure 10.

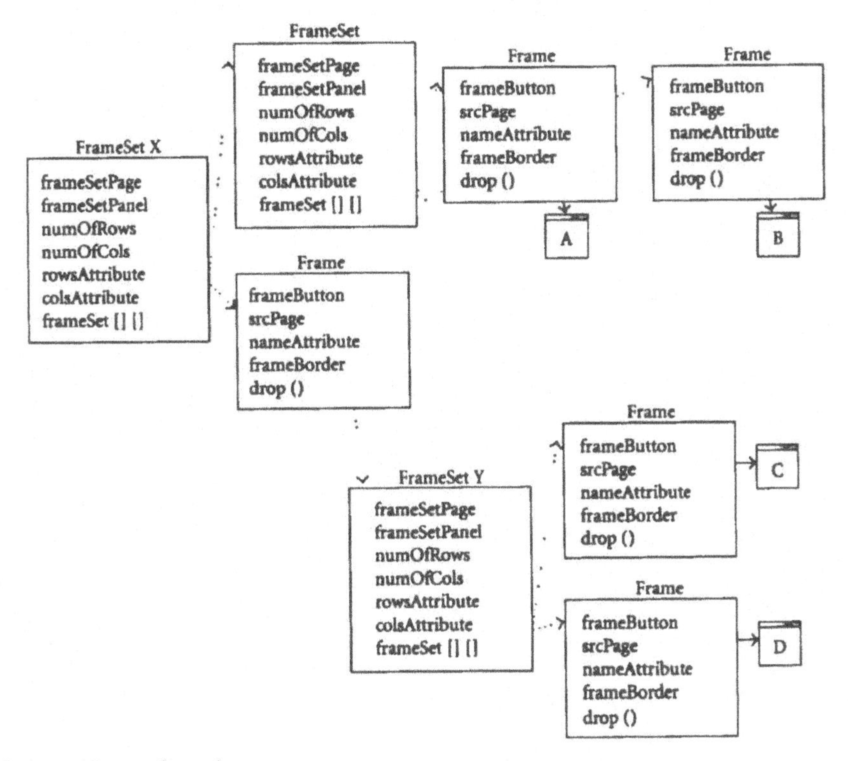

Figure 11. Frameset hierarchy.

Page-Transition Framework

Each Web page is designed using visual Web components. The reference relationships of Web components inside the same Web pages or between Web pages are created by the drag and drop operations of the components, and represented by the arrows whose colors show the types of the references. When a request is submitted from a Web page S, as a result, if a Web page T is displayed, we call page S a source page, and page T a target page. When a source page transfers to a target page, some of the Web components in the source page automatically submit some data with the request. The target page automatically analyzes those submitted data, generates some variables, and stores the analyzed results in the generated variables. In this target page, by using those generated variables, developers can easily write actions to be executed when the target page comes from each one of its source pages.

As indicated in the Struts framework [7], the Web application controlled by JavaServlet containers is composed of a sequence of display of a JSP page, receipt of a request, execution of an action, and forwarding the request to the next JSP page. The system provides the page-transition framework for Web applications, where form data can be automatically submitted and analyzed, and actions to be executed in a target page can be written for each one of the source pages of the target page. Figure 12 illustrates how form data are automatically submitted and analyzed. In the system, Web pages are composed f a variety of Web components such as text fields, field references from program and DB tables, submit buttons, HTML tables, and pivot tables. When a request is submitted from a source page A to a target page B, some of Web components included in source page A automatically submit their data. Target page B automatically receives these submitted data, analyzes the contents of the data, transforms them into appropriate values, and generates some variables to store those values. For example, a text field Web component whose name is "name" submits the text that is input in this field, and the target page receives this submitted text, generates a variable whose name is "name," and stores the text in this variable.

For example, in the Web-based chat system shown in Figure 4, a text field Web component whose name is "user" submits the text that is input in this field, and the target page receives this submitted text, generates a variable whose name is "user," and stores the text in this variable. In the Web source window of chatFrame Web page, we can refer to variables user, enter, exit, and textarea as predefined variables, which are generated from the visual design of the Web pages by the system. For example, variable user has the value of the name text field. We do not need to care about the inconsistency between the parameter names of a sender and a receiver. Using these predefined variables, we write necessary actions for the processes of entering and exiting the room. In the Web source window of utterFrame Web page, we write the process for adding the submitted message by referring to the contents of the message as a predefined variable.

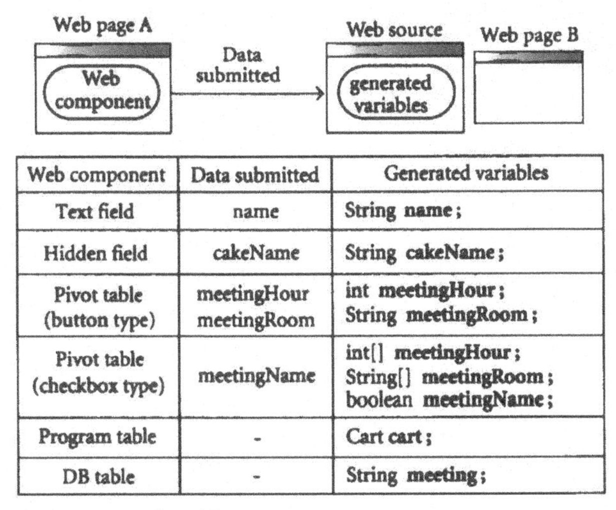

Web component	Data submitted	Generated variables
Text field	name	String **name**;
Hidden field	cakeName	String **cakeName**;
Pivot table (button type)	meetingHour meetingRoom	int **meetingHour**; String **meetingRoom**;
Pivot table (checkbox type)	meetingName	int[] **meetingHour**; String[] **meetingRoom**; boolean **meetingName**;
Program table	-	Cart **cart**;
DB table	-	String **meeting**;

Figure 12. Automatic generation of variables.

The pivot table Web component that sums up data in two dimensions (shown in Figure 7) submits the data that identify which data cells are selected or clicked. The target page automatically receives these submitted data and analyzes the contents of the data. When the data cells of the pivot table are displayed as buttons, as shown in Figure 13, the pivot table Web component assigns the row value (hour) that corresponds to the selected cell to parameter meetingHour, and the column value (room) that corresponds to the selected cell to parameter meetingRoom, and sends these parameters. The target page receives these parameters and automatically generates variables "int meetingHour" and "String meetingRoom" to store them. When the data cells of the pivot table are displayed as checkboxes, the pivot table Web component assigns the selected cell number (col + columns.length * row) to parameter meetingName, and sends this parameter. The target page receives a sequence of the selected cell numbers, obtains the number of the selected cells, and automatically generates two variables "int[] meetingHour" and "String[] meetingRoom." Then, from the selected cell numbers, it obtains the row numbers and the column numbers of the selected cells, and stores their row and column field values in meetingHour[i] and meetingRoom[i], respectively. When no fields are assigned to the row and column cells of the pivot table, the pivot-table Web component generates variable "boolean meetingName" that indicates whether or not the data cell is selected. This mechanism enables the target page to know the cells of the pivot table selected in its source page.

	Web page A (contains Pivot table) ----→	Web page B (comes from A)
Submit Button	<input type=“hidden” name=“**meetingHour**” value=“<%= rowValue %>”> <input type=“hidden” name=“**meetingRoom**” value=“<%= columnValue %>”>	**meetingHour** = Integer.parseInt(request.getParameter (“meetingHour”)); **meetingRoom** = request.getParameter (“meetingRoom”);
Checkbox	<input type=“checkbox” name=“**meetingName**” value=“<%= cellNo %>”>	**meetingHour**[i] = rows[cellNo / columns.length]; **meetingRoom**[i] = columns[cellNo % columns.length];

Figure 13. Automatic interpretation of submitted data for pivot tables.

The automatically generated variables consist of not only the variables that are generated by analyzing received requests. In addition to these variables, the target page automatically generates the variables that refer to Program tables and DB tables when the target page refers to some fields of those tables. When the target page refers to a field of a Program table Cart, it generates the variable cart that refers to the Cart object taken out of the session. When the target page refers to a field of a DB table meeting, it generates the variable meeting whose initial value is also “meeting.” When the database table to be dealt with is dynamically changed at the execution time of the application (see Figure 6(a)), we have only to assign the real database table's name to this variable meeting. The field references of the DB table display their field values according to the database table variable meeting points to.

Actions Defined in the Web-Based Chat System

In the Web source window that corresponds to chatFrame Web page, we can here define actions to be executed when control transfers to this page. The “Predefined vars:” column of the Web source window shows some variables that contain submitted data and these are automatically generated by the system. Figure 14 shows the Web source window that corresponds to utterFrame Web page. Although a typical page transition (a default transition) is specified as an arc from one Web page to another Web page, the other transitions (as when a failure occurs) can be specified in this Web source window as an action. The actions defined here will be synthesized with designed Web components to generate the program code of the Web application (see Figure 17).

Store | Window | Close | Insert web page output method | Update display

Session: Continue

Web page this page comes from: Direct transfer / Web page this page comes from: utterFrame / Page output

Predefined vars: String utter;

Fields / Methods

Page Transfers

Process when this page "utterFrame" transfers directly

Process when it transfers from page "utterFrame"

```
Vector talks = (Vector) application.getAttribute("talks");
String user = (String) session.getAttribute("user");
if (user != null) talks.add(user + ": " + utter);
```

Output process for page "utterFrame"

```
b.oOutputWebPage();
```

Figure 14. Web source window for utterFrame page.

Figure 15 illustrates how predefined variables are generated for chatFrame and utterFrame Web source windows. When we click on "enter" or "exit" button in chatFrame Web page, a request is submitted to chatFrame Web page itself. Therefore, variables enter, exit, textarea, user are generated as predefined variables in chatFrame Web source window. When we click on "utter" button in utterFrame Web page, a request is submitted to utterFrame Web page itself. Therefore, variable utter for the text field, whose name attribute is "utter," is generated as a predefined variable in utterFrame Web source window as shown in "Predefined vars:" column of Figure 14. Although utterFrame Web page contains "utter" button, no variable for that button is generated because this button's name attribute has not been specified in this Web application.

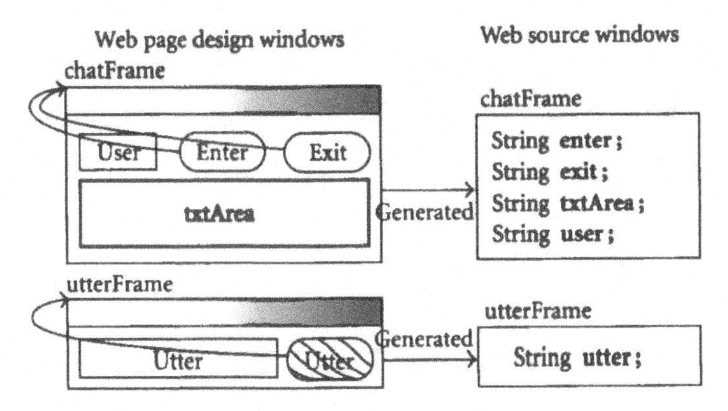

Figure 15. Predefined variables for chatFrame and utterFrame Web source windows.

Actions Defined in the Reservation System

As shown in Figure 7, the reservation system for meeting rooms consists of three Web pages, the reserveCheckbox page that displays the reservation, the input-Name page that prompts customers to enter their names, and the doubleBooking page that refuses their requests. Figure 16 illustrates an example of action code written in the Web source window that corresponds to the reserveCheckbox page. The reserveCheckbox page is accessed directly and it also comes from the reserveCheckbox page itself.

(1) When this page is accessed directly, we do nothing, that is, no action needs to be written. In this case, the text fields for entering the date and the customer's name and password are displayed.

(2) When the reserveCheckbox page comes from itself, we have the following two cases:

(2.1) the customer enters the date and his or her name and then clicks on the Apply button;

(2.2) the customer selects some of the checkboxes displayed and then clicks on the Reserve/Cancel button.

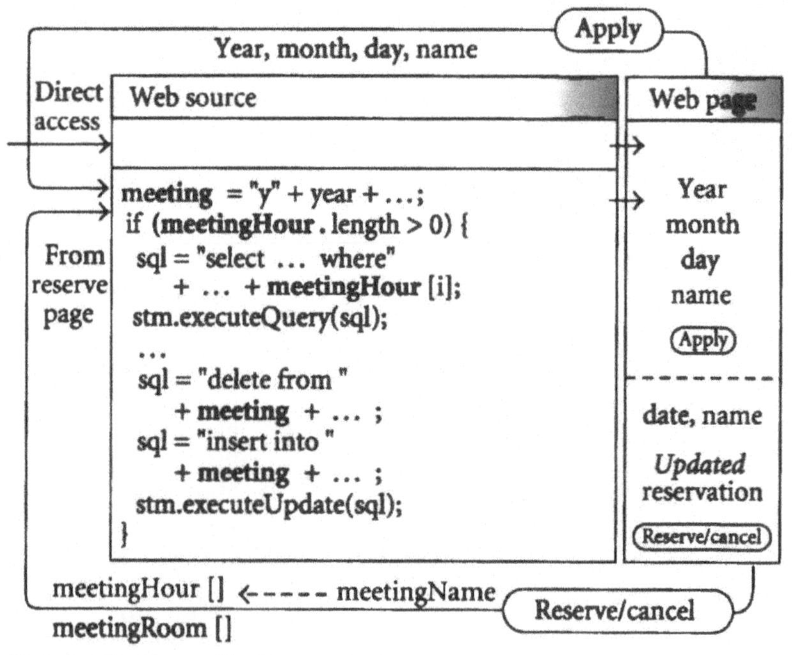

Figure 16. Actions for the reserveCheckbox page.

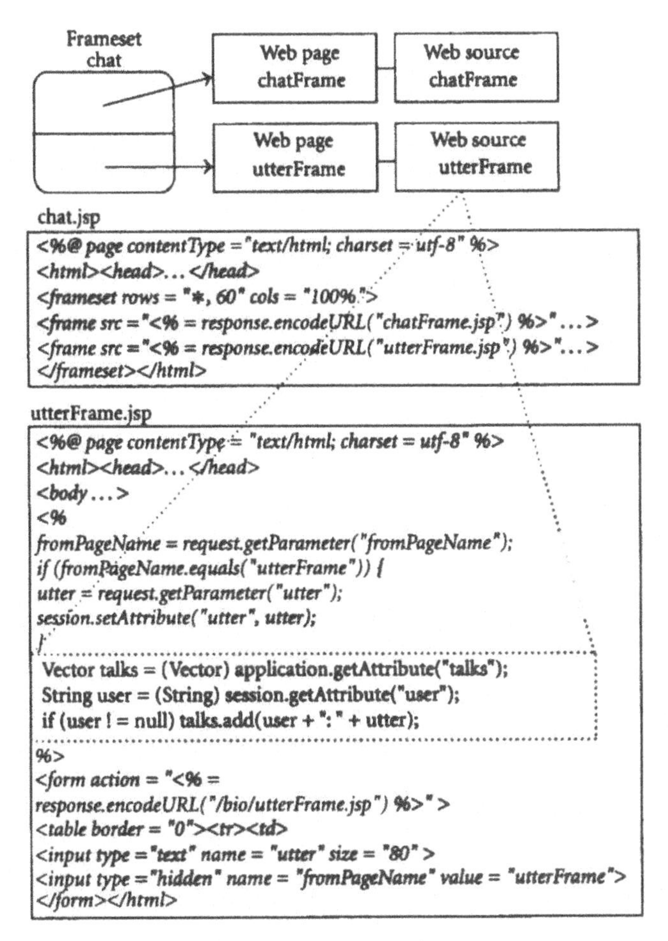

Figure 17. Automatic generation of Web application program code.

In the case of (2-1), the date and the customer's name are sent, and variables year, month, day, and name are automatically generated. Variable meeting that points to the DB table, which is used in the pivot table, is also generated. We construct a real DB table name from variables year, month, and day, and assign the real DB table name to variable meeting. To display the reservation for that date, the system dynamically expands the pivot table by retrieving the DB table this variable meeting points to.

In the case of (2-2), parameter meetingName that indicates which checkboxes are selected is sent, and variables meetingHour and meetingRoom are automatically generated that store the field values corresponding to the selected cells. In this case, the condition that meeting Hour.length is greater than zero becomes true. We write some code to update the database. First, when the customer's name has not yet been entered, we display the inputName page as shown in Figure 7. Next, we retrieve the database using the values of variables meetingHour[i] and meetingRoom[i]. When the customer tries to cancel others' reservations, we display the doubleBooking page. Finally, we execute delete SQL statements for cancellations and insert SQL statements for reservations to update the database.

From a Web page, code for displaying the corresponding Web page will be automatically generated. If it refers to some fields of a DB table, code for retrieving records from the corresponding DB table will be automatically generated. Submitted data can be automatically received and analyzed, and then the variables that contain received data will be automatically generated. The other logic such as updating a DB table needs to be written manually as an action in a Web source window.

System Configuration

The system exists on a client machine, and a Web server/Servlet container and a database server exist on server machines. The system generates JSP pages and Java class files from designed Web pages, Program tables, DB tables, and Web page source files. It then uploads them to the Web server. To run a Web application, the system invokes a Web browser so that it will display the first JSP page, which is either automatically determined or chosen by a user. To display the preview of a Web page, it generates a JSP page for preview, and uploads the JSP page to the Web server. Then, the system itself accesses the JSP page, and shows its output result in a Web preview window.

The BioPro system is not based on any of the existing frameworks. It proposes a visual programming framework, where a generated application only uses Servlets and JSPs. To connect to a PostgreSQL database server, the generated application uses JDBC. The system generates Web application program code from the visual design of a Web application. As shown in Figure 2, we first visually design the contents of each of Web pages in their Web page design windows. When we use database tables, we first visually design those database tables in DB table design windows. Instead, we may specify the names of existing database tables to display them in DB table design windows. We drag and drop the fields of a database table from a DB table design window to a Web page design window so that these fields will be displayed in the corresponding Web page. When we use a table in the program that exists only during the execution of the program, we visually design the contents of this table in a Program table design window. For example, we design a table of an online shop cart in this Program table design window. The system automatically generates JavaBeans code from these Program tables. We next write the actions that are executed when control transfers to a Web page in the Web source window that corresponds to the Web page. Control may transfer to one Web page from multiple Web pages. In the Web source window of a Web page, for each Web page control transfers from, we can write a necessary action, which is executed when control transfers from the Web page to this Web page. From these resources, the BioPro system automatically generates Servlets, JSP pages [8], and Java classes that compose a Web application in a client side, and uploads them to the Web server together with other resources such as image files, and customized Java classes. To start the Web application, the system then runs a Web browser to make it send a request to the entry Web page of the Web application.

The system generates code for connecting a database server, retrieving records, receiving submitted data, and displaying Web components designed in a Web page. These processes will be run as threads. After these threads complete, it will start a Web browser to access the entry Web page of the Web application. The entry Web page is automatically determined as a Web page that does not have its preceding page.

Figure 17 shows a method to automatically generate the program code of the Web-based chat system from the visual design created in Section 3.3 "Visual programming for frames and framesets." The part of the program code printed in italics in Figure 17 represents the code that was automatically generated by the BioPro system. We first designed a Frameset page chat in the Frameset page design window, and designed two Web pages that are displayed inside its frames in the Web page design windows. In the Web source windows of the corresponding Web pages, we then wrote the necessary actions that would be executed when control transfers to each of the Web pages by referring to predefined variables (e.g., "utter" as shown in Figure 14) that were automatically generated by the BioPro system.

The BioPro system generates the program code that composes the Web application from these pieces of design information. It generates a JSP page "chat.jsp" from the Frameset page chat, and as shown in Figure 17, it generates a JSP page "utterFrame.jsp" by synthesizing the utterFrame Web page and its Web source.

Observations on the Proposed System

Applications of Functional Web Components

As shown in Figure 4, the structure of the Web application is displayed visually, and this makes it easier to understand the program and efficient to modify and debug the program. When we design framesets and the frames that the framesets contain, we can easily create them, and divide, delete, and exchange frames by clicking or drag-and-dropping the mouse. In addition, we can create another frameset in a different Frameset page design window, and make the parent frame refer to the created child frameset. This simplifies the design of framesets and the frames that the framesets contain. We can also easily see the hierarchy of the frames in the Frameset page design windows, and see the contents of the frames in the Web page design windows that are pointed at from the corresponding frames in the Frameset page design windows.

The pivot tables can sum up a various kinds of data in two dimensions. Developers can customize the form for displaying the summed up data that correspond to each row and column. The information of selected cells in the pivot table is automatically sent to a target page, and in the target page, developers can easily obtain this information through automatically generated variables. We think that the pivot tables can be applied to various Web applications. The application favoriteCake that has been shown in Figure 1 is also one example of the applications that use pivot tables. Using a pivot table, the loveCake page displays the number of the customers who like each kind of cake. This application only uses two of the three cells in the pivot table. The column cell is expanded to display the images of the cakes that are stored in the DB table cake. The data cell is expanded to display the number of the customers who like each kind of cake using the display form Counter (see Section 4.1 "Visual programming for pivot tables"). To do this, we click on the Join button of the pivot table and enter a condition "cake.name = favoriteCake.cake" to join two DB tables cake and favoriteCake. The system automatically executes the following SQL statement to expand the pivot table:

select cake.image, favoriteCake.cake from cake, favoriteCake where cake.name = favorite- Cake.cake.

For each kind of cake (cake.image), the number of the values the field favoriteCake.cake has is equivalent to the number of the customers who like that cake.

Web source windows show some automatically generated variables in their predefined variable column. These automatically generated variables that have the values of the query data submitted to the server enable developers to easily write actions in thefields/methods columns and the page transfers columns of the Web source windows. This avoids a mismatch problem between variable names that are written in the form tags of JSP pages and in the methods of Servlets.

Comparison of Facilities with other Tools

Web application development tools are broadly classified into two groups, text-oriented and visual-oriented. BioPro is a kind of visual-oriented tool with high-level functional Web components. For eample, NetBeans [9] is a text-oriented tool, and Sun Java Studio Creator [10] is a visual-oriented tool. When we choose a Web component from a palette, NetBeans will generate and display its corresponding HTML code in a source window while Sun Java Studio Creator will paste its corresponding visual component in a form window. Table 2 shows the comparison of the BioPro system with a conventional text-oriented IDE, NetBeans [9] and a conventional visual-oriented IDE, Sun Java Studio Creator [10] concerning the facilities that assist in the development of Web applications. IDEs are integrated development environments that include editors, compilers, debuggers, project management, various kinds of source code templates, refactoring facilities, and application servers. On the other hand, unlike commercial software tools, the proposed system BioPro is not a comprehensive Web development tool. It has been developed to evaluate visual programming for high-level functional Web components. These high-level functional Web components can reduce the amount of codes required to display the components in Web pages, and perform actions. We think that it will be much easier to develop Web applications if such customizable high-level functional Web components are available even in any type of software development environment.

Table 2. Comparison with other tools.

Facilities	IDE	Visual IDE	BioPro
Code-based HTML components	O	O	O
Visual HTML components	—	O	O
Customizable visual high-level Web components	—	—	O
Visual DB tables	—	O	O
Visual Program tables	—	—	O
Assist in writing actions	—	—	O
Form data generation	—	—	O
Test/Preview Web pages	O	O	O
Debugging support	O	O	—
Refactoring support	O	O	—

Code Generation Efficiency of Functional Web Components

To evaluate the efficiency and the ease of the development with the system, we developed several typical Web applications using the system, and compared it with the development using an existing integrated development environment IDE [11] and Struts [7]. Each of four programmers developed several sample programs with IDE, BioPro, and Struts in this order. Those sample programs include Web applications (1) selectFruit (selection of fruits), (2) onlineShop (online shopping using a Program table), (3) reserveRoom (reservation for meeting rooms using one pivot and two DB tables), (4) favoriteCake (a questionnaire program about favorite cakes using a pivot table), and (5) webChat (Web-based chatting using a frameset). Figure 18 shows the time and the lines of code required to develop these applications on average. The time indicates how many hours it took to make an application, test it, and make sure of its execution result. The lines of code indicate the total lines of JSP and Java code required to develop each application. Because of the learning bias, the time required might have been advantageous to the subsequent tools, BioPro and Struts. However, by using visual programming of functional Web components, we were able to greatly reduce the code that was required to develop Web applications.

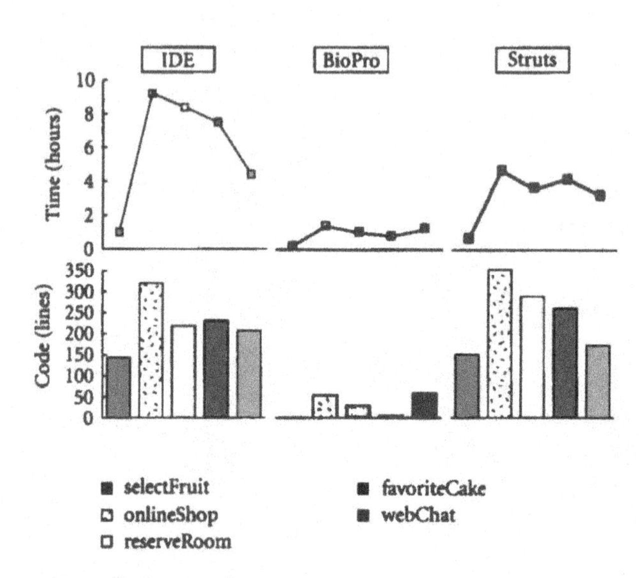

Figure 18. Time and lines of code required.

The application "webChat" is a Web-based chatting program that uses a frame set. In the Web-based chat system which is a typical example of conventional development, it required 212 lines of code (10 for chat.jsp, 118 for chatFrame.jsp, and 84 for utterFrame.jsp). This method reduced it to 63 lines of code, which is about one-third. Talking about the efficiency, it only took about fifteen minutes to design the chat Frameset, chatFrame Web page, and utterFrame Web page, which are shown in Figure 4(a).

The application "reserveRoom" is a meeting-room reservation system that makes use of a pivot table, where the data cells are displayed as either of checkboxes and buttons. The application "favoriteCake" shown in Figure 1 is a questionnaire program of favorite cakes that also uses a pivot table.

The applications that used pivot tables reduced the amount of code greatly. Pivot table components automatically retrieve the database to expand the pivot tables, and submit the information of selected cells. Target pages can automatically receive and interpret this information and generate some variables to store it.

Related Work

This section gives the outlines and brief discussions of text-oriented IDEs, Web application development frameworks, visual development environments for client programs, visual-oriented IDEs, and model-based approaches.

Text-Oriented Ides

As for world-widely used systems that assist in the development of Web applications, there are several IDEs such as Sun One Studio [11], IntelliJ IDEA [12], CodeWarrior [13], Eclipse [14], NetBeans [9]. IDE is an integrated development environment that includes editors, compilers, debuggers, project managements, various source code templates, and application servers. It assists in the development of software using object-oriented programming languages like Java. Besides these systems, Zope [15] is an application server with which users can easily develop Web applications using a Web browser that is connected to a Zope server. FAR [16] is an end-user visual language to assist in the development of the Web applications that use spreadsheets. DENIM [17] is a sketch-based visual language to assist in the early stages of Web site design using the graph representation that consists of Web pages as nodes and the dependencies between the Web pages as arcs. JWIG [18] provides a session model and a flexible mechanism for dynamic construction of XHTML documents. With JWIG, a Web application can be written as a single thread using an extension of Java. PageGen [19] provides a scheme for dynamic generation of Web pages.

These text-oriented IDEs have a variety of functions. However, they only provide fundamental Web components such as textfields, buttons, checkboxes, anchors, and tables, and Web applications are developed using text-based languages such as XHTML, JSP, JSP tag libraries, and Java. On the other hand, the proposed system provides visual high-level functional Web components, and this paper has also presented how to implement and customize these components in a flexible manner.

Web Application Development Frameworks

Several frameworks for efficiently developing Web applications have been proposed. Struts [7] provides a framework for building Web applications that consists of such components as views, controllers, and actions. Separately from business processes, users can easily write code for verifying form data and can specify target actions to which requests are forwarded. Tiles is a framework for creating Web pages that separates Web page layouts and their contents. It is used together with the Struts framework to create JSP pages to which requests are forwarded. Tiles makes it easier to change the look and feel of a Web site. JavaServer Faces [20] simplifies building user interfaces for Webapplications. It wires client-generated events to server-side event handlers. Tapestry [21] is a framework for creating Web applications in Java, where a Web application is composed of a combination of a specification file in XML, an XHTML template and a Java class. The template defines the XHTML document that includes dynamic contents, and the page components written in Java define the representation of the dynamic contents.

These Web application development frameworks make it easier to develop Web applications because they standardize various processes such as the receipt of requests, the validation of form data, and Web page transfers, and these processes become independent from the others. The BioPro system does not use these frameworks. However, it is based on the MVC architecture, where data access layer (model), presentation layer (view), and business-logic layer (controller) can be independent from the other.

Visual Development Environments for Client Programs

In the development of client programs, a variety of graphical components have been used to create their graphical user interfaces. This has made the software development easier [11]. There are several researches on software development that makes use of graphics, which include rapid development of visual applications [22], the visualization of software requirements using multimedia [23–25], assistance for object-oriented programming using UML [26, 27], the development of language processors using the graphical representation of their behaviors [28, 29], automatic form generationby the combination of graphical components [30, 31], and visual software development environments [32, 33]. To assist in database accesses using graphics, visual retrieval of structured Web information [34], and the visualization of the contents of a database [35] have been researched.

In the BioPro system, we can design database tables in the same way as visual programming tools, and the fields of the tables can be pasted in appropriate places of Web pages by mouse dragging.

Visual-Oriented Ide's

As for the development of server-side programs using graphics, there are some Web design tools such as IronSpeed [36], Sun Java Studio Creator [10], Visual Studio.NET [37], Web Sphere Studio [38], and Dreamweaver [39]. With these tools, users can design the contents of Web pages using a variety of graphical components. By connecting the pages to databases, they can create the dynamic contents of the pages as well. These tools make it easy to design Web pages, and program code is generated from those designed Web pages.

However, to develop a complete Web application, users need to write code to define business processes and add it to the generated code. On the other hand, the system this paper presents provides some high-level functional Web components such as frames and framesets [40], pivot tables [41], and customizable visual Web components. The pivot tables sum up various kinds of data in two dimensions as in Microsoft Excel. The differences between Microsoft Excel and this system is as follows:

(1) this system generates some Web components such as radio buttons and checkboxes in the data cells of a pivot table to send a request to the server;

(2) this system generates some variables that record the values indicating which radio button or checkboxes are checked to automatically receive those values on the sever side.

Model-Based Approaches

The Object-Oriented Hypermedia Design Method (OOHDM) [42–46] is a model-based approach for building hypermedia applications. It comprises four different activities: conceptual design, navigational design, abstract interface design, and implementation. It models a Web application so that the navigation model can be separated from the conceptual model. UWAT+ [47, 48] makes it possible to design Web application transactions according to the user's perspective and to integrate the Web transaction design with the information and navigation design of the Web application. Web Modeling Language (WebML) [49] is a visual notation for specifying complex Web sites at the conceptual level. WebML enables the high-level description of a Web site under distinct orthogonal dimensions: its data content (structural model), the pages that compose it (composition model), the topology of links between pages (navigation model), the layout and graphic requirements for page rendering (presentation model), and the customization features for one-to-one content delivery (personalization model). Comprehension of Web applications is a complex task, since several concerns coexist in their implementation, among which the business logic, the navigation structure (as supported by hyperlinks and form submission), and persistent data storage. Conallen's stereotypes [50] are a set of UML stereotypes designed with Web applications in mind. They add information on such things as navigation structure, page generation, and form submission that UML diagrams do not normally contain explicitly. OPM/Web [51] introduces hierarchical state expressing and suppressing to model both structure and dynamics of Web applications. WAST [52] specifies a navigational structure of Web applications and detects the inconsistency of parameter names between JSP pages and actions during the test execution.

Model-based approaches mainly support the conceptual design of Web applications. The BioPro system can assist in their implementation based on the conceptual design. In addition, the inconsistency problem of parameter names as described above will hardly occur. When users need to write code, the BioPro system shows these parameter names as automatically generated variables in Web source windows. Even if they use a wrong parameter name, this error can be detected during the compilation time because that wrong variable is not declared.

Conclusion

This paper has presented a method that makes it possible to visually design and program Web applications that use frame facilities and pivot tables. Image-oriented design using such graphical Web components and action writing for each source page of a target page, which is based on automatic interpretation of submitted data, have important roles to develop Web applications. They have made the development of Web applications easy, especially in the design of the presentation layer and action writing of the business-logic layer.

Existing tools such as Homepage Builder [4], and Dreamweaver & Fireworks [39] provide a variety of GUI components and have sufficient facilities for editing Web pages. On the other hand, the proposed system has facilitated the development of Web applications by providing Web components such as Frameset pages, Web pages, pivot tables, DB tables, Program tables, and Web source windows, where each Web component can easily refer to the definitions of the other components. As the next step, we are going to investigate how to visually incorporate rich components such as Flex, Flash, JavaScript, and Applets in the design phase. In the future, we intend to develop an end-user programming environment based on the BioPro system, where typical business patterns will be shown by using a sequence of functional components, and users will be guided and taught what to do next to develop a Web application they have in mind.

References

1. S. Pemberton, D. Austin, J. Axelsson, et al., "XHTMLTM 1.0 the extensible hypertext markup language (Second Edition)," 2002, http://www.w3.org/TR/xhtml1.
2. P. van Schaik and J. Ling, "The effects of frame layout and differential background contrast on visual search performance in Web pages," Interacting with Computers, vol. 13, no. 5, pp. 513–525, 2001.

3. T. Comber and J. Maltby, "Layout complexity: does it measure usability?," in Proceedings of the International Conference on Human-Computer Interaction (INTERACT '97), pp. 623–626, Sydney, Australia, July 1997.
4. IBM, "WebSphere Studio Homepage Builder," 2007, http://www-306.ibm.com/software/awdtools/hpbuilder.
5. T. Shimomura, "Visual design and programming for Web applications," Journal of Visual Languages & Computing, vol. 16, no. 3, pp. 213–230, 2005.
6. A. Leff and J. Rayfield, "Web-application development using the model/view/controller design pattern," in Proceedings of the 5th IEEE International Conference on Enterprise Distributed Object Computing (EDOC '01), pp. 118–127, Seattle, Wash, USA, September 2001.
7. J. Goodwill, Mastering Jakarta Struts, John Wiley & Sons, New York, NY, USA, 2002.
8. Sun Microsystems, Inc., JavaServer Pages Technology, 2006, http://java.sun.com/products/jsp.
9. NetBeans, 2008, http://www.netbeans.org.
10. Sun Microsystems, Inc., Sun Java Studio Creator, 2004, http://wwws.sun.com/software/products/jscreator.
11. R. Mogha and R. Bhargava, Sun One Studio Programming, John Wiley & Sons, New York, NY, USA, 2002.
12. JetBrains: IntelliJ IDEA, 2007, http://www.jetbrains.com/idea.
13. "CodeWarrior Development Tools," 2008, http://www.freescale.com/codewarrior.
14. S. Shavor, J. D'Anjou, S. Fairbrother, D. Kehn, J. Kellerman, and P. McCarthy, The JavaTM Developer's Guide to Eclipse, Addison-Wesley, Reading, Mass, USA, 2003.
15. A. Latteier and M. Pelletier, The Zope Book, Macmillan Computer, New York, NY, USA, 2001.
16. M. Burnett, S. K. Chekka, and R. Pandey, "FAR: an end-user language to support cottage e-services," in Proceedings of the IEEE Symposia on Human-Centric Computing Languages and Environments, pp. 195–202, Stresa, Italy, September 2001.
17. J. Lin, M. Thomsen, and J. A. Landay, "A visual language for sketching large and complex interactive designs," CHI Letters, vol. 4, no. 1, pp. 307–314, 2002.

18. A. S. Christensen, A. Møller, and M. I. Schwartzbach, "Extending Java for high-level Web service construction," ACM Transactions on Programming Languages and Systems, vol. 25, no. 6, pp. 814–875, 2003.
19. N. Al-Darwish, "PageGen: an effective scheme for dynamic generation of Web pages," Information and Software Technology, vol. 45, no. 10, pp. 651–662, 2003.
20. Sun Microsystems, Inc., JavaServer Faces, 2003, http://java.sun.com/j2ee/javaserverfaces.
21. Apache Software Foundation: Tapestry, 2003, http://jakarta.apache.org/tapestry.
22. G. D. Penna, B. Intrigila, and S. Orefice, "An environment for the design and implementation of visual applications," Journal of Visual Languages & Computing, vol. 15, no. 6, pp. 439–461, 2004.
23. D. C. Kung, "An executable visual formalism for object-oriented conceptual modeling," Journal of Systems and Software, vol. 31, no. 1, pp. 33–43, 1995.
24. D.-J. Chen, W.-C. Chen, and K. M. Kavi, "Visual requirement representation," Journal of Systems and Software, vol. 61, no. 2, pp. 129–143, 2002.
25. R. Castelló, R. Mili, and I. G. Tollis, "ViSta: a tool suite for the visualization of behavioral requirements," Journal of Systems and Software, vol. 62, no. 3, pp. 141–159, 2002.
26. S. J. Mellor and M. J. Balcer, Executable UML: A Foundation for Model-Driven Architecture, Addison-Wesley, Reading, Mass, USA, 2002.
27. C. Nentwich, W. Emmerich, A. Finkelstein, and A. Zisman, "BOX: browsing objects in XML," Software: Practice and Experience, vol. 30, no. 15, pp. 1661–1676, 2000.
28. K. Zhang, D.-Q. Zhang, and J. Cao, "Design, construction, and application of a generic visual language generation environment," IEEE Transactions on Software Engineering, vol. 27, no. 4, pp. 289–307, 2001.
29. S. Glass, D. Ince, and E. Fergus, "Llun—a high-level debugger for generated parsers," Software: Practice and Experience, vol. 31, no. 10, pp. 983–1001, 2001.
30. S. Stoecklin and C. Allen, "Creating a reusable GUI component," Software: Practice and Experience, vol. 32, no. 5, pp. 403–416, 2002.
31. S. A. Mamrak and S. Pole, "Automatic form generation," Software: Practice and Experience, vol. 32, no. 11, pp. 1051–1063, 2002.

32. K. L. Mills and H. Gomaa, "A knowledge-based method for inferring semantic concepts from visual models of system behavior," ACM Transactions on Software Engineering and Methodology, vol. 9, no. 3, pp. 306–337, 2000.

33. A. F. Blackwell, "See what you need: helping end-users to build abstractions," Journal of Visual Languages & Computing, vol. 12, no. 5, pp. 475–499, 2001.

34. W.-S. Li, J. Shim, and K. S. Candan, "WebDB: a system for querying semi-structured data on the Web," Journal of Visual Languages & Computing, vol. 13, no. 1, pp. 3–33, 2002.

35. I. F. Cruz and P. S. Leveille, "As you like it: personalized database visualization using a visual language," Journal of Visual Languages & Computing, vol. 12, no. 5, pp. 525–549, 2001.

36. Iron Speed, Inc., "Iron Speed Designer," 2004, http://www.ironspeed.com.

37. D. D. Loveh, D. Maharry, B. Sempf, and D. Xie, Effective Visual Studio .Net, Springer, New York, NY, USA, 2002.

38. I. Redbooks, Ejb 2.0 Development with Websphere Studio Application Developer, Vervante, Rolling Hls Ests, Calif, USA, 2003.

39. Adobe Systems Incorporated, "Adobe Dreamweaver and Fireworks," 2007, http://www.adobe.com/products/dreamweaver.

40. T. Shimomura, K. Ikeda, Q. L. Chen, N. S. Lang, and M. Takahashi, "Visual programming of hierarchical frames for Web applications," in Proceedings of the International Conference on Computer Engineering and Applications (CEA '07), pp. 384–389, Gold Coast, Australia, January 2007.

41. T. Shimomura, K. Ikeda, Q. L. Chen, N. S. Lang, and M. Takahashi, "Visual pivot-table components for Web application development," in Proceedings of the 3rd IASTED International Conference on Advances in Computer Science and Technology (ACST '07), pp. 90–95, Phuket, Thailand, April 2007.

42. G. Rossi and D. Schwabe, "Object-oriented design structures in Web application models," Annals of Software Engineering, vol. 13, no. 1–4, pp. 97–110, 2002.

43. D. Schwabe, L. Esmeraldo, G. Rossi, and F. Lyardet, "Engineering Web applications for reuse," IEEE Multimedia, vol. 8, no. 1, pp. 20–31, 2001.

44. D. Schwabe and G. Rossi, "From domain models to hypermedia applications: an object-oriented approach," in Proceedings of the International Workshop on Methodologies for Designing and Developing Hypermedia Applications, Edinburgh, UK, September 1994.

45. D. Schwabe and G. Rossi, "Building hypermedia applications as navigational views of information models," in Proceedings of the 28th Hawaii International Conference on System Sciences (HICSS '95), p. 231, Maui, Hawaii, USA, January 1995.
46. D. Schwabe and G. Rossi, "The object oriented hypermedia design model," Communications of the ACM, vol. 38, no. 8, pp. 45–46, 1995.
47. D. Distante, G. Rossi, G. Canfora, and S. Tilley, "A comprehensive design model for integrating business processes in Web applications," International Journal of Web Engineering and Technology, vol. 3, no. 1, pp. 43–72, 2007.
48. D. Distante, G. Canfora, S. Tilley, and S. Huang, "Redesigning legacy applications for the Web with UWAT+: a case study," in Proceedings of the 28th International Conference on Software Engineering (ICSE '06), pp. 482–491, Shanghai, China, May 2006.
49. S. Ceri, P. Fraternali, and A. Bongio, "Web modeling language (WebML): a modeling language for designing Web sites," Computer Networks, vol. 33, no. 1–6, pp. 137–157, 2000.
50. F. Ricca, M. D. Penta, M. Torchiano, P. Tonella, and M. Ceccato, "An empirical study on the usefulness of Conallen's stereotypes in Web application comprehension," in Proceedings of the 8th IEEE International Symposium on Web Site Evolution (WSE '06), pp. 58–68, Philadelphia, Pa, USA, September 2006.
51. I. Reinhartz-Berger, D. Dori, and S. Katz, "OPM/Web—object-process methodology for developing Web applications," Annals of Software Engineering, vol. 13, no. 1–4, pp. 141–161, 2002.
52. H. Tai, T. Nerome, M. Abe, and M. Hori, "A model-driven development support environment for Web applications," Transactions of Information Processing Society of Japan, vol. 44, no. 6, pp. 1498–1508, 2003.

Challenges and Improvements in Distributed Software Development: A Systematic Review

Miguel Jiménez, Mario Piattini and Aurora Vizcaíno

ABSTRACT

Distributed Software Development (DSD) has recently evolved, resulting in an increase in the available literature. Organizations now have a tendency to make greater development efforts in more attractive zones. The main advantage of this lies in a greater availability of human resources in decentralized zones at less cost. There are, however, some disadvantages which are caused by the distance that separates the development teams. Coordination and communication become more difficult as the software components are sourced from different places, thus affecting project organization, project control, and product quality. New processes and tools are consequently necessary. This work presents the findings of a systematic review of the literature related to the challenges concerning Distributed Software Development, whose purpose is to identify the solutions and improvements proposed up to the present day.

Introduction

Recent years have seen the geographic distribution of software development. The software industry now tends to relocate its production units in decentralized zones in which a skilled workforce is more readily available, thus taking advantage of political and economic factors [1]. The main objective of this is to optimize resources in order to develop higher quality products at a lower cost than that of colocated developments. Software Factories [2] are therefore organizational structures which automate parts of software development by imitating those industrial processes that were originally linked to more traditional sectors such as those of the automobile and aviation industries, decentralize production units, and promote the reusability of architectures, knowledge and components.

Distributed Software Development (DSD) allows team members to be located in various remote sites during the software lifecycle, thus making up a network of distant sub-teams. In some cases, these teams may be members of the same organization; in other cases, collaboration or outsourcing involving different organizations may exist. Traditional face-to-face meetings are, therefore, no longer common, and interaction between members requires the use of technology to facilitate communication and coordination. Although this phenomenon began in the 90s, only during the last ten years has its strategic importance been recognized [3], and related studies are very recent [4].

The distance between the different teams can vary from a few meters (when the teams work in adjacent buildings) to different continents [5]. The situation in which the teams are distributed beyond the limits of a nation is called Global Software Development (GSD) [6]. This kind of scenario is interesting for several reasons, mainly because it enables organizations to abstract themselves from geographical distance, whilst having qualified human resources and minimizing cost [7], thus increasing the market area by producing software for remote clients and obtaining a longer workday by taking advantage of time differences [8]. However, a number of problems [9], caused mainly by distance, time, and cultural differences [10], must be confronted, and these depend largely on the specific characteristics of each organization.

In this context, "offshoring" refers to the transfer of an organizational function to another country, usually one in which human resources are cheaper. We refer to "nearshoring" when jobs are transferred to geographically closer countries, thus avoiding cultural and time differences between members and saving travel and communication costs. Outsourcing is a mean to contract an external organization, independently of its location, rather than developing in-house [11].

The aforementioned development practices have as a common factor both the problems arising from distance that directly affect the processes of communication and coordination, and control activities [12]. In these environments, communication is less fluid than in colocalized development groups, and problems related to coordination, collaboration, or group awareness therefore appear which negatively affect productivity and, consequently, software quality. These factors all influence the way in which software is defined, built, tested, and delivered to customers, thus affecting the corresponding stages of the software life cycle.

In order to mitigate these effects, and with the aim of achieving higher levels of productivity, organizations require new technologies, processes, and methods [13] through improvements related to the software life cycle, project planning, estimations, risks management, quality assurance, infrastructures, team skills, and the division of responsibilities with the aim of supporting collaboration, coordination, and communication among developers [14]. Iterative approaches are commonly used in contrast to traditional waterfall or sequential methods but these become more difficult to use consistently when teams are geographically distributed [15].

The Model Driven Development (MDD) approach is currently emerging in this field, providing reusability, maintainability, interoperability, and adaptability through different languages and platforms, and improving software quality and developers' productivity. Model Driven Architecture (MDA) [16] is the most frequently adopted MDD standard and provides concepts of separation in individual models and transformation techniques.

Reference [17] discusses the main ideas with regard to how MDA can be used within a collaborative environment to assist interenterprise business processes by using tools that are able to take several input models and produce different kinds of outputs. One representative example of the application of this approach is presented in [18] with a proposal for modeling enterprise organization and developing groupware applications under a concrete MDA-based development process, thus improving communication, collaboration, and coordination between distributed actors. Tools such as InterDOC [19] also exists, which serve as an example of the power of the approach to enable the authoring process when interoperability among different collaborative applications is necessary.

This work presents a systematic review of the literature dealing with efforts related to DSD and GSD with the purpose of discovering the aspects upon which researchers have focused until this moment, thus allowing us to analyze the issues and the solutions which have been contributed up to the present through information of a highly scientific and practical value.

The paper is organized as follows. Section 2 describes the systematic review procedure applied and the results obtained. Section 3 presents an analysis of the results presented in the previous section. The issues and solutions found relating to DSD and GSD are explained in Section 4. The main success factors necessary to carry out a distributed development are listed in Section 5. Finally, Section 6 provides some concluding remarks.

Systematic Review Procedure

A systematic review of literature [20] permits the identification, evaluation, and interpretation of all the available relevant studies related to a particular research question, topic area or phenomenon, thus providing results of a high scientific value by classifying studies into primary studies and secondary or relevant studies, by means of synthesizing existing work according to a predefined strategy.

This systematic review has been carried out within the context of the FAB-RUM project, whose main objective is the development of a process with which to manage the relationships between a planning and design center and a software production factory, this work serves as a starting point upon which to focus future research.

We have followed the systematic search procedure provided by Kitchenham [20], and the selection of primary studies method followed in [21].

Question Formularization

The research question that guided this systematic review was: What are the initiatives carried out in relation to the improvement of DSD processes?

The keywords that guided the search to answer the research question were: distributed, software, development, global, enterprise, organization, company, team, offshore, offshoring, outsource, outsourcing, nearshore, nearshoring, model, strategy, and technique.

The ultimate goal of this systematic review consists of identifying the best procedures, models, and strategies employed, and to determine the most important improvement factors for the main problems found. The population will be composed of publications found in the selected sources which apply procedures or strategies related to DSD.

Sources Selection

The search strings (shown in Table 1) were established by combining the keyword list from the previous section through the logical connectors "AND" and "OR."

Table 1. Basic search strings.

	Basic search strings
1	(*"distributed software development"* OR *"global software development"*) AND ((*enterprise* OR *organization* OR *company* OR *team*) AND (*offshore* OR *offshoring* OR *outsource* OR *outsourcing* OR *nearshore* OR *nearshoring*)
2	(*"distributed software development"* OR *"global software development"*) AND (*model* OR *strategy* OR *technique*)

The studies were obtained from the following search sources: Science@Direct (http://www.sciencedirect.com/), Wiley Interscience (http://www.interscience.wiley.com/), IEEE Digital Library (http://www.computer.org/), and ACM Digital Library (http://portal.acm.org/dl.cfm). The quality of these sources guarantees the quality of the studies. The basic search chains had to be adapted to the search engines of each source.

Studies Selection

The inclusion criteria for determining whether a study should be considered relevant (a potential candidate to become a primary study) were based on analyzing the title, abstract, and keywords from the studies retrieved by the search to determine whether they dealt with DSD as regards being orientated towards process improvement, quality, coordination, collaboration, communication, and related issues that carry out any improvement concerning the subject in question. In some cases it was necessary to read the entire document to determine its relevance.

After analyzing the results of the first iteration of the systematic review, we applied exclusion criteria to obtain the primary studies, excluding those studies which, despite addressing the issue of DSD, did not contribute to any significant improvement method. We also dismissed those studies which focused solely upon social issues, cultural or time differences or focused solely upon free software, although other papers that address these topics in a secondary manner have been taken into consideration.

The search procedure produced 768 initial studies, of which 497 were not repeated. 170 of these were selected as being relevant, and 78 were selected as primary studies (the complete list of primary studies is shown in the appendix. Table 2 shows the distribution of studies found according to the sources used.

Table 2. Distribution of studies found.

		Studies				
Sources	**Search date**	**Found**	**Not repeated**	**Relevant**	**Primaries**	**%**
Science@Direct	15/08/2008	175	143	53	19	23,8
Wiley InterScience	27/09/2008	30	20	17	13	16,3
IEEE Digital Library	17/08/2008	66	49	19	14	18,8
ACM Digital Library	16/08/2008	497	355	80	32	41,3
Total	—	**768**	**567**	**170**	**78**	**100,0**

Information Extraction

The process of extracting information from the primary studies followed an inclusion criterion based on obtaining information concerning the key success factors, improvement strategies employed, processes improved and the most important ideas in each study, thus establishing a categorization between objective and subjective results. All articles were categorized by paying close attention to the methodological study followed according to the models presented in [22]; these categorizations are as follows:

(i) case studies,

(ii) literature reviews,

(iii) experiments,

(iv) simulations,

(v) surveys.

The nonexperimental model for studies (which makes a proposal without testing it or performing experiments) was also applied.

Information corresponding to a specific template (including the type of study, methodology employed, affected processes, and a description of the approach) was extracted from each paper selected for analysis, with particular attention being paid to the problems dealt with and the solutions contributed.

Trends in Distributed Software Development Research

This section analyzes and discusses the content of the primary studies found in order to extract relevant information.

Figure 1(a) shows that the majority of the primary studies analyzed are case studies and experimental papers. Nonexperimental studies and surveys in which members involved in the development take part in outlining their difficulties have a significant representation.

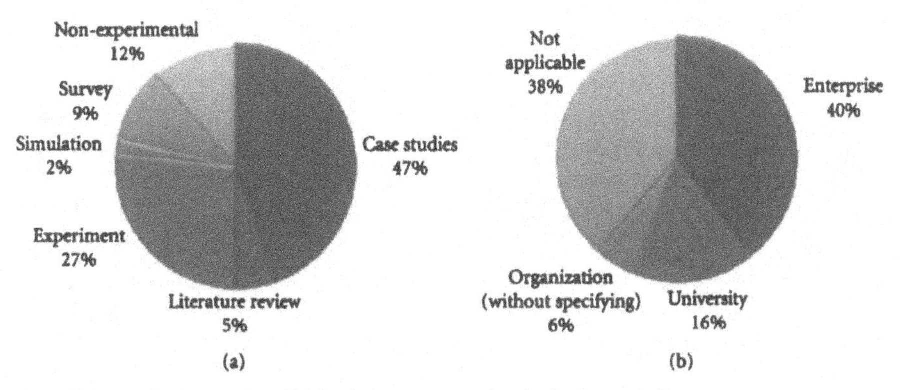

Figure 1. Type of articles analyzed (a), and environments of study development (b).

However, as Figure 1(b) shows, the majority of primary studies are focused upon the business field but studies in the university environment also appear in which groups of students carried out developments in different locations. 38% of the studies did not indicate their field of work or their classification was not applicable owing to the nature of the study, while 6% were from organizations which did not specify their corporate or university environment.

Publications Tendency

After concentrating on the number of relevant studies found through the systematic search carried out, it can be concluded that the subject of DSD is evidently an area which was not widely studied until a few years ago, and that it is only recently that a greater number of publications have appeared; thus, as Figure 2 shows, 2006 is the year in which by far the greatest number of studies was published, bearing in mind that the data shown for 2008 only reflects the studies found before September of that year.

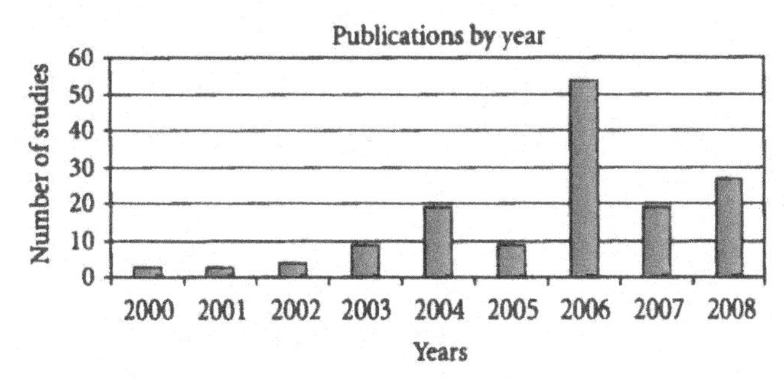

Figure 2. Trends in publications concerning DSD.

Standards Employed

Figure 3 presents the standards addressed by the articles analyzed. Based on the available data, it may be inferred that few studies indicate the use of specific standards. In part, this is attributable to the fact that the vast majority of studies deal with issues such as communication difficulties in which the standard used is not of importance. The standards supported by most primary studies are CMM, CMMI, and ISO 9001; it is common to jointly apply both. The majority of the studies which applied CMM and CMMI employed a maturity level of 2.

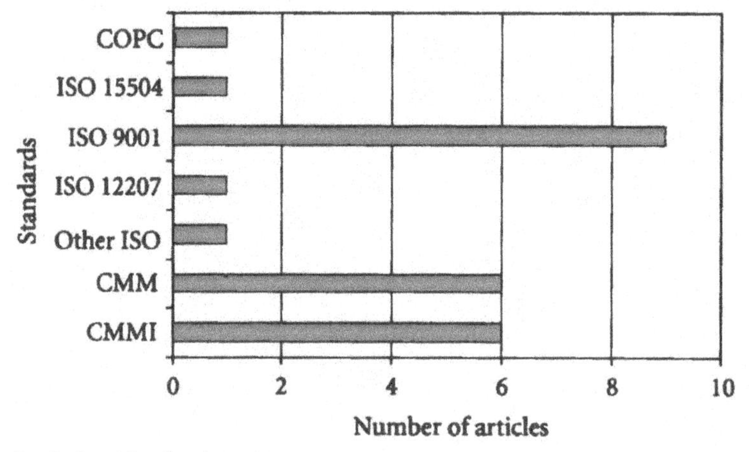

Figure 3. Standards employed in the studies.

Improved or Analyzed Processes

Taking the primary studies analyzed as a reference, we carried out a classification in terms of processes in the software life cycle to which improvements were proposed or success factors or areas to be improved related to DSD were discussed. Primary studies were classified according to the improved or studied processes, in each case based on the ISO/IEC 12207 standard [23], with the aim of obtaining a vision of the process life cycle that requires special attention when working in a distributed environment, and discovering the improvement efforts carried out until that moment.

The ISO 12207 standard establishes the activities that may be carried out during the software life cycle, which are grouped into main processes, support processes, and general processes. The results are presented graphically in Figure 4, which indicates frequency in function of the number of studies that address each process.

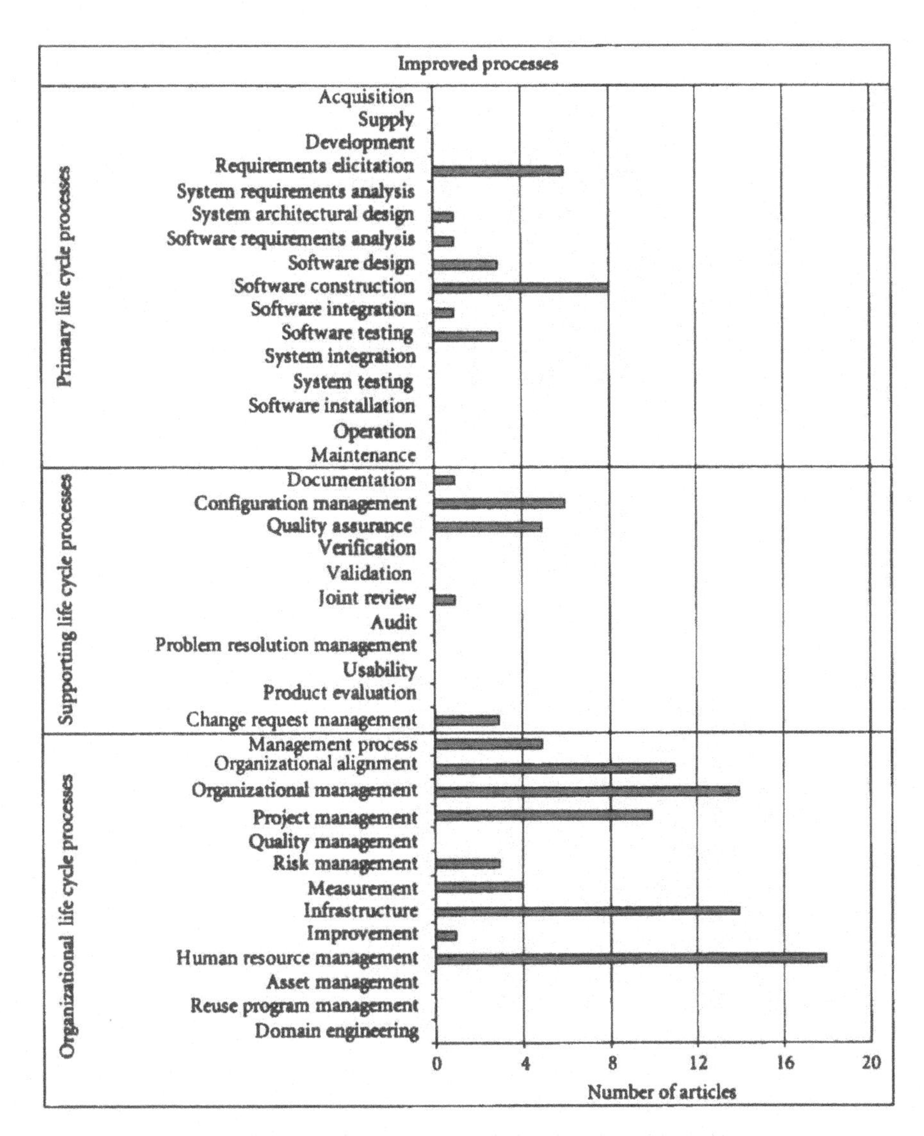

Figure 4. Processes improved or analyzed by the primary studies adjusted to ISO 12207.

The results obtained indicate that greater efforts are focused on human resources, organizational management, infrastructure, organizational alignment, and project management. From these data we can infer that communication between team members is a critical factor. Most of the studies are centered on the organizational processes, and we thus believe that there is a need for more studies focused on the level of projects and technical aspects.

Contents of the Studies

Table 3 provides a schematic representation of the lines towards which the primary studies have focused. Most of the works study tools or models designed specifically for DSD which attempt to improve certain aspects related to development and coordination. Another large part of the studies are related to communication processes and the integration of collaborative tools, combining tools such as e-mail or instant messaging, and studying their application by means of different strategies. Most of the studies address the subject of communication difficulties in at least a secondary manner, presenting this aspect as being one of the most important in relation to the problematic nature of DSD.

Table 3. Thematic areas dealt with in the primary studies.

Thematic areas	Studies (%)
Process control, task scheduling, and project coordination	43.5
Collaborative tools, techniques, and frameworks	35.9
Configuration management	5.4
Multiagent systems	4.3
Knowledge management	7.6
Defects detection	2.2
Test management	1.1

Challenges and Improvements

In this section, we synthesize the challenges and proposed improvements identified through the systematic review, discussing the main subjects.

Communication

The software life cycle requires a great deal of communication between those members involved in the development who exchange a large amount of information through different tools and different formats without following communication standards, and who thus face misunderstandings and high response times. These drawbacks, combined with the complex infrastructure and the great size of personal networks which change over time, are summarized in a decrease in communication frequency and quality, which directly affects productivity. In order to decrease these effects, both methodologies and processes must be supported by collaborative tools, which are a means of avoiding ambiguity and face-to-face meetings without comprising the quality of the results, as is proposed by M. A. Babar et al. [PS56]. K. Mohan and B. Ramesh [PS40] discuss the need for user-friendly tools, and integrate collaborative tools and agents to improve knowledge integration. M. R. Thissen et al. [PS70] examine communication tools and describe collaboration processes, dealing with techniques such as conference calls and e-mail.

Cultural differences imply different terminologies which may cause mistakes in messages and translation errors. Different levels of understanding of the problem domain also exist, as do different levels of knowledge, skills, and training between teams. The use of translation processes and codification guidelines is therefore useful [PS6].

Requirements should also be clearly defined and modeled in order to make them easily understood, and dependencies among modules should be identified in the architecture. G. N. Aranda et al. [PS34] propose a technique with which to reduce communication problems in the process of requirements elicitation by selecting a suite of groupware tools and techniques from the field of cognitive psychology.

The security of communications must also be taken into account. All the members involved must be able to work with several tools, and the human factor takes on more importance; the team members' communication skills are a critical factor.

Group Awareness

Members of a virtual team tend to be less productive due to feelings of isolation and indifference. Literature deals with the poor socialization and sociocultural differences which cause a lack of trust [PS39]. Developers need to have as much information as possible at their disposal, and to know the full status of the project and its past history, which will in turn allow them to create realistic assumptions about the project. Frequent changes in processes, lack of continuity in communications, and lack of collaborative tool integration cause remote groups to be unaware of what is important because they do not know what other people are working on. As a consequence, they cannot find the right person and/or timely information which will enable them to work together efficiently, resulting in misalignment, replanning, redesign, and rework.

M. A. D. Storey et al. [PS65] propose a framework for the comparison and understanding of visualization tools that provides awareness of software development activities, giving a solid grounding to the existing theoretical foundation of the field. Augur [PS14] similarly describes a visualization tool which supports

DSD processes by creating visual representations of both software artifacts and software development activities, thus allowing developers to explore the relationships between them.

J. D. Herbsleb et al. [PS26] present a tool which provides a visualization of the changing management system, thus making it easy to discover who has experience in working on which parts of the code, and to obtain contact information for that person. In the same line, R. Holmes and R. J. Walker [PS25] present the YooHoo awareness system to help developers to keep apprised of code changes, providing notifications in a flexible manner.

Apart from using these tools, the development process must also be adapted to provide the team members with a better awareness of the project status. It must therefore be automated to provide notifications of actions and decisions to the roles involved.

Software Configuration Management

Distributed environments present problems derived from conflicts related to source code control. Coordination and synchronization become more complex as the degree of distribution of the team grows, and traceability is a critical factor. Source control systems must support access through Internet, thus confronting its unreliable and insecure nature and the higher response times.

To reduce these drawbacks, S. E. Dossick and G. E. Kaiser [PS11] propose CHIME, an Internet- and Intranet-based application which allows users to be placed in a 3D virtual world representing the software system. Users interact with project artifacts by "walking around" the virtual world, in which they collaborate with other users through a feasible architecture. B. Al-Ani et al. [PS12] present a similar tool which visualizes the developers and artifacts in a project using a 3D metaphor and give managers an overview of ongoing activities in the project. With the same purpose in mind, J. T. Biehl et al. [PS2] present FASTDash, a user-friendly tool that uses a spatial representation of the shared code base which highlights team members' current activities, allowing a developer to rapidly determine which team members have source files checked out, which files are being viewed, and what methods and classes are currently being changed, providing immediate awareness of potential conflict situations, such as two programmers editing the same source file.

B. Bruegge et al. [PS5] present ADAMS, an artifact-based process support system, supporting permissions definition, quality management and storing traceability links between artifacts.

Knowledge Management

The team members' experiences, methods, decisions, and skills must be accumulated during the development process through effective information-sharing mechanisms, so that each team member can use the experience of his/her predecessor and the experience of the team accumulated during development, thus saving costs and time by avoiding redundant work. Distributed environments must facilitate knowledge sharing by maintaining a product/process repository focused on well-understood functionality by linking content from sources such as e-mail and online discussions, and sharing metadata information among several tools.

To solve the drawbacks caused by distribution, M. A. Babar [PS23] proposes the application of an electronic workspace paradigm to capture and share knowledge to support the software architecture processes.

H. Zhuge [PS76] presents an approach that works with a knowledge repository in which information related to each project is saved by using internet-based communication tools, thus enabling a new team member to become quickly experienced by learning the knowledge stored.

K. Mohan and B. Ramesh [PS40] present an approach based on a traceability framework that identifies the key knowledge elements which are to be integrated, and a prototype system that supports the acquisition, integration, and use of knowledge elements, allowing the knowledge fragments stored in diverse environments to be integrated and used by various stakeholders in order to facilitate a common understanding.

Change cannot be limited solely to tools, but must also take place in the organization and role distribution. Documentation must always be updated and structured to prevent assumptions and ambiguity, therefore facilitating the maintainability of the software developed.

Coordination

Coordination in multisite developments becomes more difficult in terms of articulation work, as problems derived from communication, lack of group awareness, and the complexity of the organization appear which influence the way in which the work must be structured and managed [PS3]. J. D. Herbsleb et al. [PS22] suggest that multisite communication and coordination require more people to participate which causes delays. Large changes involve multiple sites and greater implementation times. Changes in multiple distributed sites involve a large number of people. More progress reports, project reviews, conference calls, and regular meetings to take corrective action are therefore needed, thus minimizing task dependencies with other locations. Collaborative tools must support analysis, design and development to permit monitoring activities and managing dependencies, notifications, and implementation of corrective measures. P. Ovaska et al. [PS47] study the coordination of interdependencies between activities, including the figure of a chief architect to coordinate the work and maintain the conceptual integrity of the system.

S. Setamanit et al. [PS59] describe a simulation model to study different ways in which to configure global software development processes. Such models, based on empirical data, allow research into and calculation of the impact of coordination efficiency and its effects on productivity.

C. R. de Souza et al. [PS63] present the Ariadne tool which analyzes software projects for dependencies and helps to find coordination problems through a visual environment.

Collaboration

Software development is a collaborative activity in which business analysts, customers, system engineers, architects, and developers interact. The concurrent edition of models and processes requires synchronous collaboration between architects and developers who cannot be physically present at a common location. Software modeling requires concurrency control in real time, thus enabling geographically dispersed developers to edit and discuss the same diagrams, and improving productivity by providing a means through which to easily capture and model difficult concepts through virtual workspaces and the collaborative edition of artifacts by means of tools which permit synchronized interactions. S. Liu et al. [PS35] present an interesting approach which can support real-time collaborative UML-based modeling.

B. Bruegge et al. [PS4] describe SYSIPHUS, a distributed environment which provides a uniform framework for system models, collaboration artifacts, and organizational models, with services for exploring, searching, filtering, and analyzing the models.

A further approach is presented by J. Suzuki and Y. Yamamoto [PS16], [PS67] with the SoftDock framework which solves the issues related to software component modeling and their relationships, describing and sharing component models information, and ensuring the integrity of these models. Developers can therefore work by analyzing, designing, and developing software from component models and transfer them by using an exchange format, thus permitting communication between team members. S. Sarkar et al. [PS57] describe CollabDev, a human assisted collaborative knowledge tool with which to analyze applications in multiple languages and render various structural, architectural, and functional insights to the members involved in maintenance.

J. T. Biehl et al. [PS78] present IMPROMPTU, a framework for collaboration in multiple display environments, which allows users to share task information through displays via off-the-shelf applications.

In another direction, X. WenPeng et al. [PS75] study Galaxy Wiki, an online collaborative tool based on the wiki concept which permits the existence of a collaborative authoring system for documentation and coordination purposes, thus allowing developers to compile, execute, and debug programs in wiki pages.

The most valuable characteristics of these kinds of tools for an organization are their simplicity, usability, accessibility, adaptability, and broadband requirements. We therefore believe that proposals based on the wiki concept and Intranet web-based environments are more generic and easier to apply.

Project and Process Management

High organizational complexity, scheduling, task assignment, and cost estimation become more problematic in distributed environments as a result of volatile requirements, changing specifications, cultural diversity, and the lack of informal communication [PS7]. Managers must control the overall development process, improving it during the enactment and minimizing any factors that may decrease productivity, taking into account the possible impact of diverse cultures, identifying interrelated tasks, and minimizing dependencies among distributed groups.

The maturity of the process becomes a key success factor. M. Passivaara and C. Lassenius [PS48] propose incremental integration and frequent deliveries by following informing and monitoring practices.

H. Spanjers et al. [PS64] present SoftFab, an infrastructure which enables projects to automate the building and test process, and which manages all the tasks remotely though a control center.

G. Gousios et al. [PS17] propose a model for evaluating developers' contributions by combining traditional metrics with data mined from software repositories to extract contribution indicators. In the same line, N. Nagappan et al. [PS43] present a metric scheme to quantify organizational complexity.

R. J. Madachy [PS38] deals with economic issues, presenting a set of cost models to estimate distributed teams' work, and taking into account different environmental characteristics of the teams, localized labor categories, calendars, compensation rates, and currencies for costing.

The automation of the process through an adaptable tool is consequently necessary in order to manage tasks and metrics through customizable reports managed by a central server and ensuring the application of the development processes in compliance with a predefined standard.

Process Support

Processes should reflect the direct responsibilities and dependencies between tasks, notifying the people involved of the changes that concern them, thus avoiding the information overload of team members. Process modeling and enactment should support the intersite coordination and cooperation of the working teams, offering automated support to distributed project management. Problems derived from process evolution, mobility, and tool integration appear within this context. Process engines have to support changes during enactment. Furthermore, distributed environments usually involve a large network of heterogeneous, autonomous and distributed models, and process engines, which requires the provision of a framework for process system interoperability.

In relation to these problems, A. Fernández et al. [PS13] present the SPEARMINT process modeling environment, which supports extensive capabilities for multiview modeling and analysis, and XCHIPS for Web-based process support which permits enactment and simulation functionalities.

S. Setamanit et al. [PS59] describe a hybrid computer simulation model of software development processes to study alternative ways in which to configure GSD projects in order to confront communication problems, control and coordination problems, process management, and time and cultural differences.

Quality and Measurement

The quality of products is highly influenced by the quality of the processes that support them. In DSD projects the impact of issues can be magnified when a problem is discovered, and it is more difficult to recover from this than in collocated projects. Organizations should introduce new quality assurance models and measures to obtain information which can be adapted to the distributed scenarios, thus ensuring that the requirements reflect the customer's needs. One of the most frequently recommended practices is that of automated code inspections [PS4] and the application of coding standards. With this aim, K. V. Siakas and B. Balstrup [PS27] propose the capability model eSCM-SP, which has many similarities with other capability-assessment models such as CMMI, Bootstrap or SPICE, and the SQM-CODE model, and considers the factors that influence software quality management systems from a cultural and organizational perspective.

J. D. Herbsleb et al. [PS21] work with several interesting measures, such as the interdependence measure which allows the degree of dispersion of work among sites to be determined by looking up the locations of all the individuals. F. Lanubile et al. [PS30] similarly propose metrics associated with products and processes oriented towards software defects such as: discovery effort, reported defects, defects density, fixed defects or unfixed defects.

Furthermore, software architecture evaluation usually involves a large number of stakeholders who need face-to-face evaluation meetings, and adequate collaborative tools are therefore needed, such as that proposed by M. A. Babar et al. [PS56].

We observed a lack of empirical studies that allow us to enumerate reliable measures, and more articles related to tests in distributed environments, which are directly related to software quality, are also necessary.

Risk Management

Risk management is a critical project management activity. In addition to all the known traditional issues connected with collocated environments [PS7], DSD development includes issues related to coordination, problem resolution, evolving requirements, knowledge, sharing and risk identification [14]. Software defects become more frequent due to the added complexity, and in most cases, this is related to communication problems and a lack of group awareness. Defects control must be adapted by making a greater effort in risk management activities. The use of adequate measures and the requirements definition is important key factors.

In an attempt to minimize these problems, F. Lanubile et al. [PS30] define a process, specifying roles, guidelines, forms and templates, and describe a web-based tool that adopts a re-engineered inspection process in order to minimize synchronous activities and coordination problems and thus support geographically dispersed teams.

R. Kuni and Navneet Bhushan [PS29] propose the WOOM methodology to provide measures and facilitate decision making, taking into account both the risks during various lifecycle phases and mitigation plans.

Rules and guidelines with which to organize the teams and their interactions become necessary. Teams must be continuously controlled in order to detect problems and take corrective actions.

Success Factors

From the experimental studies analyzed, we have extracted the following success factors of DSD. The primary studies referenced are listed in the appendix.

i. Intervention of human resources by participating in surveys [PS56], [PS21].
ii. Carrying out improvements based on the needs of the company, taking into account the technologies and methodologies used [PS1]. The tools employed at the present must be adapted and integrated [PS58].
iii. Training of human resources in the tools and processes introduced [PS22].
iv. Registration of activities with information on pending issues, errors and people in charge [PS2], and the provision of awareness of software development activities [PS65].
v. Establishment of an efficient communication mechanism between the members of the organization, allowing a developer to discover the status and changes made within each project [PS67], [PS2].
vi. Using a version control tool in order to control conflictive situations [PS49].
vii. There must be a manner in which to permit the planning and scheduling of distributed tasks, taking into account costs and dependencies between projects, and the application of corrective measures and notifications [PS14], [PS38].
viii. Application of maturity models and agile methodologies [PS32] based on incremental integration and frequent deliveries.
ix. Application of MDD approaches to automate development tasks [PS64], [PS72].
x. Systematic use of metrics tailored to the organization [PS22].

Conclusions

In this work we have applied a systematic review method in order to analyze the literature related to the topic of DSD within the FABRUM project context whose main objective is to create a new DSD model with which to manage the relationships between a planning and design center and a software production factory. This work serves as a starting point from which to establish the issues upon which subsequent research will be focused.

The results obtained from this systematic review have allowed us to obtain a global vision of a relatively new topic which should be investigated in detail. However, every organization has concrete needs which basically depend on its distribution characteristics, its activity and the tools it employs. These factors therefore cause this subject to be extremely wide-ranging, and lead to the necessity of adapting both the technical and organizational procedures, according to each organization's specific needs.

The proposals found in the analyzed studies were, in general, mainly concerned with improvements related to the use of collaborative tools, the integration of existing tools, source code control, or the use of collaborative agents. Moreover, it should be stressed that the evaluation of the results obtained from the proposed improvements are often based on studies in a single organization, and sometimes only takes into account the developers' subjective perception.

On the other hand, it should be noted that maturity models such as CMM, CMMI, or ISO, which would be of particular relevance to the present investigation, represent only 17% of all analyzed works. The fact that almost all the experimental studies that employed CMMI and CMM applied a maturity level of 2 suggests that the cost of implementing higher maturity levels in distributed environments might be too high. However, there is a need for more studies related to the application of maturity models and metrics to quantify issues related to the process areas. The application of agile methodologies based on incremental integration and frequent deliveries, and frequent reviews of problems to adjust the process become important success factors. We also found an increasing interest in modeling in software development, and MDA approaches as a means to improve productivity, quality and understanding among members involved in the development process.

Finally, we must emphasize that the search was reduced to a limited number of search engines and excluded studies which addressed the subject of DSD but did not contribute any significant method or improvement in this research context. However, since this is such a wide area, some of these works present interesting parallel subjects for the development of this investigation, and their study would, therefore, be important in a future work. We also have found studies related to the business perspective or focused on the customer which may be useful for related works. Furthermore, many studies mainly related to tools which are not included in the context of DSD but are useful in fields related to communications or source control also exist.

Appendix

A. Primary Studies Selected

The selected primary studies in the systematic review are presented in Table 4.

Table 4. Primary studies selected in the systematic review.

Number	Year	Source	Type of study	Reference
PS1	2004	Science Direct	Use Case	[24]
PS2	2007	ACM	Use Case	[25]
PS3	2006	ACM	Use Case	[26]
PS4	2006	IEEE	Experimental	[27]
PS5	2006	IEEE	Experimental	[28]
PS6	1998	Science Direct	Literature review	[29]
PS7	2006	IEEE	Use Case	[30]
PS8	2008	ACM	Experimental	[31]
PS9	2007	Science Direct	Use Case	[32]
PS10	2007	Science Direct	Use Case	[33]
PS11	1999	ACM	Nonexperimental	[34]
PS12	2008	ACM	Experimental	[35]
PS13	2004	Wiley Interscience	Use Case	[36]
PS14	2004	ACM	Use Case	[37]
PS15	2008	Wiley Interscience	Use Case	[38]
PS16	1996	Science Direct	Experimental	[39]
PS17	2008	ACM	Experimental	[40]
PS18	2006	ACM	Use Case	[41]
PS19	2008	Science Direct	Experimental	[42]
PS20	2006	IEEE	Experimental	[43]
PS21	2000	ACM	Survey	[44]
PS22	2001	ACM	Survey	[45]
PS23	2008	ACM	Experimental	[46]
PS24	2005	ACM	Survey	[47]
PS25	2008	ACM	Experimental	[48]
PS26	2006	IEEE	Use Case, Survey	[49]
PS27	2006	Wiley Interscience	Nonexperimental	[50]
PS28	2008	Science Direct	Use Case	[51]
PS29	2006	IEEE	Experimental	[52]
PS30	2003	Wiley Interscience	Use Case	[53]
PS31	2006	Science Direct	Use Case	[54]
PS32	2006	ACM	Survey	[55]
PS33	2006	ACM	Survey	[56]
PS34	2006	IEEE	Use Case	[57]
PS35	2006	IEEE	Experimental	[58]
PS36	2002	IEEE	Use Case	[59]
PS37	2008	Wiley Interscience	Survey	[60]
PS38	2008	Wiley Interscience	Nonexperimental	[61]
PS39	2008	Wiley Interscience	Use Case	[62]
PS40	2007	Science Direct	Use Case	[63]
PS41	2004	Science Direct	Use Case	[64]
PS42	2007	Science Direct	Use Case	[65]
PS43	2008	ACM	Use Case	[66]
PS44	1991	Science Direct	Use Case	[67]
PS45	2007	Wiley Interscience	Use Case	[68]
PS46	2008	ACM	Experimental	[69]
PS47	2003	Wiley Interscience	Use Case	[70]
PS48	2003	Wiley Interscience	Use Case, Survey	[71]
PS49	2006	ACM	Use Case	[72]

Table 4. *(Continued)*

Number	Year	Source	Type of study	Reference
PS50	2004	ACM	Literature review	[73]
PS51	2003	Wiley Interscience	Use Case	[74]
PS52	2006	IEEE	Use Case	[75]
PS53	2007	ACM	Use Case	[76]
PS54	2005	Science Direct	Literature review	[77]
PS55	2008	ACM	Experimental	[78]
PS56	2006	Science Direct	Experimental	[79]
PS57	2008	ACM	Experimental	[80]
PS58	2003	ACM	Experimental	[81]
PS59	2007	Wiley Interscience	Simulation	[82]
PS60	2004	ACM	Experimental	[83]
PS61	2006	IEEE	Literature review	[84]
PS62	2006	Wiley Interscience	Use Case	[85]
PS63	2007	ACM	Experimental	[86]
PS64	2006	IEEE	Use Case	[87]
PS65	2005	ACM	Experimental	[88]
PS66	1999	IEEE	Nonexperimental	[89]
PS67	1995	Science Direct	Nonexperimental	[90]
PS68	2006	Science Direct	Experimental	[91]
PS69	2006	Science Direct	Use Case	[92]
PS70	2007	ACM	Use Case	[93]
PS71	2005	IEEE	Nonexperimental	[94]
PS72	2008	ACM	Experimental	[95]
PS73	2004	ACM	Use Case	[96]
PS74	2006	Science Direct	Nonexperimental	[97]
PS75	2007	ACM	Experimental	[98]
PS76	2002	Science Direct	Experimental	[99]
PS77	2006	ACM	Use Case	[100]
PS78	2008	ACM	Use Case	[101]

Acknowledgements

The authors acknowledge the assistance of MELISA Project (PAC08-0142-3315), financed by the "Junta de Comunidades de Castilla-La Mancha" of Spain. This work is part of FABRUM Project (PPT-430000-2008-63), financed by "Ministerio de Ciencia e Innovación" of Spain and by Alhambra-Eidos (http://www.alhambra-eidos.es/).

References

1. W. Aspray, F. Mayadas, and M. Y. Vardi, "Globalization and offshoring of software," Report of the ACM Job Migration Task Force, Association for Computing Machinery, New York, NY, USA, 2006.
2. J. Greenfield, K. Short, S. Cook, S. Kent, and J. Crupi, Software Factories: Assembling Applications with Patterns, Models, Frameworks, and Tools, John Wiley & Sons, New York, NY, USA, 2004.

3. R. Davison, "Offshoring information technology: sourcing and outsourcing to a global workforce," Information Technology for Development, vol. 13, no. 1, pp. 101–102, 2007.
4. R. Prikladnicki, D. Damian, and J. L. N. Audy, "Patterns of evolution in the practice of distributed software development: quantitative results from a systematic review," in Proceedings of the 12th Conference on Evaluation and Assessment in Software Engineering (EASE '08), Bari, Italy, June 2008.
5. R. Prikladnicki, J. L. N. Audy, and J. R. Evaristo, "Distributed software development: toward an understanding of the relationship between project team, users and customers," in Proceedings of the 5th International Conference on Enterprise Information Systems (ICEIS '03), pp. 417–423, Angers, France, April 2003.
6. J. D. Herbsleb and D. Moitra, "Global software development," IEEE Software, vol. 18, no. 2, pp. 16–20, 2001.
7. W. Kobitzsch, D. Rombach, and R. L. Feldmann, "Outsourcing in India," IEEE Software, vol. 18, no. 2, pp. 78–86, 2001.
8. C. Ebert and P. De Neve, "Surviving global software development," IEEE Software, vol. 18, no. 2, pp. 62–69, 2001.
9. L. Layman, L. Williams, D. Damian, and H. Bures, "Essential communication practices for extreme programming in a global software development team," Information and Software Technology, vol. 48, no. 9, pp. 781–794, 2006.
10. S. Krishna, S. Sahay, and G. Walsham, "Managing cross-cultural issues in global software outsourcing," Communications of the ACM, vol. 47, no. 4, pp. 62–66, 2004.
11. S. McConnell, Rapid Development: Taming Wild Software Schedules, Microsoft Press, Redmond, Wash, USA, 1996.
12. D. Damian, F. Lanubile, and H. L. Oppenheimer, "Addressing the challenges of software industry globalization: the workshop on global software development," in Proceedings of the 25th International Conference on Software Engineering, pp. 793–794, Portland, Ore, USA, May 2003.
13. D. Damian and F. Lanubile, "The 3rd international workshop on global software development," in Proceedings of the 26th International Conference on Software Engineering (ICSE '04), pp. 756–757, Edinburgh, UK, May 2004.
14. R. Sangwan, M. Bass, N. Mullick, D. J. Paulish, and J. Kazmeier, Global Software Development Handbook, Auerbach Series on Applied Software Engineering Series, Auerbach, Boston, Mass, USA, 2006.

15. M. A. Cusumano, "Managing software development in globally distributed teams," Communications of the ACM, vol. 51, no. 2, pp. 15–17, 2008.
16. OMG, "MDA guide version 1.0.1," Object Management Group, Needham, Mass, USA, June 2003.
17. L. Kutvonen, "Relating MDA and inter-enterprise collaboration management," in Proceedings of the 2nd European Workshop on Model Driven Architecture (MDA) with an Emphasis on Methodologies and Transformations (EWMDA '04), pp. 84–88, University of Kent, Canterbury, UK, September 2004.
18. J. L. Garrido, M. Noguera, M. González, M. V. Hurtado, and M. L. Rodríguez, "Definition and use of computation independent models in an MDA-based groupware development process," Science of Computer Programming, vol. 66, no. 1, pp. 25–43, 2007.
19. R. S. P. Maciel, C. G. Ferraz, and N. S. Rosa, "An MDA domain specific architecture to provide interoperability among collaborative environments," in Proceedings of the 19th Brazilian Symposium on Software Engineering (SBES '05), pp. 1–16, Uberlandia, Brazil, October 2005.
20. B. Kitchenham and S. Charters, "Guidelines for performing systematic literature reviews in software engineering," Keele University & Durham University Joint Report, Staffordshire, UK, 2007.
21. F. J. Pino, F. García, and M. Piattini, "Software process improvement in small and medium software enterprises: a systematic review," Software Quality Journal, vol. 16, no. 2, pp. 237–261, 2008.
22. M. V. Zelkowitz and D. R. Wallace, "Experimental models for validating technology," Computer, vol. 31, no. 5, pp. 23–31, 1998.
23. ISO/IEC 12207:2002, "AMENDMENT 1: Information technology—Software life cycle processes," International Organization for Standardization, 2002.
24. M. Akmanligil and P. C. Palvia, "Strategies for global information systems development," Information & Management, vol. 42, no. 1, pp. 45–59, 2004.
25. J. T. Biehl, M. Czerwinski, G. Smith, and G. G. Robertson, "FASTDash: a visual dashboard for fostering awareness in software teams," in Proceedings of the 25th SIGCHI Conference on Human Factors in Computing Systems (CHI '07), pp. 1313–1322, San Jose, Calif, USA, April 2007.
26. B. Brian, "Impact of organizational structure on distributed requirements engineering processes: lessons learned," in Proceedings of the International Workshop on Global Software Development for the Practitioner (GSD '06), Shanghai, China, May 2006.

27. B. Bruegge, A. H. Dutoit, and T. Wolf, "Sysiphus: enabling informal collaboration in global software development," in Proceedings of the IEEE International Conference on Global Software Engineering (ICGSE '06), pp. 139–148, Florianopolis, Brazil, October 2006.
28. B. Bruegge, A. De Lucia, F. Fasano, and G. Tortora, "Supporting distributed software development with fine-grained artifact management," in Proceedings of the IEEE International Conference on Global Software Engineering (ICGSE '06), pp. 213–222, Florianopolis, Brazil, October 2006.
29. J. M. Carey, "Creating global software: a conspectus and review," Interacting with Computers, vol. 9, no. 4, pp. 449–465, 1998.
30. V. Casey and I. Richardson, "Project management within virtual software teams," in Proceedings of the IEEE International Conference on Global Software Engineering (ICGSE '06), pp. 33–42, Florianopolis, Brazil, October 2006.
31. V. Clerc, "Towards architectural knowledge management practices for global software development," in Proceedings of the 3rd International Workshop on Sharing and Reusing Architectural Knowledge (SHARK '08), Leipzig, Germany, May 2008.
32. K. Crowston, Q. Li, K. Wei, U. Y. Eseryel, and J. Howison, "Self-organization of teams for free/libre open source software development," Information and Software Technology, vol. 49, no. 6, pp. 564–575, 2007.
33. A. De Lucia, F. Fasano, G. Scanniello, and G. Tortora, "Enhancing collaborative synchronous UML modelling with fine-grained versioning of software artifacts," Journal of Visual Languages and Computing, vol. 18, no. 5, pp. 492–503, 2007.
34. S. E. Dossick and G. E. Kaiser, "CHIME: a metadata-based distributed software development environment," in Proceedings of the 7th European Software Engineering Conference, held jointly with the 7th ACM SIGSOFT International Symposium on the Foundations of Software Engineering, Toulouse, France, September 1999.
35. B. Al-Ani, E. Trainer, R. Ripley, A. Sarma, A. van der Hoek, and D. Redmiles, "Continuous coordination within the context of cooperative and human aspects of software engineering," in Proceedings of the International Workshop on Cooperative and Human Aspects of Software Engineering (CHASE '08), Leipzig, Germany, May 2008.
36. A. Fernández, B. Garzaldeen, I. Grützner, and J. Münch, "Guided support for collaborative modeling, enactment and simulation of software development

processes," Software Process: Improvement and Practice, vol. 9, no. 2, pp. 95–106, 2004.

37. J. Froehlich and P. Dourish, "Unifying artifacts and activities in a visual tool for distributed software development teams," in Proceedings of the 26th International Conference on Software Engineering (ICSE '04), vol. 26, pp. 387–396, Edinburgh, UK, May 2004.
38. P. J. Gomes and N. R. Joglekar, "Linking modularity with problem solving and coordination efforts," Managerial and Decision Economics, vol. 29, no. 5, pp. 443–457, 2008.
39. I. Gorton and S. Motwani, "Issues in co-operative software engineering using globally distributed teams," Information and Software Technology, vol. 38, no. 10, pp. 647–655, 1996.
40. G. Gousios, E. Kalliamvakou, and D. Spinellis, "Measuring developer contribution from software repository data," in Proceedings of the International Working Conference on Mining Software Repositories, pp. 129–132, Leipzig, Germany, 2008.
41. C. A. Halverson, J. B. Ellis, C. Danis, and W. A. Kellogg, "Designing task visualizations to support the coordination of work in software development," in Proceedings of the 20th Anniversary ACM Conference on Computer Supported Cooperative Work (CSCW '06), pp. 39–48, Banff, Canada, November 2006.
42. B. Hanks, "Empirical evaluation of distributed pair programming," International Journal of Human Computer Studies, vol. 66, no. 7, pp. 530–544, 2008.
43. T. Heistracher, T. Kurz, G. Marcon, and C. Masuch, "Collaborative software engineering with a digital ecosystem," in Proceedings of the IEEE International Conference on Global Software Engineering (ICGSE '06), pp. 119–123, Florianopolis, Brazil, October 2006.
44. J. D. Herbsleb, A. Mockus, T. A. Finholt, and R. E. Grinter, "Distance, dependencies, and delay in a global collaboration," in Proceedings of the ACM Conference on Computer Supported Cooperative Work, pp. 319–328, Philadelphia, Pa, USA, December 2000.
45. J. D. Herbsleb, A. Mockus, T. A. Finholt, and R. E. Grinter, "An empirical study of global software development: distance and speed," in Proceedings of the 23rd International Conference on Software Engineering, pp. 81–90, Toronto, Canada, May 2001.
46. M. Ali-Babar, "The application of knowledge-sharing workspace paradigm for software architecture processes," in Proceedings of the 3rd International

Workshop on Sharing and Reusing Architectural Knowledge (SHARK '08), Leipzig, Germany, May 2008.

47. J. D. Herbsleb, D. J. Paulish, and M. Bass, "Global software development at Siemens: experience from nine project," in Proceedings of the 27th International Conference on Software Engineering (ICSE '05), pp. 524–533, St. Louis, Mo, USA, May 2005.
48. R. Holmes and R. J. Walker, "Promoting developer-specific awareness," in Proceedings of the International Workshop on Cooperative and Human Aspects of Software Engineering (CHASE '08), Leipzig, Germany, May 2008.
49. H. Holmstrom, E. Ó. Conchúir, P. J. Ågerfalk, and B. Fitzgerald, "Global software development challenges: a case study on temporal, geographical and socio-cultural distance," in Proceedings of the IEEE International Conference on Global Software Engineering (ICGSE '06), pp. 3–11, Florianopolis, Brazil, October 2006.
50. K. V. Siakas and B. Balstrup, "Software outsourcing quality achieved by global virtual collaboration," Software Process: Improvement and Practice, vol. 11, no. 3, pp. 319–328, 2006.
51. J. Kotlarsky, P. C. van Fenema, and L. P. Willcocks, "Developing a knowledge-based perspective on coordination: the case of global software projects," Information and Management, vol. 45, no. 2, pp. 96–108, 2008.
52. R. Kuni and N. Bhushan, "IT application assessment model for global software development," in Proceedings of the IEEE International Conference on Global Software Engineering (ICGSE '06), pp. 92–100, Florianopolis, Brazil, October 2006.
53. F. Lanubile, T. Mallardo, and F. Calefato, "Tool support for geographically dispersed inspection teams," Software Process: Improvement and Practice, vol. 8, no. 4, pp. 217–231, 2003.
54. L. Layman, L. Williams, D. Damian, and H. Bures, "Essential communication practices for Extreme Programming in a global software development team," Information and Software Technology, vol. 48, no. 9, pp. 781–794, 2006.
55. G. Lee, W. DeLone, and J. A. Espinosa, "Ambidextrous coping strategies in globally distributed software development projects," Communications of the ACM, vol. 49, no. 10, pp. 35–40, 2006.
56. E. Lindqvist, B. Lundell, and B. Lings, "Distributed development in an intranational, intra-organisational context: an experience report," in Proceedings of the International Workshop on Global Software Development for the Practitioner, Shanghai, China, May 2006.

57. G. N. Aranda, A. Vizcaíno, A. Cechich, M. Piattini, and J. J. Castro-Sáchez, "Cognitive-based rules as a means to select suitable groupware tools," in Proceedings of the 5th IEEE International Conference on Cognitive Informatics, vol. 1, pp. 418–423, Beijing, China, July 2006.

58. S. Liu, Y. Zheng, H. Shen, S. Xia, and C. Sun, "Real-time collaborative software modeling using UML with rational software architect," in Proceedings of the International Conference on Collaborative Computing: Networking, Applications and Worksharing (CollaborateCom '06), Atlanta, Ga, USA, November 2006.

59. W. J. Lloyd, M. B. Rosson, and J. D. Arthur, "Effectiveness of elicitation techniques in distributed requirements engineering," in Proceedings of the 10th Anniversary Joint IEEE International Requirements Engineering Conference (RE '02), Essen, Germany, September 2002.

60. J. Ma, J. Li, W. Chen, R. Conradi, J. Ji, and C. Liu, "A state-of-the-practice study on communication and coordination between Chinese software suppliers and their global outsourcers," Software Process: Improvement and Practice, vol. 13, no. 3, pp. 233–247, 2008.

61. R. J. Madachy, "Cost modeling of distributed team processes for global development and software-intensive systems of systems," Software Process: Improvement and Practice, vol. 13, no. 1, pp. 51–61, 2008.

62. N. B. Moe and D. Šmite, "Understanding a lack of trust in global software teams: a multiple-case study," Software Process: Improvement and Practice, vol. 13, no. 3, pp. 217–231, 2008.

63. K. Mohan and B. Ramesh, "Traceability-based knowledge integration in group decision and negotiation activities," Decision Support Systems, vol. 43, no. 3, pp. 968–989, 2007.

64. J. Van Moll, J. Jacobs, R. Kusters, and J. Trienekens, "Defect detection oriented lifecycle modeling in complex product development," Information and Software Technology, vol. 46, no. 10, pp. 665–675, 2004.

65. B. E. Munkvold and I. Zigurs, "Process and technology challenges in swift-starting virtual teams," Information and Management, vol. 44, no. 3, pp. 287–299, 2007.

66. N. Nagappan, B. Murphy, and V. R. Basili, "The influence of organizational structure on software quality: an empirical case study," in Proceedings of the 30th International Conference on Software Engineering (ICSE '08), pp. 521–530, Leipzig, Germany, May 2008.

67. K. Narayanaswamy and N. M. Goldman, "A flexible framework for cooperative distributed software development," The Journal of Systems and Software, vol. 16, no. 2, pp. 97–105, 1991.
68. R. M. De Araujo and M. R. S. Borges, "The role of collaborative support to promote participation and commitment in software development teams," Software Process: Improvement and Practice, vol. 12, no. 3, pp. 229–246, 2007.
69. R. J. Ocker and J. Fjermestad, "Communication differences in virtual design teams: findings from a multi-method analysis of high and low performing experimental teams," ACM SIGMIS Database, vol. 39, no. 1, pp. 51–67, 2008.
70. P. Ovaska, M. Rossi, and P. Marttiin, "Architecture as a coordination tool in multi-site software development," Software Process: Improvement and Practice, vol. 8, no. 4, pp. 233–247, 2003.
71. M. Paasivaara and C. Lassenius, "Collaboration practices in global inter-organizational software development projects," Software Process: Improvement and Practice, vol. 8, no. 4, pp. 183–199, 2003.
72. L. Pilatti, J. L. N. Audy, and R. Prikladnicki, "Software configuration management over a global software development environment: lessons learned from a case study," in Proceedings of the International Workshop on Global Software Development for the Practitioner (GSD '06), Shanghai, China, May 2006.
73. A. Powell, G. Piccoli, and B. Ives, "Virtual teams: a review of current literature and directions for future research," ACM SIGMIS Database, vol. 35, no. 1, pp. 6–23, 2004.
74. R. Prikladnicki, J. L. N. Audy, and R. Evaristo, "Global software development in practice lessons learned," Software Process: Improvement and Practice, vol. 8, no. 4, pp. 267–281, 2003.
75. R. Prikladnicki, J. L. N. Audy, and R. Evaristo, "A reference model for global software development: findings from a case study," in Proceedings of the IEEE International Conference on Global Software Engineering (ICGSE '06), pp. 18–28, Florianopolis, Brazil, October 2006.
76. N. Ramasubbu and R. K. Balan, "Globally distributed software development project performance: an empirical analysis," in Proceedings of the 6th Joint Meeting of the European Software Engineering Conference and the ACM SIGSOFT Symposium on the Foundations of Software Engineering (ESEC/FSE '07), pp. 125–134, Dubrovnik, Yugoslavia, September 2007.
77. S. Sakthivel, "Virtual workgroups in offshore systems development," Information and Software Technology, vol. 47, no. 5, pp. 305–318, 2005.
78. R. S. Sangwan and J. Ros, "Architecture leadership and management in globally distributed software development," in Proceedings of the 1st International

Workshop on Leadership and Management in Software Architecture, pp. 17–21, Leipzig, Germany, May 2008.

79. M. A. Babar, B. Kitchenham, L. Zhu, I. Gorton, and R. Jeffery, "An empirical study of groupware support for distributed software architecture evaluation process," Journal of Systems and Software, vol. 79, no. 7, pp. 912–925, 2006.
80. S. Sarkar, R. Sindhgatta, and K. Pooloth, "A collaborative platform for application knowledge management in software maintenance projects," in Proceedings of the 1st Bangalore Annual Compute Conference, Bangalore, India, January 2008.
81. A. Sarma, Z. Noroozi, and A. Van der Hoek, "Palantír: raising awareness among configuration management workspaces," in Proceedings of the 25th International Conference on Software Engineering, pp. 444–454, Portland, Ore, USA, May 2003.
82. S.-O. Setamanit, W. Wakeland, and D. Raffo, "Using simulation to evaluate global software development task allocation strategies," Software Process: Improvement and Practice, vol. 12, no. 5, pp. 491–503, 2007.
83. N. S. Shami, N. Bos, Z. Wright, et al., "An experimental simulation of multi-site software development," in Proceedings of the Conference of the Centre for Advanced Studies on Collaborative Research, Markham, Canada, October 2004.
84. B. Sengupta, S. Chandra, and V. Sinha, "A research agenda for distributed software development," in Proceedings of the 28th International Conference on Software Engineering (ICSE '06), pp. 731–740, Shanghai, China, May 2006.
85. D. Šmite, "Global software development projects in one of the biggest companies in Latvia: is geographical distribution a problem?," Software Process: Improvement and Practice, vol. 11, no. 1, pp. 61–76, 2006.
86. C. R. de Souza, S. Quirk, E. Trainer, and D. F. Redmiles, "Supporting collaborative software development through the visualization of socio-technical dependencies," in Proceedings of the International ACM Conference on Supporting Group Work, pp. 147–156, Sanibel Island, Fla, USA, 2007.
87. H. Spanjers, M. ter Huurne, B. Graaf, M. Lormans, D. Bendas, and R. van Solingen, "Tool support for distributed software engineering," in Proceedings of the IEEE International Conference on Global Software Engineering (ICGSE '06), pp. 187–198, Florianopolis, Brazil, October 2006.
88. M.-A. D. Storey, D. Čubranić, and D. M. German, "On the use of visualization to support awareness of human activities in software development: a survey and a framework," in Proceedings of the ACM Symposium on Software Visualization (SoftVis '05), pp. 193–202, St. Louis, Mo, USA, May 2005.

89. J. Suzuki and Y. Yamamoto, "SoftDock: a distributed collaborative platform for model-based software development," in Proceedings of the 10th International Workshop on Database and Expert Systems Applications (DEXA '99), Florence, Italy, September 1999.
90. M. Baentsch, G. Molter, and P. Sturm, "WebMake: integrating distributed software development in a structure-enhanced Web," Computer Networks and ISDN Systems, vol. 27, no. 6, pp. 789–800, 1995.
91. Y. Tamura, S. Yamada, and M. Kimura, "A reliability assessment tool for distributed software development environment based on Java and J/Link," European Journal of Operational Research, vol. 175, no. 1, pp. 435–445, 2006.
92. L. Taxén, "An integration centric approach for the coordination of distributed software development projects," Information and Software Technology, vol. 48, no. 9, pp. 767–780, 2006.
93. M. R. Thissen, J. M. Page, M. C. Bharathi, and T. L. Austin, "Communication tools for distributed software development teams," in Proceedings of the ACM SIGMIS CPR Conference: The Global Information Technology Workforce, pp. 28–35, Saint Louis, Mo, USA, April 2007.
94. P. F. Tiako, "Collaborative approach for modeling and performing mobile software process components," in Proceedings of the International Symposium on Collaborative Technologies and Systems, pp. 40–47, Saint Louis, Mo, USA, May 2005.
95. S. Vale and S. Hammoudi, "Towards context independence in distributed context-aware applications by the model driven approach," in Proceedings of the 3rd International Workshop on Services Integration in Pervasive Environments, Sorrento, Italy, July 2008.
96. P. Wongthongtham, E. Chang, and T. S. Dillon, "Ontology-based multi-agent system to multi-site software development," in Proceedings of the Workshop on Quantitative Techniques for Software Agile Process, Newport Beach, Calif, USA, November 2004.
97. P. Wongthongtham, E. Chang, T. S. Dillon, and I. Sommerville, "Ontology-based multi-site software development methodology and tools," Journal of Systems Architecture, vol. 52, no. 11, pp. 640–653, 2006.
98. W. Xiao, C. Chi, and M. Yang, "On-line collaborative software development via wiki," in Proceedings of the International Symposium on Wikis, pp. 177–183, Montreal, Canada, October 2007.
99. H. Zhuge, "Knowledge flow management for distributed team software development," Knowledge-Based Systems, vol. 15, no. 8, pp. 465–471, 2002.

100. B. Ramesh, L. Cao, K. Mohan, and P. Xu, "Can distributed software development be agile?," Communications of the ACM, vol. 49, no. 10, pp. 41–46, 2006.

101. J. T. Biehl, W. T. Baker, B. P. Bailey, D. S. Tan, K. M. Inkpen, and M. Czerwinski, "IMPROMPTU: a new interaction framework for supporting collaboration in multiple display environments and its field evaluation for co-located software development," in Proceedings of the 26th Annual SIGCHI Conference on Human Factors in Computing Systems, pp. 939–948, Florence, Italy, April 2008.

Vertical Mining of Frequent Patterns from Uncertain Data

Laila A. Abd-Elmegid, Mohamed E. El-Sharkawi,
Laila M. El-Fangary and Yehia K. Helmy

ABSTRACT

Efficient algorithms have been developed for mining frequent patterns in traditional data where the content of each transaction is definitely known. There are many applications that deal with real data sets where the contents of the transactions are uncertain. Limited research work has been dedicated for mining frequent patterns from uncertain data. This is done by extending the state of art horizontal algorithms proposed for mining precise data to be suitable with the uncertainty environment. Vertical mining is a promising approach that is experimentally proved to be more efficient than the horizontal mining. In this paper we extend the state-of-art vertical mining algorithm Eclat for mining frequent patterns from uncertain data producing the proposed UEclat algorithm. In addition, we compared the proposed UEclat algorithm with the UF-growth algorithm. Our experimental results show that the proposed algorithm outperforms the UF-growth algorithm by at least one order of magnitude.

Keywords: Frequent patterns, Uncertain data, Vertical mining, Tidset, Diffset, Association rules, Data mining

Introduction

Frequent pattern mining has been a focused theme in data mining research for over a decade. It is a core technique used in many mining tasks like sequential pattern mining, structured pattern mining, correlation mining, associative classification, and frequent pattern-based clustering (C. Zhu, X. Zhang, J. Sun, and B. Huang, 2009), as well as their broad applications (H. Kriegel, P. Kroger and A. Zimek, 2009) (Y. Koh, N. Rountree, R. O'Keefe, 2008) (A.Ceglar and J. Roddick, 2006). So, a great effort has been dedicated to this research and tremendous progress has been made to develop efficient and scalable algorithms for frequent pattern mining (A. Ceglar and J. Roddick, 2006) (Z. Zheng, R. Kohavi, and L. Mason, 2001) (J. Han, H. Cheng, D. Xin, and X. Yan, 2007). All these algorithms deal with precise data sets (J. Han, J. Pei, Y. Yin and R. Mao, 2004) (M. Zaki, 2000) (P. Shenoy, J. Haritsa, S. Sudarshan, G. Bhalotia, M. Bawa and D. Shah, 2000) (M. Zaki and K. Gouda, 2003)(W. Consue, and W. Kurutach, 2003) (B. Goethals, 2004)(M. Song, S. Rajasekaran, 2006). Such data is characterized by known and definite existence of the items or events in the transactions. However, there are datasets where the exact existence of items in the transactions cannot be gained. These datasets are called uncertain data. The existence of an item in a transaction is best captured by a likelihood measure or a probability (Chui, C.-K., Kao, B., Hung, E.). As an example, a medical dataset may contain a table of patients' records, each of which contains a set of diseases that a patient suffers. In such case the physician may highly suspect (but cannot guarantee) that a patient suffers from a specific disease. So he expresses his suspection by a probability of the existence of such disease (H., Li, H., Yang, Q. 2007). Another example of uncertain dataset is pattern recognition applications. Given a satellite picture, image processing techniques can be applied to extract features that indicate the presence or absence of certain target objects (such as bunkers). Due to noises and limited resolution, the presence of a feature in a spatial area is often uncertain and expressed as a probability (Dai, X., Yiu, M.L., et al. 2005). Figure 1 shows an example of precise and uncertain data sets. Few algorithms have been dedicated for mining frequent patterns from uncertain data. All these algorithms follow the horizontal data representation.

TID	Items
1	B, C
2	A, D, E
3	B, C, D
4	A, D

Precise data set

Patient ID	Depression	Eating Disorder
1	90%	80%
2	40%	70%

Uncertain data set

Figure 1. Example of precise and uncertain data sets

Although vertical data representation is a promising approach no published research work has been studied this issue. In this paper we study the problem of mining frequent patterns from uncertain data using the vertical data representation Tidset. We extend the state-of-art vertical mining algorithm Eclat to be suitable with the uncertain environment. During such extension we propose the Utidset structure for vertical representation of uncertain data. A comparative study between the proposed UEclat algorithm and the well known UF-growth algorithm is conducted and showed that the proposed algorithm outperforms the UF-growth by at least one order of magnitude.

The rest of the paper is organized as follows: In Section two we introduce the preliminaries of mining frequent itemsets. Whereas, in Section three we list and discuss the related work. Section four explains in details the proposed UElcat algorithm. A performance study is given in Section five. Finally, a conclusion is given in Section six.

Background

The problem of mining frequent itemsets can be formulated as follows. Let I be a set of items and T a database of transactions, where each transaction has a unique transaction identifier (Tid) and contains a set of items. A set $X \subseteq I$ is called an itemset, and a set $Y \subseteq T$ is called a tidset. An itemset that contains k items is called a k-itemset. The support of an itemset X, denoted $\sigma(X)$, is the number of transactions in which X occurs. An itemset is frequent if its support is greater than or equal to a user-specified minimum support (min_ sup) value (M. Zaki and K. Gouda, 2003). Figure 2 shows the frequent itemsets for different values of min_sup on a given transactional database.

All frequent itemsets
Minimum support= 50%

TID	Items
1	ACTW
2	CDW
3	ACTW
4	ACDW
5	ACDTW
6	CDT

Support	*Itemsets*
100%	C
83%	W, CW
67%	A, D, T, AC, AW, CD, CT, ACW
50%	AT, DW, TW, ACT, ATW, CDW, CTW, ACTW

Figure 2. Illustrative example for mining vfrequent itemsets

A key difference between precise and uncertain data is that each transaction of the latter contains items and their existential probabilities. The existential probability P(x, ti) of an item x in a transaction ti indicates the likelihood of x being present in ti. Using the "possible world" interpretation of uncertain data (Leung, C.K.-S., Carmichael, C.L., Hao, B. 2007)(C. Aggarwal, 2009), there are two possible worlds for an item x and a transaction ti: (i) W1 where $x \in ti$ and (ii) W2 where $x \notin ti$. Although it is uncertain which of these two worlds be the true world, the probability of W1 be the true world is P(x, ti) and that of W2 is 1 - P(x, ti). Figure 3 shows all possible worlds for a data set contains only two transactions and two items. To a further extent, there are many items in each of many transactions in a transaction database TDB. Hence, the expected support of a pattern (or a set of items) X in TDB can be computed by summing the support of X in possible world Wj (while taking in account the probability of Wj to be the true world) over all possible worlds (Leung, C.K.-S., Carmichael, C.L., Hao, B. 2007).

The following formula in rule 1 is used to calculate the expected support of any itemset X. a summarized form of rule 1 exists in rule 2. With this setting, a pattern X is considered frequent if its expected support equals or exceeds the user-specified support threshold min_sup.

$$\exp Sup(X) = \sum_{j} \left[\sup(x) \text{ in } W_j \times \prod_{i=1}^{|TDB|} \left(\prod_{I \in t_i \text{ in } W_j} P(x,t_i) \times \prod_{y \notin t_i \text{ in } W_j} \left(1 - P(y,t_i)\right) \right) \right] \quad (1)$$

$$= \sum_{i=1}^{|TDB|} \left(\prod_{x \in X} P(x,t_i) \right). \quad (2)$$

There are two types of data representation; the horizontal and vertical representation as in Figure 4. In the horizontal representation approach, the data is organized as a set of rows. Each row has a key identifier that is the transaction identifier (TID) and a set of IIDs (Item Identifier). While in the vertical representation approach, the data is organized as a set of columns; each column has a key identifier, which is the item identifier (IID) and a set of TIDs (M. Zaki and K. Gouda, 2003). There are many variations of vertical and horizontal representations presented in (P. Shenoy, J. Haritsa, S. Sudarshan, G. Bhalotia, M. Bawa and D. Shah, 2000).

Most of the previous work on mining frequent patterns is based on the horizontal representation. However, recently a number of vertical mining algorithms have been proposed for mining frequent itemsets. Mining algorithms using the vertical representation have shown to be effective and usually outperform horizontal approaches (M. Song, S. Rajasekaran, 2006). This advantage stems from the fact that frequent patterns can be counted via tidset intersections, instead of using complex internal data structures like the hash/search trees that the horizontal algorithms require (M. Zaki and K. Gouda, 2003).

Also in the vertical mining, the candidate generation and counting phases are done in a single step. This is done because vertical mining offers natural pruning of irrelevant transactions as a result of an intersection. Another feature of vertical mining is the utilization of the independence of classes, where each frequent item is a class that contains a set of frequent k-itemsets (where k > 1) (M. Zaki, 2000).

Related Work

Limited research work has been dedicated for mining frequent patterns from uncertain data. Several studies show that broad classes of algorithms can be extended to the uncertain data setting. To the best of our knowledge no research work has been done to study the feasibility of extending vertical mining algorithms for mining uncertain data. The following paragraphs describe the horizontal algorithms proposed for mining frequent patterns from uncertain data.

Chui et al. proposed the U-Apriori algorithm, which is a modification of the Apriori algorithm. Specifically, instead of incrementing the support counts of candidate patterns by their actual support, U-Apriori increments the support counts of candidate patterns by their expected support (using Equation (2)). However, U-Apriori suffers from the following problems: (i) Inherited from the Apriori algorithm, U-Apriori does not scale well when handling large amounts because it also follows a level-wise generate-and-test framework. (ii) If the existential probabilities of most items within a pattern X are small, increments for each transaction can be insignificantly small. Consequently, many candidates would not be recognized as infrequent until most (if not all) transactions were processed.

Leung et al. proposed a UF-tree which is a variant of the FP-tree. Each node in the UF-tree stores (i) an item, (ii) its expected support, and (iii) the number of occurrence of such expected support for such an item. The proposed UF-growth algorithm constructs the UF-tree as follows. It scans the database once and accumulates the expected support of each item. Hence, it finds all frequent items (i.e. items having expected support ≥ minsup). It sorts these frequent items in descending order of accumulated expected support. The algorithm then scans the database the second time and inserts each transaction into the UF-tree in a similar fashion as in the construction of an FP-tree except that the new transaction is merged with a child (or descendant) node of the root of the UF-tree (at the highest support level) only if the same item and the same expected support exist in both the transaction and the child (or descendant) nodes.

Recently, Aggarwal (C. Aggarwal, 2009) extended several existing classical frequent item set mining algorithms for deterministic datasets, and compared their relative performance in terms of efficiency and memory usage. The study focused on candidate generate-and-test algorithms, hyper-structure algorithms and pattern growth based algorithms. According to the experiments in the study, the hyper-structure and the candidate generate-and-test algorithms are proved to perform much better than tree-based algorithms.

Vertical Mining of Frequent Patterns from Uncertain Data

In this section we propose the UEclat algorithm for vertical mining of frequent patterns from uncertain data. First we introduce the proposed Utidset structure that is used in the mining process. Second, we explain in details the UEclat algorithm provided by an illustrative example.

Construction of the Utidset Mining Structure

According to the special nature of uncertain data, a key challenge in its mining is how to represent and store this data. In tidset vertical representation of precise data, each item is associated with a set of transactions identifiers (Tids) where this item appears. The case is different in uncertain data as the item's appearance in the transaction is represented by an existential probability ranging from a positive value close to 0 (indicating that the item has insignificantly low chance to be present in the transaction) to value of 1 (indicating that the item is definitely present).

To effectively represent uncertain data in vertical representation, we propose the Utidset structure which is a variant of the tidset. In the Utidset structure, each node stores (i) an item, (ii) its existential probability in every transaction. It scans the database once and accumulates the expected support of each item. Hence, it finds all frequent items. The following example illustrates the construction process of the Utidset structure.

Table 1. Uncertain transactional data

TID	Items
T1	A (0.9), B(0.8), C(0.7), D(0.6), F(0.8)
T2	A(0.9), C(0.7), D(0.6), F(0.1)
T3	B(0.9), C(O.5), E(0.4)
T4	B(0.9), E(0.2)
T5	A(0.9), C(0.7), D(0.6), E(0.3)

Example 1

Consider the following uncertain transactional data set in Table 1 and construct the Utidset for all items respecting to minimum support equals 1.

Here, each transaction contains items and their corresponding existential probability. For example the existential probability of item A is 0.9 in all transactions (T1, T2, and T5). However, the case is different for item E, for example, where there are different existential probabilities in different transactions.

The Utidset can be constructed as follows. First, the UEclat algorithm scans the data only once and accumulates the expected support of each item. The expected support is calculating by summing the probabilities of the current processed item in all its transactions. Table 2 shows all the items with their corresponding Utidset and expected support. A pruning step is done for removing all items with expected support less than the minimum expected support. At such step, both items E and F are removed. Table 3 shows the Utidset vertical representation of frequent items.

Calculating the Support of k-Itemsets where (k > 1)

After representing all the frequent items using the Utidset structure, we need to move for mining other frequent k-itemsets where k>1. The main issue here is how to calculate the support of k-itemsets. In the main Eclat algorithm that is used for mining precise data, the support of any given k-itemset is calculated simply by counting the number of transactions result from intersecting the subsets of the k-itemset. So, for any two subsets Y, Z $\subseteq$ X, such that Y$\cup$ Z=X the support of X is calculated by intersecting both tidsets of Y and Z. However, the case is different in uncertain data because the item Y may have high existential probability in a specific transaction and at the same transaction item Z may have low existential probability which will affect the real probability of item X. So, in vertical mining of uncertain data it is not enough to count the common transactions between any two subsets to calculate the support of their superset, we also need to consider the existential probabilities of both subsets in each common transaction. According to rule 2 described in Section two, the expected support of any k-itemset is the multiplication of the existential probabilities of its subsets in all transactions. For simplification we can conclude that: For any k-itemset X with subset itemsets Y and Z

$$Exp\_sup(X) = \sum^{i=n} ((p(Y, t_i) * p(Z, t_i))$$

For example the expected support of itemset BC is calculated by ((0.6*o.8) + (0.9*0.7)). The first bracket (0.6*o.8) is the support of BC in transaction T1. Whereas the second bracket (0.9*0.7) is the support of BC in transaction T3. Only transactions T1 and T3 are considered because they are the common transactions between the two subsets B and C. The total expected support of itemset BC is calculated by summing all the transactional supports that will result in value of 1.11.

Mining Frequent k-Itemsets

Once the Utidsets of all frequent items are constructed, the proposed UEclat algorithm recursively mines frequent itemsets from this Utidset structure. At the first step each frequent item is added to the output set. After that, for every such frequent itemset i, the i-projected database D_i is created. This is done by first finding every item j that frequently occurs together with i. the support of this set {i, j} is computed using the previous rule. If {i, j} is frequent, then j is inserted into D_i. The algorithm is called recursively to find all frequent itemsets in the new database D_i. Figure 4 shows the pseudo code of the UEclat algorithm.

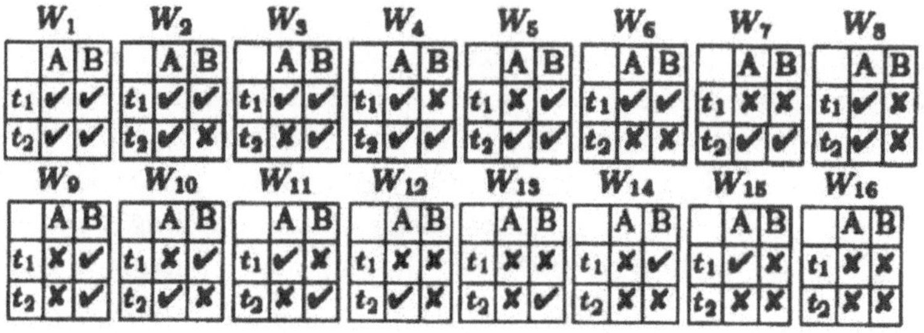

World	t_1 A	t_1 B	t_2 A	t_2 B
W_1	✔	✔	✔	✔
W_2	✔	✔	✔	✘
W_3	✔	✔	✘	✔
W_4	✔	✘	✔	✔
W_5	✘	✔	✔	✔
W_6	✔	✔	✘	✘
W_7	✘	✘	✔	✔
W_8	✔	✘	✔	✘
W_9	✘	✔	✘	✔
W_{10}	✘	✔	✔	✘
W_{11}	✔	✘	✘	✔
W_{12}	✘	✘	✔	✘
W_{13}	✘	✘	✘	✔
W_{14}	✘	✔	✘	✘
W_{15}	✔	✘	✘	✘
W_{16}	✘	✘	✘	✘

Figure 3. Possible worlds from dataset with two transactions and two items

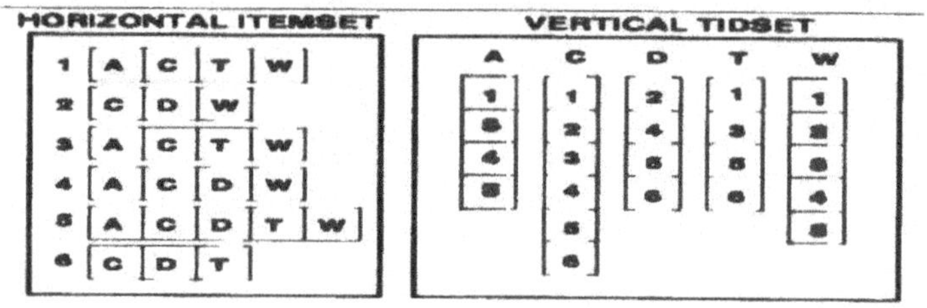

HORIZONTAL ITEMSET

1	A	C	T	W	
2	C	D	W		
3	A	C	T	W	
4	A	C	D	W	
5	A	C	D	T	W
6	C	D	T		

VERTICAL TIDSET

A	C	D	T	W
1	1	2	1	1
3	2	4	3	2
4	3	5	5	3
5	4	6	6	4
	5			5
	6			

Figure 4. Horizontal and vertical representation of data

The following example illustrates how the UEclat algorithm mines all frequent k-itemsets from the Utidset structure.

Example 2

Once the Utidset structure is constructed as in Table 2, the proposed UEclat algorithm recursively mines frequent itemsets from the structure with minimum expected support equals to 1 as follows. At the beginning, the UEclat algorithm starts to mine 2-itemsets. For each 2-itemset the expected support is calculated according to the proposed rule. A pruning process is done for all itemsets with expected support less than 1. In this example, the itemsets AB and BD are removed. Table 4 shows result of this step. For clarification purpose, we associate with each transaction the values of its two existential probabilities. Note that the columns of the infrequent itemsets AB and BD are highlighted by a grey color.

Based on the frequent 2-itemsets in Table 4, the same recursive process is done for mining frequent 3-itemsets. Table 5 shows the Utidset of frequent 3-itemsets. Here there is only frequent 3-itemset which is ACD with expected support 1.13 and as a result there is no further processing.

Table 2. Utidset vertical representation of the all items

ITEM	A	B	C	D	E	F
TIDSET	T1(0.9)	T1(0.8)	T1(0.7)	T1(0.6)	T3(0.4)	T1(0.8)
	T2(0.9)	T3(0.9)	T2(0.7)	T2(0.6)	T4(0.2)	T2(0.1)
	T5(0.9)	T4(0.9)	T3(0.5)	T5(0.6)	T5(0.3)	
			T5(0.7)			
Exp_sup	**2.7**	**2.6**	**2.6**	**1.8**	**0.9**	**0.9**

Table 3. Tidset vertical representation of frequent items

ITEM	A	B	C	D
TIDSET	T1(0.9)	T1(0.8)	T1(0.7)	T1(0.6)
	T2(0.9)	T3(0.9)	T2(0.7)	T2(0.6)
	T5(0.9)	T4(0.9)	T3(0.5)	T5(0.6)
			T5(0.7)	
Exp_sup	**2.7**	**2.6**	**2.6**	**1.8**

Table 4. Utidset vertical representation of 2-itemsets

ITEM	ACD
TIDSET	T1(0.9 * 0.7 * 0.6)
	T2(0.9 * 0.7 * 0.6)
	T5(0.9 * 0.7 * 0.6)
Exp_sup	**1.13**

Table 5. Tidset vertical representation of frequent 3-itemsets

ITEM	AB	AC	AD	BC	BD	CD
TIDSET	T1(0.9 * 0.8)	T1(0.9 * 0.7)	T1(0.9 * 0.6)	T1(0.6 * 0.8)	T1 (0.8 * 0.6)	T1(0.7 * 0.6)
		T2(0.9 * 0.7)	T2(0.9 * 0.6)	T3(0.9 * 0.7)		T2 (0.7 * 0.6)
		T5(0.9 * 0.7)	T5(0.9 * 0.6)			T5 (0.7 * 0.6)
Exp_sup	**0.72**	**1.89**	**1.62**	**1.11**	**0.48**	**1.26**

Performance Study

In this section we measure the performance of the proposed UEclat algorithm and also compare its performance of with the most recent algorithm UF-growth. Datasets used in the experiments are downloaded from http://kdd09.crowdvine.com/talks/show/4894. Two data sets are used in the experiments, T40I10D100K and T25I15D320k, are generated using the IBM synthetic data set generator. These data sets contain 100k records with an average transaction length of 10 items and a domain of 1,000 items. These data sets are used in the performance study of the UF-growth algorithm. All programs are implemented and compiled with Microsoft Visual C# Net 2005. All experiments are performed on an Intel processor 2GHz Core 2 Due with 2G of memory, running Windows Vista. The accumulated time is measured from the beginning of reading the data set and converting it to its structure to the end of the frequent pattern mining process. In Figure 6 and Figure 7, a comparison between the UEclat algorithm and the UF-growth algorithm is conducted for the T40I10D100K data set for varying minimal support thresholds. Whereas the performance regarding the T25I15D320k data set is shown in Figure 8 and Figure 9. All expermiental results confirmed that, when minsup increased, fewer patterns had expected support ≥ min_sup, and thus shorter runtimes were required. One can see that the execution time of UEclat algorithm is better than UF-growth in general. However, when the minimum support is lower, the performance of our method gets better than UF-growth. The utilization of the simple data representation used in the UTidset structure and fast counting mechanism accelerate the process of mining large number of frequent patterns and thus result in less processing time rather than the one required by the UF-growth algorithm. These experiments show that vertical mining of uncertain data is a promising approach that can achieve efficient performance regarding its features as been proved in traditional precise data.

```
UEclat([P]):
For all Xᵢ ∈   [P] do
        For all Xⱼ∈ [P], with j > i
                R = Xi ∪ Xj
                t(R) = t(Xi ) ∩ t(Xj)
                 σ(R) = ∑ (∏ p (R, tᵢ))
                if σ(R) >= min sup then
                Dᵢ = Dᵢ ∪ {R}        //Dᵢ initially empty
                Next j
                if Dᵢ≠ ∅ ; then UEclat(Di)
                Next i
```

Figure 5. UEclat algoirthm

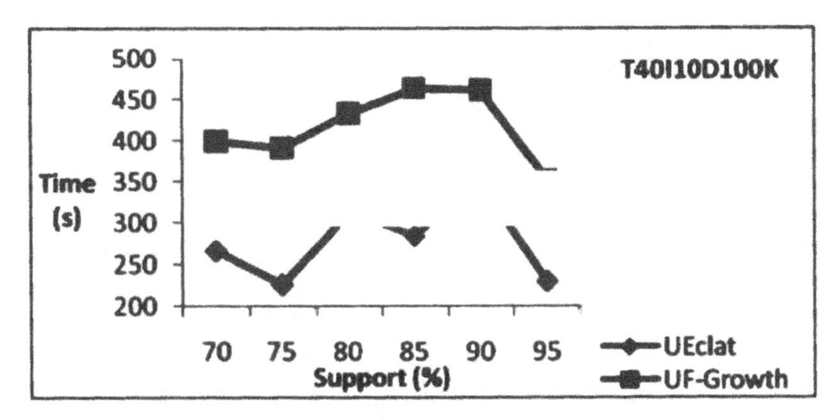

Figure 6. Run time for T40I10D100K Data set for support from 70% to 95%

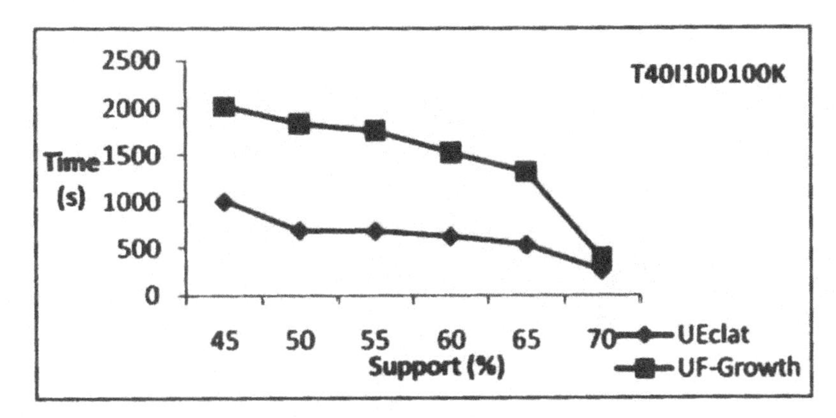

Figure 7. Run time for T40I10D100K Data set for support from 45% to 70%

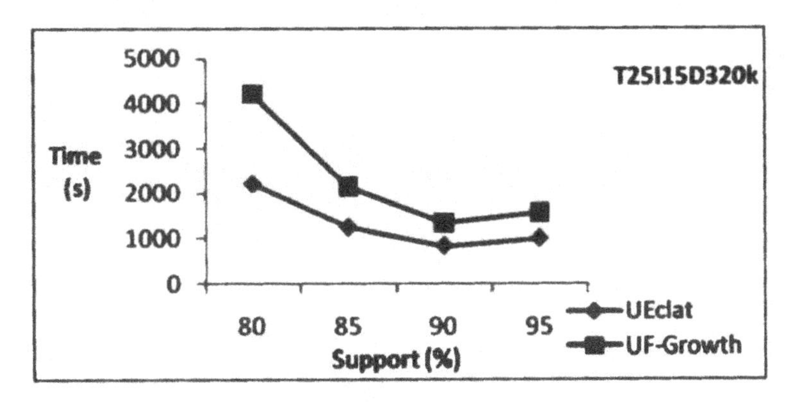

Figure 8. Run time for T25I15D320k Data set for support from 80% to 95%

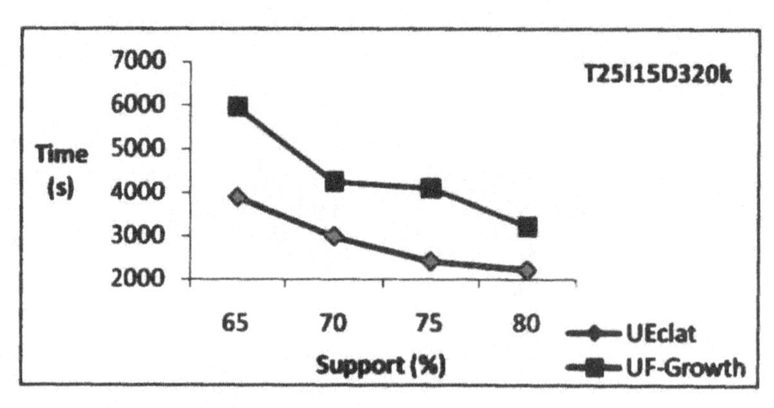

Figure 9. Run time for T25I15D320k Data set for support from 65% to 80%

Conclusion

Most existing algorithms mine frequent patterns from traditional transaction databases that contain precise data. In these databases, users definitely know whether an item (or an event) is present in, or is absent from, a transaction in the databases. However, there are many real-life situations in which one needs to deal with uncertain data. In such data users are uncertain about the presence or absence of some items or events. For example, a physician may highly suspect (but cannot guarantee) that a patient suffers from a specific disease. The uncertainty of such suspicion can be expressed in terms of existential probability. Since there are many real-life situations in which data are uncertain, efficient algorithms for mining uncertain data are in demand. Two algorithms have been proposed for mining frequent patterns from uncertain data. The previous two algorithms follow the horizontal data representation. In this paper we studied the problem of mining frequent itemsets from existential uncertain data using the Tidset vertical data representation. We introduced the U-Eclat algorithm, which is a modified version of the Eclat algorithm, to work on such datasets. A performance study is conducted to highlight the efficiency of the proposed algorithm also a comparative study between the proposed algorithm and the well known algorithm UF-growth is conducted and showed that the proposed algorithm outperforms the UF-growth.

References

1. A.Ceglar and J. Roddick. (2006). "Association Mining," In ACM Computing Surveys, Vol.38, No.2, Article no. 5, July 2006.

2. B. Goethals. (2004). "Memory Issues in Frequent Itemset Mining," In Proceedings of the ACM Symposium on Applied Computing (SAC), pp.530 –534, March 2004.
3. C. Aggarwal. (2009). "Managing and Mining Uncertain Data," Springer, 2009.
3. C. Zhu, X. Zhang, J. Sun, and B. Huang. (2009). "Algorithm for Mining Sequential Pattern in Time Series Data," International Conference on Communications and Mobile Computing, pp. 258–262, January 2009.
4. Chui, C.-K., Kao, B., Hung, E. Mining frequent itemsets from uncertain data. In: Zhou, Z.- Dai, X., Yiu, M.L., et al. (2005). Probabilistic spatial queries on existentially uncertain data. In: Bauzer Medeiros, C., Egenhofer, M.J., Bertino, E. (eds.) SSTD 2005. LNCS, vol. 3633, pp. 400–417. Springer, Heidelberg (2005)
5. H. Kriegel, P. Kroger and A. Zimek, (2009). "Clustering high-dimensional data: A survey on subspace clustering, pattern-based clustering, and correlation clustering," ACM Transactions on Knowledge Discovery from Data (TKDD), Vol.3, No.1, march 2009.
6. H., Li, H., Yang, Q. (2007). (eds.) PAKDD 2007. LNCS (LNAI), vol. 4426, pp. 47–58. Springer, Heidelberg (2007) http://kdd09.crowdvine.com/talks/show/4894
7. J. Han, H. Cheng, D. Xin, and X. Yan. (2007). "Frequent Pattern Mining: Current Status and Future Directions," Data Mining and Knowledge Discovery, Vol.15, No.1, pp. 55–86, 2007.
8. J. Han, J. Pei, Y. Yin and R. Mao. (2004). "Mining Frequent Patterns without Candidate Generation: A Frequent-Pattern Tree Approach," Data Mining and Knowledge Discovery, Vol.8, No.1, pp. 53–87, 2004.
9. Leung, C.K.-S., Carmichael, C.L., Hao, B. (2007). Efficient mining of frequent patterns from uncertain data. In: Proc. IEEE ICDM Workshops, pp. 489–494 (2007)
10. M. Song, S. Rajasekaran. (2006). "A Transaction Mapping Algorithm for Frequent Itemsets Mining" , IEEE Transactions on Knowledge and Data Engineering , Vol.18, No.4, pp. 472–481, April 2006.
11. M. Zaki and K. Gouda. (2003). "Fast Vertical Mining Using Diffsets," In Knowledge Discovery and Data Mining (KDD), pp. 326–335, 2003.
12. M. Zaki. (2000). "Scalable Algorithms for Association Mining," IEEE Transactions on Knowledge and Data Engineering , Vol.12, No.3, pp. 372–390, May-June 2000.

13. P. Shenoy, J. Haritsa, S. Sudarshan, G. Bhalotia, M. Bawa and D. Shah. (2000). "Turbo-Charging Vertical Mining of Large Databases," In ACM Special Interest Group on Management of Data (SIGMOD), Vol.29, No.2, June 2000.
14. W. Consue, and W. Kurutach. (2003). "Novel Vertical Mining on Diffsets Structure," In Proceedings of the IEEE/WIC International Conference on Intelligent Agent Technology (IAT), pp. 343–349, October 2003.
15. Y. Koh, N. Rountree, R. (2008). O'Keefe, "Mining Interesting Imperfectly Sporadic Rules," Knowledge and Information Systems ,Vol. 14 , No. 2, pp: 179–196, January 2008.
16. Z. Zheng, R. Kohavi, and L. Mason. (2001). "Real world performance of association rule algorithms," In ACM International Conference on Knowledge Discovery and Data Mining (SIGKDD), pp: 401–406, 2001.

An Open-Source Representation for 2-DE-Centric Proteomics and Support Infrastructure for Data Storage and Analysis

Romesh Stanislaus, John M. Arthur, Balaji Rajagopalan, Rick Moerschell, Brian McGlothlen and Jonas S. Almeida

ABSTRACT

Background

In spite of two-dimensional gel electrophoresis (2-DE) being an effective and widely used method to screen the proteome, its data standardization has still not matured to the level of microarray genomics data or mass spectrometry approaches. The trend toward identifying encompassing data standards has been expanding from genomics to transcriptomics, and more recently to proteomics.

The relative success of genomic and transcriptomic data standardization has enabled the development of central repositories such as GenBank and Gene Expression Omnibus. An equivalent 2-DE-centric data structure would similarly have to include a balance among raw data, basic feature detection results, sufficiency in the description of the experimental context and methods, and an overall structure that facilitates a diversity of usages, from central reposition to local data representation in LIMs systems.

Results & Conclusion

Achieving such a balance can only be accomplished through several iterations involving bioinformaticians, bench molecular biologists, and the manufacturers of the equipment and commercial software from which the data is primarily generated. Such an encompassing data structure is described here, developed as the mature successor to the well established and broadly used earlier version. A public repository, AGML Central, is configured with a suite of tools for the conversion from a variety of popular formats, web-based visualization, and interoperation with other tools and repositories, and is particularly mass-spectrometry oriented with I/O for annotation and data analysis.

Background

The post genomic era has seen an increasing effort put into systematic surveys of various proteomes. Consequently, proteomics is rapidly evolving into a high throughput experimental approach that enables the identification, for example, of differentially expressed proteins as biomarkers for disease and pathogenesis. Similarly, there is a critical need for central repositories and common data formats to make the most of the copious amounts of data generated by the different screening initiatives. The higher methodological complexity of proteomics makes data integration a challenge, greatly complicated by the fact that there are no comprehensive data structures in many proteomic fields. In spite of the fact that separation by 2-dimensional gel electrophoresis (2-DE) followed by spot identification by mass spectrometry has been a major workhorse and a versatile tool in discovery proteomics [1,2], it remains under-supported by stable data formats and repositories. High resolution 2-DE provides a powerful tool for the reproducible separation, visualization, and quantification of thousands of proteins in a single gel. The increasing variety and amount of proteins being separated and the number of researchers using the 2-DE method has generated an immense diversity of datasets produced by different laboratories and using different instruments.

The lack of common formats has had an even more pernicious effect at the level of centralized data reposition, as well as in the development of incipient publicly available open source software that would enable the experimental biologist to analyze 2-DE data. The field relies on a fragmented collection of proprietary tools associated with specific instruments. A consequence of the lack of stable open formats and the piecemeal processing by instrument specific data analysis tools is that the experimental context for the generation of specific datasets is rarely stored with the raw data. Interestingly, the availability of tools for MS-based proteomics screening is far better, with two standards having emerged under the patronage of different organizations, mzXML [3] and mzData [4]. The emergence of a stable data standard, or format, coupled with a public repository, catalyzes the subsequent establishment of additional specialized or more abstract formats, as well as analysis and visualization tools (e.g., for mzXML [5] and mzData [6]). The key ingredient for this process is a consistent data format, which gives the tool creator a stable platform from which to work.

However, this is being changed by the work recently undertaken by HUPO-PSI-GEL. Specifically, the Gel Markup Language (gelML; currently in its 2nd milestone; [7]) and GelInfoML (currently in its 1st milestone; [8]) are hoping to fill the lack of gel standards. gelML captures a gel electrophoresis experiment from experimental procedures up to sample processing. It makes use of the Functional Genomics Experimental Object Model [9,10], which defines common components found in many biological experiments and extensions thereof according to the authors. Along with gelML, HUPO-PSI is working on guidelines for the documentation of proteomics experiments known as the minimum information about proteomics experiment, or MIAPE [11]. GelML, along with GelInfoML, which is based on MIAPE guidelines, hopes to be a comprehensive data standard for 2-D gel electrophoresis.

There is a clear need for stable open data structures, data analysis tools, and data repositories in 2-DE-centric proteomics that enable the comprehensive representation of the data from raw image pixel values to the experimental methodology used to generate it. Several groups, including ours, have recently made incremental advances toward that goal [12-14]. This report describes resources built around the annotated gel markup language (AGML) format [13], which has been improved along with converters and analysis and management tools. The public repository, known as AGML Central [15], and its conversion utilities that allow data upload in a variety of formats were correspondingly upgraded. The emphasis on interoperation and cross-reference with other open formats was also reflected by the extended support for spot identification by Mass Spectrometry, which is addressed by a number of stable, open formats.

Results

XML Representation of 2-Dimensional Gel Electrophoresis Experiments

Concept

The creation of a comprehensive representation for 2-D gel electrophoresis was based on two criteria: independence of data, and an adequate minimum amount of data. Fulfillment of these criteria allows another scientist in the same field to replicate a given experiment. Thus, AGML 2.0 includes a substructure describing the protocol used in running the 2-DE experiment. This was achieved by incorporating other open data formats developed to describe gel-centric experimental protocols into AGML 1.0 [13] (see below).

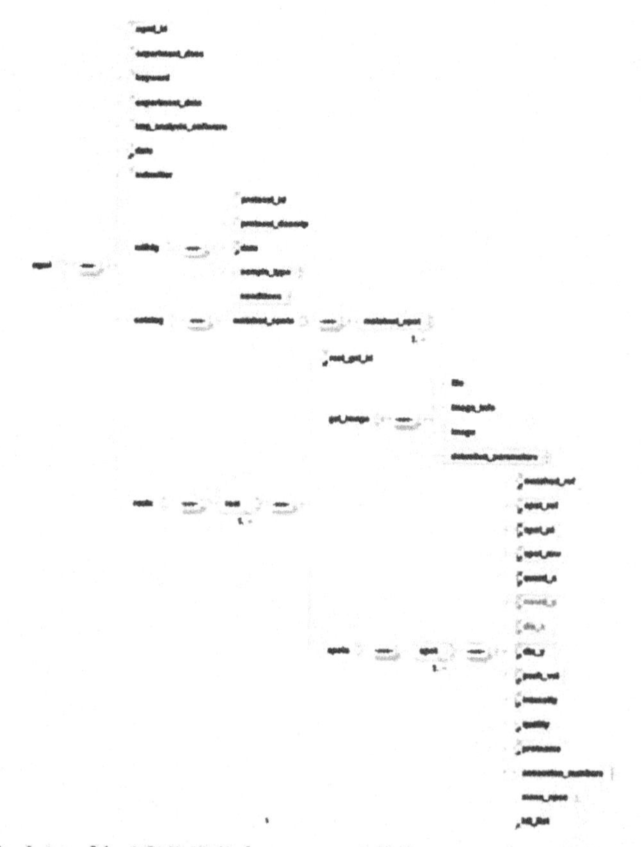

Figure 1. Top level view of the AGML XML data structure. AGML represented as a UML class diagram can be found at the project website [25]. The element 'detection_parameters' (a child of 'gel_image') enables AGML 2.0 to handle DIGE gel images. Also, the fact that AGML stores the raw gel image enables their reanalysis by other means. Please note that, for clarity, not all elements are shown in the figure.

Similarly, AGML 2.0 also includes a dedicated MS data substructure that provides the option of using already established MS data structures [3,12] or provides the link to a proteomics identification database such as the proteomics identifications (PRIDE) database [16].

AGML 2.0 Data Structure

An AGML document can be broadly divided into a) an identification section, b) a protocol section, and c) a gel section (Figure 1).

(a) The identification section consists of information that identifies the experiment and the instruments used in the analysis of data. Two identifiers deserve particular notice. The experiment_desc element can be used to describe the experiment with keywords or a descriptive sentence. The agml_id element is a unique identifier generated automatically for data submitted to AGML Central.

(b) The protocol section, also known as the minimum information about 2-D gel electrophoresis (MI2DG) [17], is intended to put the obtained data in context with the methodological and biological information. We have abstracted this from the spot information for several reasons. The main reason is that the sample processing method is not directly relevant to the image analysis; however, it is important in the final analysis of data. Furthermore, separating the protocol section and the spot information into different subsections allows for their independent description. Additionally, the protocol section includes a number of elements describing the experimental protocol, and provides important covariates for data analysis and semantic searching of the AGML data structure. It contains several elements that identify the sample, protocol type, and conditions used for the electrophoresis run.

In the AGML Central infrastructure, the MI2DG information has a dedicated management web portal so that protocols can be created and modified based on an existing protocol. This eliminates the need for the researcher to repeatedly key in all of the information, and allows researchers to easily modify a protocol based on previously published protocols. An additional advantage of the autonomous protocol database [17] is that researchers have archived catalogs of all of the protocols used in their labs. MI2DG entries have their own unique identifiers (mi2dg_id) and can be independently referenced.

(c) The gel information section consists of spot information divided into two sections: catalog and reals. The catalog section describes the aligned spots in all of the real gels. The subelement protname describes the most likely protein ID out of the possible candidates described in the hit_list. The reals element describes an individual gel (real) that comprises one or many spots. Each spot element describes an individual spot in a gel by defining many subelements that, taken together, uniquely identify the spot. The matched_ref element identifies a catalogued gel and is therefore present in both the catalog and real sections. On the other hand, spot_ref describes an ID given by the acquisition system (e.g., PDQUEST identifies spots with ssp string). This enables linkage of the data stored in the acquiring machine to AGML for auditing purposes.

The gel_image element contains all of the details of the image file uploaded by the user. The image subelement contains the whole image file as base64-encoded binary data. This element can store the raw image as well as the processed image. Additionally, the file and image_info elements contain image specific information that is useful in further analysis of the image. The detection_parameters element and its subelements in the gel_image section enable the inclusion of images scanned at different wavelengths, such as Differential In-Gel Electrophoresis (DIGE) gel images, into AGML. This additional tag gives AGML the ability to store any 2-DE gel, regardless of the number of different wavelengths used in the analysis. Inclusion of the gel images results in the document being very large; however, having the image in the document provides the user with immediate access to the original data.

AGML version 1.0 [13] provided elements to include mass-to-charge and intensity pairs to describe the mass spectrometry results obtained for individual 2-DE spots. Version 2.0 extends this by giving different options to store the mass-spectrometry data, which underscores the fact that 2-DE experiments are tightly coupled to the mass spectrometric identification (Figure 2). This element acts only as a place holder for connecting established mass spectrometry-centric XML data structures such as mzXML [3] and mzData [12], and for describing mass spectrometric information for the respective spot. Extension to accommodate other mass spectrometric schemas could easily be achieved by referencing those using standard XML namespace rules [18]. Additionally, using the 'link' element, one can identify whether the proteomic data has also been submitted or identified and placed in a repository such as PRIDE [16]. The support for 2-DE and MS experimental designs is also extended to the manipulation of the gels, a specific requirement to accommodate AGML-centric laboratory management systems. Additional elements under mass_spec can also identify the location (location) on the plate used for mass spectrometry analysis where the sample was applied. The element pooledwith can identify whether the sample spotted on the plate came from a pooled sample of spots from many gels.

AGML Central: Data Repository and Analysis Framework

AGML Central [15], a web-based analysis pipeline created around the AGML format, was expanded into a 2-DE data warehouse (Figure 3). The first set of programs developed were converters that would translate native data files into AGML data files. Currently, there are converters for PDQUEST (Bio-Rad, CA, USA), Phoretix2D (Non-linear Dynamics), Melanie (GenBio SA), and DeCyder DIGE (GE Healthcare). The software programs described below were developed based on the AGML structure. Thus, there is no need to re-write these applications to work with individual files generated by different analytical instruments as long as they are in AGML format.

Figure 2. View of the sub element <mass_spec> structure. The non-compulsory 'mass_spec' provides the option to a) store the mass spec data in a native format provided by AGML, b) store it in another format such as mzXML or mzData, or c) add a link to a location where the data is stored, such as PRIDE. Additionally, the elements 'location' and 'pooledwith' (children of 'mass_spec') can capture the location of the plate well where the spot was deposited or pooled respectively.

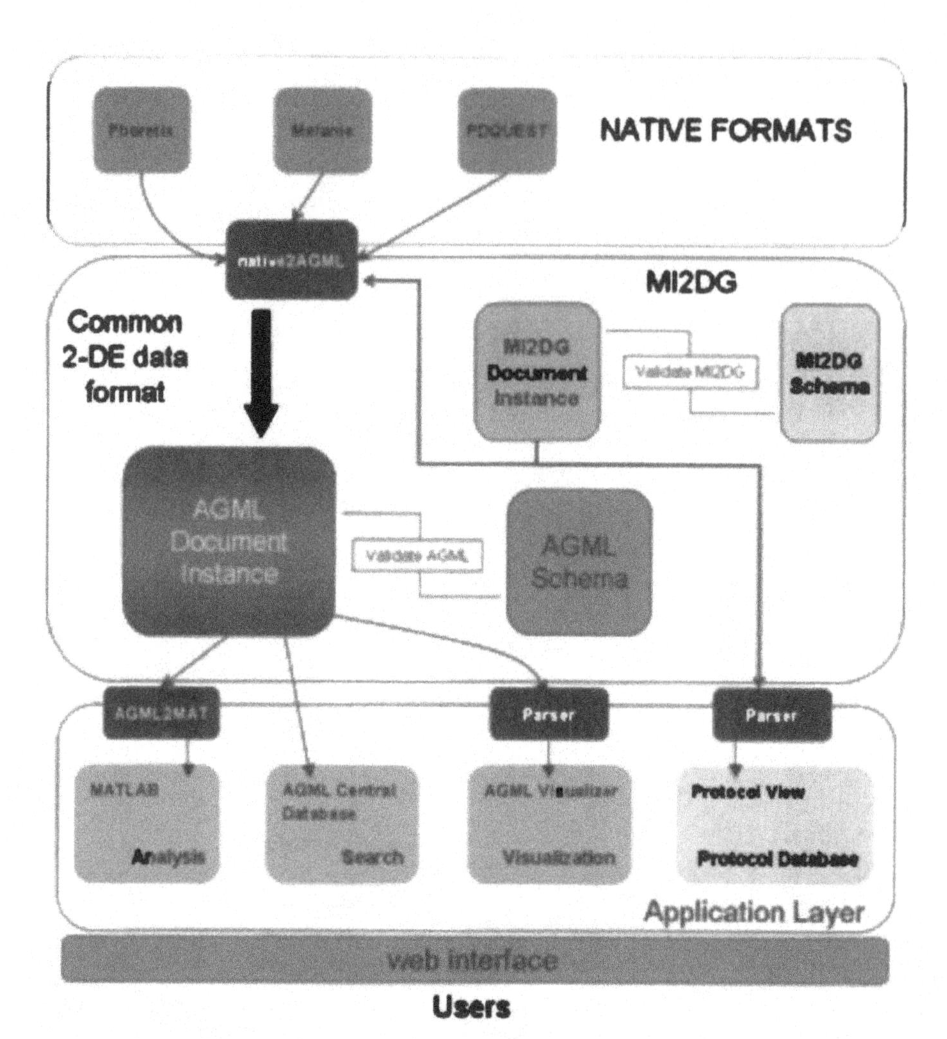

Figure 3. AGML Central web infrastructure. AGML XML format describing a 2-DE experiment is central to the AGML Central architecture. The web infrastructure is written in PHP programming language, a widely-used general-purpose scripting language, and the storage of the XML instance documents is provided by PostgreSQL, an open source object-relational database management system. The AGML document is stored as a logical unit within the database. This eliminates the need to store the document as blobs and also provides for fast retrieval of the data. Analysis software is written in MATLAB® (The MathWorks, Inc.), a technical computing environment ideal for handling high dimensional data.

AGML Central includes AGML Visualizer, an instrument-neutral Java applet that visualizes 2-DE gels. Built-in capabilities such as searching, displaying the experimental protocol, and displaying individual spot information makes the analysis of 2-DE data easy. This feature also greatly enhances the dissemination of the 2-DE data by allowing the data generated from one instrument to be viewed, even in the absence of the corresponding acquisition instrument. AGML

Visualizer can be launched in two different ways: as an applet, or through the Java Web Start infrastructure. Using the Java Web Start infrastructure, an AGML document can be opened on a local hard drive. However, the applet makes use of the AGML file that has been deposited into the AGML Central database.

Another advantage of using the AGML Central framework is the ability to make use of the tools developed for the analysis of AGML files. A number of AGML-centric data analysis tools were also developed, not just to add to the functionality of AGML Central, but also to illustrate how researchers can test novel algorithms directly on database entries. This not only illustrates the use of AGML Central as a data service, but also underscores the importance of such a service to 2-DE-centric research. For example, the tool illustrated in Figure 4 provides the sort of double clustered heat map functionality that is often used to explore microarray data.

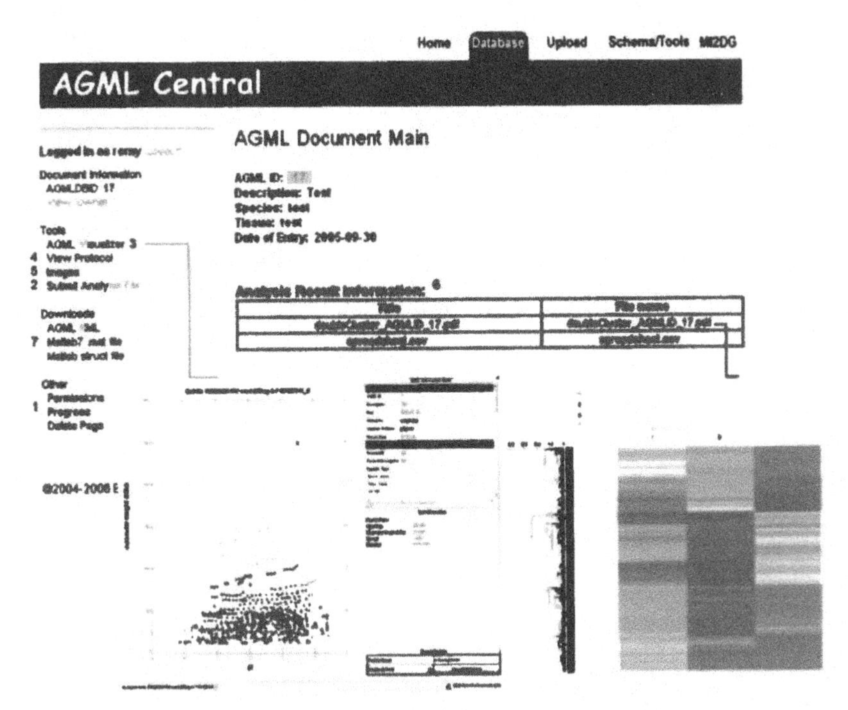

Figure 4. AGML Central web site displaying the AGML Document Main page. This page gives access to all of the information relating to a 2-DE experiment. The owner of the page can also give permissions to others to view the experiment, check progress, and delete the submitted files (1). Collaborators of the project can also submit files to the project (2) or view the experiment using AGML Visualizer (3). They can also view the 2-DE protocol used for the experiment by clicking on the view protocol link (4). Raw images can be viewed or downloaded by going to the images link (5). All of the project files are described on this page under analysis result information (6) and can be viewed or downloaded. Additionally, the AGML XML file and MATLAB® mat files are available for download from this page for the experiment (7). Thus the MATLAB® code written for this project can be used to analyze the 2-DE data directly without further manipulation.

The prototypic statistical analysis applications that use AGML Central as a data service were written in MATLAB® (The MathWorks, Inc., Natick, MA, USA) and are available at the AGML Central website [19]. The statistical tools currently available are for cluster analysis, principal component analysis, and normalization of 2-DE results [19].

Discussion

AGML was developed through close interaction with bioinformaticians and experimentalists to create a common data format that is open, accessible, and encompassing of all aspects of 2-DE experiments [13]. The AGML format accommodates a description that spans from the start of the experiment to the final identification step, and as a consequence, all of the data is placed within the experimental context. The resulting definition, AGML 2.0, allows users to establish both the provenance and relevance of a 2-DE experiment, thereby enabling the development of effective search and analysis tools.

Specialized databases exist throughout the world that focus on 2-DE data [20-22], with SWISS-2DPAGE being a major database [23]. Other research efforts have been directed toward comprehensive representation of proteomics experiments, such as PEDRo [14] and HUP-ML [24]. AGML was developed as a pragmatic representation of the 2-DE-centric subdomain and can be used to interoperate with those much larger and more encompassing representations.

In spite of being the workhorse in proteomic study, a gel-centric approach has had weak bioinformatics support due to the lack of stable gel-centric data standards and formats. In order to assist with tool development and data dissemination, a public database of AGML formatted entries, AGML Central [25], was developed. This web interface comes with a visualization plug-in and a portal for retrieval and submission by external applications. For example, MATLAB® (The MathWorks, Inc., Natick, MA, USA) GUI functions are available in the tools page for direct access to data to and from AGML Central. The ability to map AGML XML format to a MATLAB® 'struct' (agml) enables statisticians and bioinformaticians to create algorithms based on it. This MATLAB® struct can hold information extracted from an AGML XML document; hence, users of this format can, by extension, use any algorithm developed for the struct 'agml'. This feature allows the development of AGML Central-based pipelines and analysis tools. Additionally, the results of the analysis using other methods can also be submitted to AGML Central, to be appended to the corresponding entry. Although AGML Central is a public repository, all data submitted is private by default. The owner of the data can decide to make it public by providing selective access. There are currently 26 entries, of which only 2 have been made public. However, 14 of the entries have been designated for collaboration. It is our hope that as collaboration is completed all data will be made public.

The AGML concept and its implementation facilitate the management of proteomic data coming from diverse labs using different instruments and protocols, and enable the creation of much needed public 2-DE databases [26]. For this reason, the AGML format provides a wider community of developers (through the accompanying open source project) and a larger audience of users (such as bioinformaticians and statisticians) with a way to access information generated by 2-DE experiments, thus enabling them to develop comprehensive data mining algorithms that allow for exploratory and confirmatory data analysis. For example, Oates et al [27] used the AGML Central infrastructure to manage, integrate, and analyze 2-DE data to identify biomarkers that differentiate the two most common causes of acute renal failure. They used AGML Central to disseminate both their protocol and proteomic data in the AGML format to their bioinformatics collaborators. They then used the AGML data structure in the MATLAB® native format, which is provided by AGML Central, to do exploratory analysis on their proteomic data. Using AGML Central allowed the collaborators access to all of the information at any time, thus streamlining their collaborative effort to get results faster. Additionally, a nascent controlled vocabulary exists for AGML; please see Minimum Ontology for 2DE Gel Electrophoresis [28] for more information on this effort. Completion of this work will give the AGML format true agility and the ability to work with semantic web technologies.

The standards being created by HUPO-PSI-GEL for 2-D gel electrophoresis data markup, gelML and GelInfoML [7] and two analogous MIAPE modules (GE and GI; [29]), hope to encompass more details and be a comprehensive data standard for 2-D gel electrophoresis as a whole. gelML and GelInfoML are both based on the Functional Genomics Experiment [9] modeling framework. While these efforts will ultimately result in community-based data standards, AGML was created to answer this need more pragmatically. Since its inception in 2004 [13], AGML has been more interested in getting the data to tool makers. In the process AGML has acquired many of the features that are being proposed. Briefly, the <mi2dg> elements can be analogous to GelInfoML, and the <reals> can be analogous to gelML. Once stable gelML and GelInfoML standards are published, AGML documents will be made available to be translated to these standards, thereby making the data available for any tools developed for HUPO-PSI-GEL standards.

Overcoming barriers in data flow is a central theme in the route toward Systems Biology and this is especially true for rapidly expanding methodologies such as those developed for proteomics research. Rapid growth of the field has seen the emergence of high throughput instruments from different vendors that use many different proprietary data standards that, due to the lack of data interoperability, limit data integration. This fact is underscored by the formation of the Interoperable Informatics Infrastructure Consortium, whose major goal is to eliminate barriers to application interoperability, data integration, and eventually knowledge sharing [30]. Additionally, work undertaken by HUPO-PSI to advance the field of proteomics data standards also points to the need that exists in the area of data interoperability in proteomics [31].

Conclusion

The gel-centric AGML data structure is a comprehensive format for the representation of 2-DE proteomic data. It seeks to address the glaring need for a pragmatic format in which both experimental results and their experimental context can be represented. The future of 2-D electrophoresis tool development may depend on stable standards and formats being devised and used. The stability that comes with such endeavors is critical to enabling the development of open source data analysis tools that are long overdue for gel-centric proteomics.

Methods

Python, Matlab, PHP and bash Programming languages were used in the development of this project. UML diagrams were used to visualize the data structure and XML was used in creating the AGML data structure.

Authors' Contributions

RS and JSA conceived and wrote the manuscript. RS designed and implemented the object model. JAM, BR, RM and BM provided experimental expertise. All authors read and approved the final manuscript.

Acknowledgements

This work was supported by the NHLBI Proteomics initiative through contract N01-HV-28181. The authors thank Rebecca Partida for her expert assistance in preparation of the manuscript.

References

1. Fu Q, Garnham CP, Elliott ST, Bovenkamp DE, Van Eyk JE: A robust, streamlined, and reproducible method for proteomic analysis of serum by

delipidation, albumin and IgG depletion, and two-dimensional gel electrophoresis. Proteomics 2005, 5(10):2656–2664.

2. Gorg A, Weiss W, Dunn MJ: Current two-dimensional electrophoresis technology for proteomics. Proteomics 2004, 4(12):3665–3685.
3. Pedrioli PG, Eng JK, Hubley R, Vogelzang M, Deutsch EW, Raught B, Pratt B, Nilsson E, Angeletti RH, Apweiler R, Cheung K, Costello CE, Hermjakob H, Huang S, Julian RK, Kapp E, McComb ME, Oliver SG, Omenn G, Paton NW, Simpson R, Smith R, Taylor CF, Zhu W, Aebersold R: A common open representation of mass spectrometry data and its application to proteomics research. Nat Biotechnol 2004, 22(11):1459–1466.
4. mzData [http://www.psidev.info/]
5. mzXML [http://sashimi.sourceforge.net/]
6. mzData Tools [http://www.psidev.info/index.php?q=node/95]
7. Gel Markup Language [http://www.psidev.info/index.php?q=node/83]
8. GelInfoML [http://www.psidev.info/index.php?/q=node/83#miape]
9. Functional Genomics Experimental Object Model [http://fuge.sourceforge.net]
10. Jones AR, Pizarro A, Spellman P, Miller M: FuGE: Functional Genomics Experiment Object Model. Omics 2006, 10(2):179–184.
11. Taylor CF, Paton NW, Lilley KS, Binz PA, Julian RK Jr., Jones AR, Zhu W, Apweiler R, Aebersold R, Deutsch EW, Dunn MJ, Heck AJ, Leitner A, Macht M, Mann M, Martens L, Neubert TA, Patterson SD, Ping P, Seymour SL, Souda P, Tsugita A, Vandekerckhove J, Vondriska TM, Whitelegge JP, Wilkins MR, Xenarios I, Yates JR 3rd, Hermjakob H: The minimum information about a proteomics experiment (MIAPE). Nat Biotechnol 2007, 25(8):887–893.
12. Orchard S, Hermjakob H, Taylor CF, Potthast F, Jones P, Zhu W, Julian RK Jr., Apweiler R: Further steps in standardisation. Report of the second annual Proteomics Standards Initiative Spring Workshop (Siena, Italy 17–20th April 2005). Proteomics 2005, 5(14):3552–3555.
13. Stanislaus R, Jiang LH, Swartz M, Arthur J, Almeida JS: An XML standard for the dissemination of annotated 2D gel electrophoresis data complemented with mass spectrometry results. BMC Bioinformatics 2004, 5:9.
14. Taylor CF, Paton NW, Garwood KL, Kirby PD, Stead DA, Yin Z, Deutsch EW, Selway L, Walker J, Riba-Garcia I, Mohammed S, Deery MJ, Howard JA, Dunkley T, Aebersold R, Kell DB, Lilley KS, Roepstorff P, Yates JR 3rd, Brass A, Brown AJ, Cash P, Gaskell SJ, Hubbard SJ, Oliver SG: A systematic

approach to modeling, capturing, and disseminating proteomics experimental data. Nat Biotechnol 2003, 21(3):247–254.

15. Stanislaus R, Chen C, Franklin J, Arthur J, Almeida JS: AGML Central: web based gel proteomic infrastructure. Bioinformatics 2005, 21(9):1754–1757.
16. Jones P, Cote RG, Martens L, Quinn AF, Taylor CF, Derache W, Hermjakob H, Apweiler R: PRIDE: a public repository of protein and peptide identifications for the proteomics community. Nucleic Acids Res 2006, 34(Database issue):D659–63.
17. Minimum information about 2-D gel electrophoresis [http://www.agml.org/mi2dg]
18. XML namespace rules [http://www.w3.org/TR/REC-xml-names/]
19. Almeida JS, Stanislaus R, Krug E, Arthur JM: Normalization and analysis of residual variation in two-dimensional gel electrophoresis for quantitative differential proteomics. Proteomics 2005, 5(5):1242–1249.
20. Ericsson C, Petho Z, Mehlin H: An on-line two-dimensional polyacrylamide gel electrophoresis protein database of adult Drosophila melanogaster. Electrophoresis 1997, 18(3-4):484–490.
21. Li F, Li M, Xiao Z, Zhang P, Li J, Chen Z: Construction of a nasopharyngeal carcinoma 2D/MS repository with Open Source XML database--Xindice. BMC Bioinformatics 2006, 7:13.
22. Yoshida Y, Miyazaki K, Kamiie J, Sato M, Okuizumi S, Kenmochi A, Kamijo K, Nabetani T, Tsugita A, Xu B, Zhang Y, Yaoita E, Osawa T, Yamamoto T: Two-dimensional electrophoretic profiling of normal human kidney glomerulus proteome and construction of an extensible markup language (XML)-based database. Proteomics 2005, 5(4):1083–1096.
23. Hoogland C, Mostaguir K, Sanchez JC, Hochstrasser DF, Appel RD: SWISS-2DPAGE, ten years later. Proteomics 2004, 4(8):2352–2356.
24. Japan Human Proteome Organization [http://www.jhupo.org]
25. AGML Central [http://www.agml.org]
26. Prince JT, Carlson MW, Wang R, Lu P, Marcotte EM: The need for a public proteomics repository. Nat Biotechnol 2004, 22(4):471–472.
27. Oates JC, Varghese S, Bland AM, Taylor TP, Self SE, Stanislaus R, Almeida JS, Arthur JM: Prediction of urinary protein markers in lupus nephritis. Kidney Int 2005, 68(6):2588–2592.
28. Minimum Ontology for 2DE Gel Electrophoresis [http://charlestoncore.musc.edu/ont/mo2dg.html]

29. MIAPE: The Minimum Information About a Proteomics Experiment [http://www.psidev.info/miape/]

30. I3C Announces new life science protocols to simplify data exchange, knowledge sharing. [http://www.sun.com/smi/Press/sunflash/2002-06/sunflash.20020610.3.xml]

31. Orchard S, Taylor CF, Hermjakob H, Weimin Z, Julian RK Jr., Apweiler R: Advances in the development of common interchange standards for proteomic data. Proteomics 2004, 4(8):2363–2365.

Pegasys: Software for Executing and Integrating Analyses of Biological Sequences

Sohrab P. Shah, David Y. M. He, Jessica N. Sawkins, Jeffrey C. Druce, Gerald Quon, Drew Lett, Grace X. Y. Zheng, Tao Xu and B. F. Francis Ouellette

ABSTRACT

Background

We present Pegasys—a flexible, modular and customizable software system that facilitates the execution and data integration from heterogeneous biological sequence analysis tools.

Results

The Pegasys system includes numerous tools for pair-wise and multiple sequence alignment, ab initio gene prediction, RNA gene detection, masking

repetitive sequences in genomic DNA as well as filters for database formatting and processing raw output from various analysis tools. We introduce a novel data structure for creating workflows of sequence analyses and a unified data model to store its results. The software allows users to dynamically create analysis workflows at run-time by manipulating a graphical user interface. All non-serial dependent analyses are executed in parallel on a compute cluster for efficiency of data generation. The uniform data model and backend relational database management system of Pegasys allow for results of heterogeneous programs included in the workflow to be integrated and exported into General Feature Format for further analyses in GFF-dependent tools, or GAME XML for import into the Apollo genome editor. The modularity of the design allows for new tools to be added to the system with little programmer overhead. The database application programming interface allows programmatic access to the data stored in the backend through SQL queries.

Conclusions

The Pegasys system enables biologists and bioinformaticians to create and manage sequence analysis workflows. The software is released under the Open Source GNU General Public License. All source code and documentation is available for download at http://bioinformatics.ubc.ca/pegasys/.

Background

Pipelines for Biological Sequence Analysis

Large scale sequence analysis is a complex task that involves the integration of results from numerous computational tools. For high-throughput data analysis, these tools must be tied together in a coordinated system that can automate the execution of a set of analyses in sequence or in parallel. To this end, a diverse array of software systems for biological sequence analysis have emerged in recent years. For example, the Ensembl pipeline [1] automates the annotation of several eukaryotic genomes, Mungall et al [2] have created a robust pipeline for annotation and analysis of the Drosophila genome, GenDB [3] is used as an annotation system for several prokaryotic genomes and Yuan et al [4] have published resources for annotating the rice and other plant genomes. These pipelines are extensive in their scope, are well-designed and meet their objectives. In surveying these and other systems, we have identified three critical areas that are essential for building on the design of existing biological sequence analysis pipelines:

- There is a need for flexible architecture so that one software system can be used to analyse different data sets that may require different analysis tools.

- A system needs to allow for the inclusion of new tools in a modular fashion so the software architecture does not have to change with the addition of new tools.
- A system should provide the framework to facilitate data integration of analysis results from different tools that were computed on the same input.

The Need for Flexible Architecture

The systems outlined above differ substantially from each other in their design and application, but share common attributes. The diversity is naturally reflective of the varied computational tasks that biologists working on different projects need to perform in order to analyse their data. A researcher working on bacteria will need different tools for her analyses than someone working on mouse. The specificity driven by the needs of a research project makes it impossible to use a pipeline designed for a particular data set for analysis of another data set that has inherent differences such as the organism from which it was generated. As a result, numerous software pipelines have been created, many of which perform similar analyses (such as genome annotation) but on different data. For example, the concept of constructing a pipeline or 'workflows' of data processing are common to nearly all high-throughput sequence analysis projects. This shared concept provides an opportunity to harness the commonality in software so that a new system need not be designed for every new project.

Incorporating New Tools into Existing Frameworks

The bioinformatics community is faced with a challenging and dynamic environment where new computational tools and data sets for sequence analysis are constantly being generated. Capitalizing on algorithmic and computational advances is critical to discovering more about the data being analysed. For a system that has a rigid pipeline that is 'hard coded', it may require a significant programming investment to incorporate a new tool. This may discourage biologists from integrating a new tool on the basis of logistics, rather than on the basis of scientific applicability. Therefore, a system should provide a framework that is designed for flexibility and extensibility.

Facilitating Data Integration

Genome annotation requires data integration. For example ab initio prediction of gene structures on genomic sequence can be greatly enhanced by using supporting sequence similarity searches [5-7]. Concordance between different methodologies lends stronger support and gives more compelling evidence to an algorithm or a person trying to infer true biological features from computationally derived features [8]. It follows that any analysis pipeline or system should provide a design that facilitates integration of heterogeneous sources of data.

The Pegasys Biological Sequence Analysis System

To meet the challenges outlined above we have designed and implemented Pegasys: a flexible, modular and customizable framework for biological sequence analysis. The software is implemented in the Java programming language and is Open Source, released under the GNU General Public License. The features of Pegasys allow it to be used on a wide variety of tasks and data. Analysis modules for pair-wise and multiple sequence alignment, ab initio gene prediction, masking of repetitive elements, prediction of RNA sequences and eukaryotic splice site predictors have been developed. A new set of analyses is performed by first creating a new 'workflow'. We define a workflow as a set of analyses a biologist wishes to perform on a single sequence or set of sequences. Each workflow has the following qualities: a) the analyses can be linked together such that output from one analysis can be used as input to a subsequent analysis, b) analyses can accept outputs from more than one analysis as input, and c) analyses that are not serially dependent can be executed in parallel.

Analysis tools in the Pegasys system are wrapped in modules that can easily be plugged into the system. The backend database system provides a data model that abstracts the concept of a computational feature and captures data from all the different analysis tools in the same framework. We have implemented data adaptors that can export computational results in General Feature Format [9] and Genome Annotation Markup Elements (GAME) XML [10] for import into the Apollo genome editor [11]. For simple workflows where data integration is not applicable, for example one analysis on an input sequence, raw, untransformed output from the analysis can also be retrieved.

The system is fronted by a graphical user interface that allows users to create workflows at run-time and have them executed on the Pegasys server. The GUI also allows users to save their workflows for repeat execution on different input, or using different reagents.

To demonstrate the utility of Pegasys in widely different bioinformatics tasks, we present three use cases of the system: a single application workflow, a workflow designed for formatting a database for BLAST [12,13] and searching the newly formatted database, and finally a workflow designed for genome annotation of eukaryotic genomic sequence.

We are releasing this work with the intention that a wide variety of sequence analyses in the bioinformatics research community will be enabled. Full details of the availability, support and documentation of Pegasys can be found at http://bioinformatics.ubc.ca/pegasys/ .

Implementation

The design of the Pegasys system is guided by three main principles: modularity, flexibility and data integration. With these principles in mind, we designed Pegasys with the following architecture.

Architecture and Data Flow

The architecture of the system has a layered topology that uses a client/server model. The client has a graphical user interface (see Figure 4) for the creation of workflows. Once a workflow is created, it is sent to the server where it is executed. The server is made up of separate layers for job scheduling, execution, database interaction, and adaptors. The connectivity between layers is shown in Figure 1. The application layer converts the work flow rendered in XML into a directed acyclic graph (DAG) of analyses in memory. While traversing the DAG, the application schedules all of the analyses on a distributed compute cluster and facilitates the flow of data so that a particular node's program is only executed once all of its inputs are ready (i.e. all of the 'parent' analyses are complete). As each analysis completes, the results are inserted into the backend database layer. Complete reports and computational features of a sequence are inserted into relational tables. Sophisticated queries on the data, in which results from selected programs can be integrated together over a portion or all of the input sequence, can then be run to compile data for output. The data is exported from the system via the adaptor layer in various formats (currently GFF, GAME XML and raw output from each analysis tool are supported) for human interpretation or for import into other applications such as viewing tools (DAS [14]), editing tools (Apollo [11]) or statistical analysis tools such as R [15].

The Pegasys Data Structure

The core data structure of the Pegasys system is a DAG G(V, E), consisting of a set of nodes V and a set of edges connecting the nodes E (see Figure 2). The DAG data structure models a workflow created by a user of the Pegasys system. A node can take one of three forms: a) an input sequence or b) an individual run of a program in the system or c) an output node. An edge (v1, v2) where v1 and v2 are nodes in V links data flow between v1 and v2. An edge represents a serial dependency, indicating that the input of v2 is tied to the output of v1. We refer to this relationship as a parent-child relationship: node v2 is a child of node v1 and node v1 is the parent of node v2. The edge ensures that the output format from v1 is consistent with the input format of v2. A node in the DAG can have more than one parent and therefore can have heterogeneous input from multiple sources. The edges in the graph are directional and can only connect two nodes that are executed one after another. The graph therefore has a chronological axis: the child nodes are executed after their parent nodes have completed.

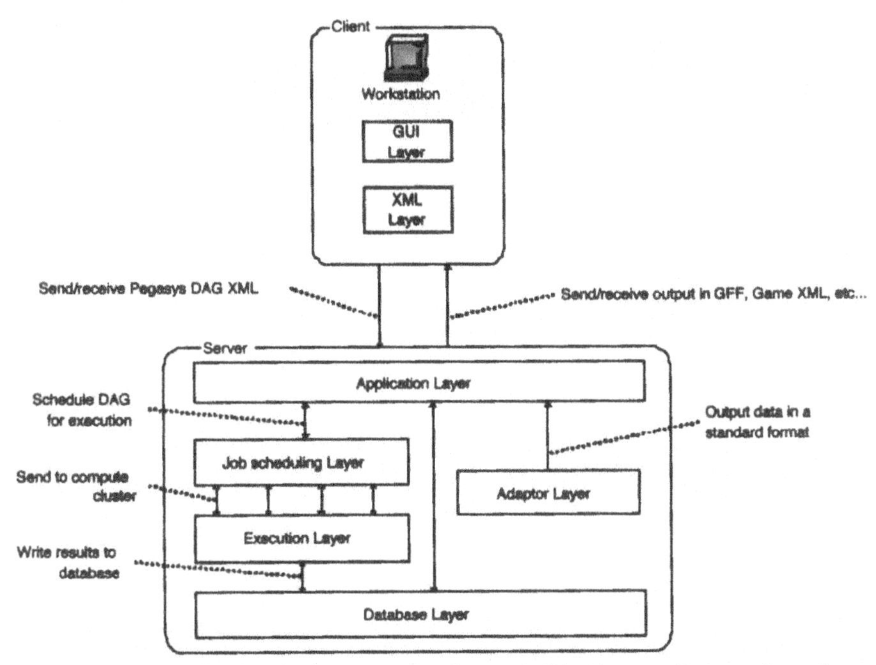

Figure 1. Diagram showing the client/server model and layering of the Pegasys architecture. Arrows between the layers indicate a transfer of data. The workflow created by manipulating the GUI in the client is sent as a Pegasys DAG XML file to the server. The application layer then processes the XML file, and sends jobs to the job scheduling layer. The analyses are then executed and the results are stored in the database. The adaptor layer takes results stored in the PegasysResultSet data structure in memory in the application layer and can create output in GFF or GAME XML format. This file is then returned to the GUI where it can be digested by the user or input into a visualization tool.

The DAG is created dynamically at run time as the user manipulates the GUI (see The Graphical User Interface section). The user can create workflows using any combination of the available programs in Pegasys by dragging/dropping and linking graphical icons that represent sequence analysis tools on a canvas together with edges in much the same way that one would use drawing tool software to create a flow diagram. Each program icon can be clicked to open a dialogue box that can take inputs for parameters that are supported by that particular program.

Once all of the parameters for all the nodes have been filled in, the information for each node and their relationships to each other are compiled into a structured XML file. This file is then used as input to the Pegasys server that executes the analyses in parallel (described in the Architecture and Data Flow section) or can be saved for later editing or distribution. During the execution of the DAG, the data structure can adjust itself to accommodate outputs generated from the nodes. Consider the edge (v3, v5) depicted in Figure 2 that connects an ab initio gene prediction program v3 with a sequence alignment program v5. In v5, the user wishes to search the coding regions from the output of v3 against a protein database. v5 cannot know how many genes will be predicted from v3 before v3 has terminated. Once v3 has terminated however, v5 will replicate itself for each 'output unit' generated from v3 (see Figure 2B). In this case, v5 replicates itself for each of the coding regions and the DAG executes each 'copy' of v5 in parallel. This built-in elasticity confers maximum parallel execution of analyses and therefore more efficient execution of the computations in the DAG.

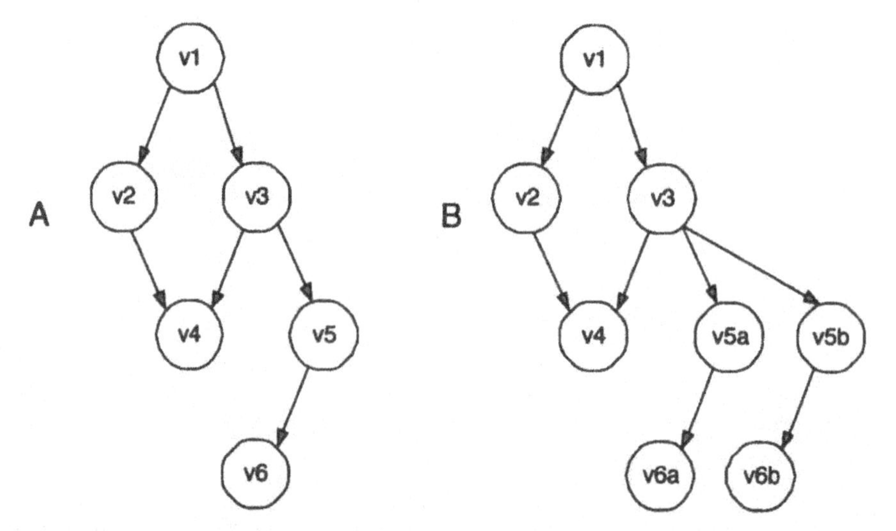

Figure 2. Diagram showing an abstract representation of a Pegasys DAG. A): Consider v1: this could be an input sequence that is used by two sequence analysis programs v2 and v3. v4 is dependent on the output of both v2 and v3 and therefore cannot execute until v2 and v3 have completed. In this diagram, v2 and v3 will be executed in parallel as will v4 and v5. B): DAG in the case where v3 produces two instances of the expected output to v5. The sub-DAG rooted at v5 replicates itself (v5a and v5b) for each instance of its input. All of the new sub-DAGs are executed in parallel.

The Program Module

The Program module is the fundamental unit of the nodes of the aforementioned DAG in the application layer of the server and is a real instance of a node $v \in V$.

'Program' is an object oriented class that abstracts the concept of a Unix program that is natively compiled. Unix programs generally have a set of input command line parameters and output that is sent to the standard output, standard error or an output file. The Program class has a data structure to store a program's command line arguments and parameters. It contains methods for setting the path to the program's location on the system, executing the program and capturing its output from a file, standard error and standard output streams. To abstract a sequence analysis program, we created a PegasysProgram class that extends Program by adding an input sequence attribute and a PegasysResultSet to store the results of the analysis. The ProgramResultSet is a hierarchical, recursive data structure that allows storage of nested analysis results. For example a BLAST output has a list of similar sequences that each in turn has a list of high scoring pairs. Similarly Genscan produces output that contains a list of predicted genes, each of which could have a promoter, a list of exons and a poly-A signal. PegasysResultSet captures the hierarchical nature of these results.

For each sequence specific analysis tool in Pegasys, we created a class that extends PegasysProgram. Each of these classes implement their own methods that load the particular output of the program and parse it into their PegasysResultSet. For example, the locations of computational evidences such as predicted exons from a gene finding tool, or a high scoring pair from an alignment algorithm are parsed along with a statistic and/or score when available. This architecture generalises a computational feature so that programmatically, results from different analysis programs can be treated equally. As mentioned earlier, this allows the user to output results from different programs in a unified format such as GFF, or GAME XML. In addition, it facilitates querying for all computational evidence computed on a segment of sequence that may be of interest to the biologist.

Creating a new PegasysProgram derivative involves writing a parser for the particular application that can extract data that is amenable to being loaded into a PegasysResultSet. The system, at the time of this writing has PegasysPrograms for RepeatMasker [16], BLAST (blastn, blastp, blastx, tblastn, tblastx) [12,13], WU BLAST [17], the EMBOSS [18] implementation of Smith-Waterman [19], Genscan [20], HMMgene [21], Mlagan [22], Sim4 [23], TrnaScan-SE [24] and GeneSplicer [25].

The Database

The backend database of the Pegasys system was created with the goal of maximizing information capture during execution of a workflow. The database tracks all parameters used for the invocations of analysis programs, all input sequences, and all output generated by computation.

The Pegasys Schema

The Pegasys schema has three main tables: 'sequence' which stores the input sequences, 'program_run' which stores the information about an individual program's process on the system and 'pegasys_result' which stores the locations of computational features on the input sequence. Peripheral to the three core tables are seventeen meta tables that store information about the data in the core tables. The full schema is presented in Figure 3.

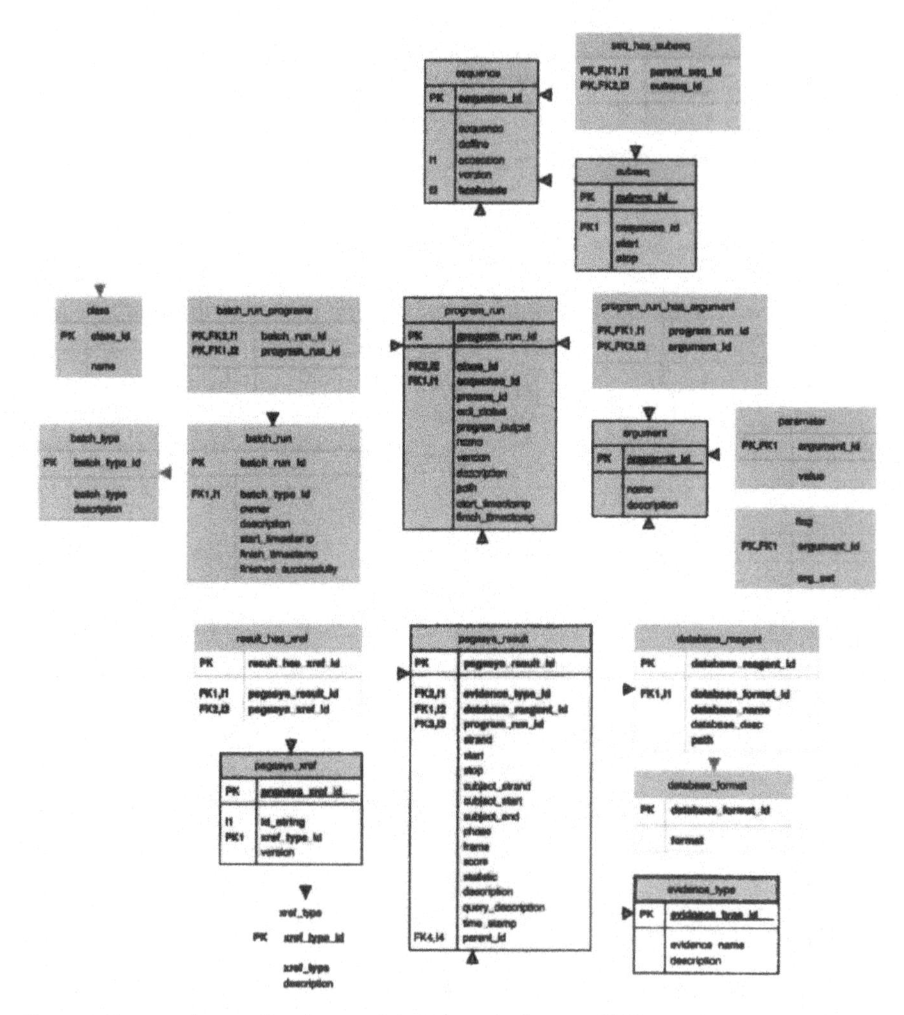

Figure 3. Diagram showing the relations of the Pegasys database model. There are three core tables to the database: sequence (shown in blue), program_run (shown in orange) and pegasys_result (shown in yellow). The meta tables for each of the three core tables are colour coded to match the corresponding core table. Foreign keys are indicated with 'FK' and indexed fields are marked with T.

The 'program_run' table is designed to store all information on an invocation of an analysis tool in order to facilitate reprocessing of results without having to recompute an analysis and can also aid in diagnosing problems that are bound to occur in the system. 'program_run' stores the class that invoked the process, the raw unprocessed output of the program, the start and end time of the process and the exit status of the process. In addition, all command line arguments used to invoke the program are stored in support tables to 'program_run' in the structured tables 'argument', 'parameter', and 'flag'. Entries into 'program_run' can be grouped into batches for selective retrieval of analysis results.

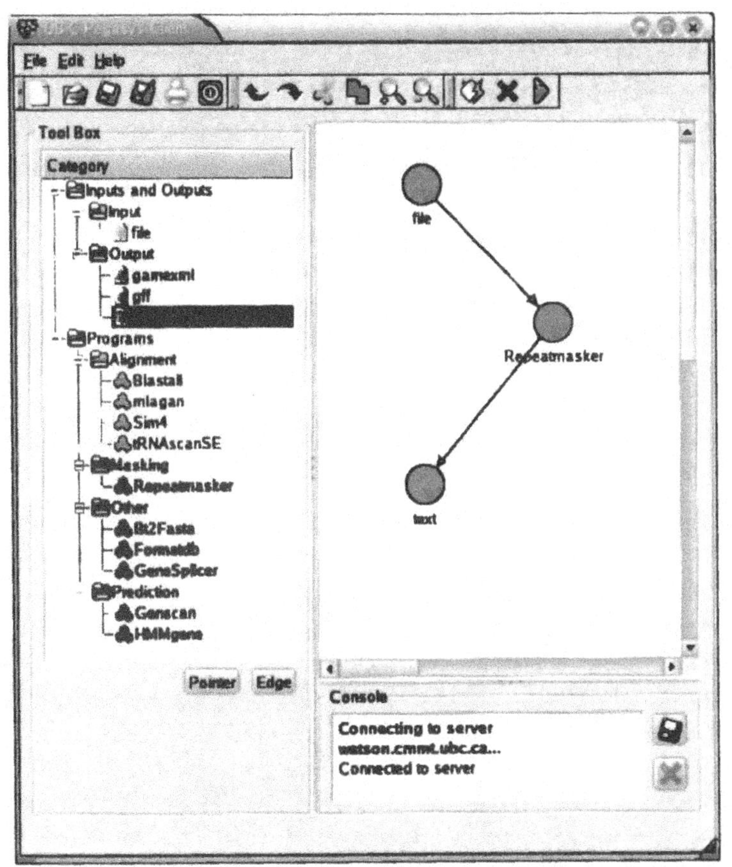

Figure 4. Screenshot of the Pegasys GUI showing the three pane design. The visible pane is the canvas pane which allows the user to create a workflow by clicking and dragging icons corresponding to the programs available to the system. The icons can be connected to each other through edges. The parameters used for the execution of each program can be set by double clicking the icon and filling in the dialogue box that appears (see Figure 5). Expected inputs and outputs for the edge can be set by double clicking the edge and filling in the dialogue (see Figure 6). This workflow will run RepeatMasker on the sequence specified in the File node and write the results to a text file whose path is specified in the text output node. The RepeatMasker analysis itself is run on the compute server and the results are communicated back to the client.

The 'sequence' table stores the raw sequence string itself, a unique hash code for the sequence string generated by the java.lang.String.hashCode() function, an identifier for the sequence (by default the GenBank accession.version number) and a description of the sequence (by default the NCBI definition line of the FASTA file). This table does not store meta data about the sequence, rather it is meant to store unique sequences used for computation. The system assumes additional information on the sequence is stored elsewhere. The uniqueness is enforced by ensuring all sequences have distinct hash codes, description and identifiers. Support tables for sequence have been created to enable the analysis of sub-sequences of a larger input sequence. The subsequence relationship to the sequence is stored in the 'subseq' and 'seq_has_subseq' relations. These tables are useful for 'sliding window' analyses or when focusing in on small regions of interest of a larger input sequence.

The 'pegasys_result' table stores the results of the computations. It has attributes for a computational evidence type, a database reagent (if the result is from similarity searches or uses a particular model in ab initio predictions), the strand, start and end positions of the computational feature, a score and a statistic for the computational feature and a free-text description of the feature. If available, the strand, start and end position on the target sequence of an alignment are also recorded. To support hierarchical computational evidences, the table has a 'parent_id' that is a self-referential foreign key. This enables relating a particular row entry in the table to another row in the table. Theoretically, the table supports infinite nesting of hierarchical data types, although in practice results are no more than 2 levels deep.

The support tables for 'pegasys_result' allow cross-referencing of ids. For example, the system models the concept of linking out an identifier from the result of a database search so that the full sequence and meta data of that sequence can be easily retrieved. This cross-referencing of a 'pegasys_result' to an identifier is stored in the 'result_has_xref' relation. The type of identifier is labeled by a controlled vocabulary so that one can query on a particular type of cross-reference (such as accession number) as well as add a new type of cross-reference to the system. Additional support tables to 'pegasys_result' are: 'database-format', 'database_reagent' and 'evidence-type'. Each of these tables stores controlled nomenclature that is referenced by 'pegasys_result'. The 'database-format' contains values such as blast, fasta, and genscan for BLAST formatted, FASTA formatted and Genscan training model respectively. The 'database_reagent' table stores the names and descriptions of sequence databases and statistical models that are used in the analysis, so that a user can query the Pegasys database for results from a particular database reagent. This structure also allows adding new database reagents into the system seamlessly. The 'evidence-type' table stores an ontology of computational evidence types, for example 'blastn_hit' or 'genscan_exon'. For each program that is part of the Pegasys system, the computational evidence(s) that it outputs must be recorded in the 'evidence-type' table prior to its use.

Database API

To communicate programmatically with the database, we have created a modular application programming interface (API). The PegasysDB class contains public methods for insertion and retrieval of sequences, analysis results and sets of results (from different programs) on a particular sequence. Application developers that wish to access data from a Pegasys database can use these high-level methods to rapidly store and access data in a straightforward manner without having to study the underlying schema of the database. The database API uses the PostgreSQL JDBC driver and so is backend relational database management system (RD-BMS) independent.

Adaptors

We have implemented several adaptors for exporting data from a PegasysProgram or set of PegasysPrograms that contain analysis results. The derived PegasysAdaptor classes all implement a print method to output data in a specific format. We currently have derived PegasysAdaptor classes for GAME XML for import into Apollo [11] and GFF [9] which can be imported into numerous tools and servers such as the Distributed Annotation System [14] (DAS) and Gbrowse [26]. The adaptor architecture is extensible and easily allows the development and inclusion of new adaptors for additional formats. The PegasysAdaptor classes serve as an important bridge from the Pegasys data structure to other well-used standards and permits interoperability between data computed using Pegasys and many other bioinformatics tools and databases.

Parallelism

Our local installation of Pegasys runs on a 28 CPU distributed memory compute cluster that runs the OpenPBS parallel batch server [27]. We have implemented 'serial' parallelism into the system meaning that each application is a serial process, but many serial processes can be run in parallel. It is important to note that this is distinct from parallelism where a single application is itself implemented using a message passing library that can use many distributed processors in a compute cluster environment. To enable serial parallelism, we implemented a Runnable thread class in the Pegasys application layer that can navigate a command line argument of a PegasysProgram, and create a script at runtime that is used to submit a job to a PBS job queue. To monitor job progress, we implemented a Java server called QstatServer, that registers each job sent to the PBS job queue. The QstatServer maintains a hash table of jobs in the queue and informs the Pegasys application layer when a particular job has terminated. This architecture enables the Pegasys application server to execute jobs in sequence or in parallel according to the structure of the DAG that was sent by the client.

Pegasys and Java

The Pegasys system is implemented in the Java programming language. Java offers robust data typing that facilitates object-oriented programming in its truest form. The principles and advantages of object-oriented design are well documented in the software engineering literature (see [28]). Java is becoming widely adopted in the bioinformatics software domain. For example, the Ensembl database has a Java API to programmatically access genome annotations [29]. The Biojava toolkit [30] is an extensive set of packages written in Java for sequence manipulation, analysis and processing. The Apollo genome editor [11], that we use with Pegasys, allows biologists and bioinformaticians to edit and create annotations in a sophisticated GUI and is written in Java. We have integrated the Biojava toolkit into Pegasys for manipulation of sequence files as well as parsing of BLAST output. Using Java also allows us to make use of the JDBC library for database connectivity that facilitates standard database interactions independent of the RDBMS engine. To enable parallelism, we made use of the robust Thread and Runnable classes that allow development of multi-threaded programs.

We have designed Pegasys in a layered architecture that consists of independent Java packages that can easily be imported into any external Java application that wishes to make use of them. These packages are well described in the Pegasys user manual, available at: http://bioinformatics.ubc.ca/pegasys/ . Implementing Pegasys in Java has brought the system strength and robustness that would not have been attainable with using a scripting language. Pegasys provides a Java alternative to existing Perl-based sequence analysis systems such as GenDB [3] and BioPipe [31].

The Graphical User Interface

The Pegasys graphical user interface (GUI) is designed for ease of use while maximizing functionality. When the client is started, the user sees a simple three pane design (see Figure 4). On left of the screen is a list of programs (the 'Tool Box') available to the user. The list is retrieved from the server as an XML configuration file when the client starts, ensuring all the programs that are available to the user from the client are available on the server. The canvas for drawing the workflow is on the upper right side of the screen, and on the bottom of the screen there is a console to view feedback from the client program.

The structure of the workflow the user creates on the canvas mirrors the structure of the DAG (see The Pegasys data structure section). The nodes of this DAG can either be input files, output files, or a program, while the edges that connect the nodes manage the flow of input and output information. For example, the Genscan program node can produce many types of outputs, a list of nucleotide FASTAs of predicted transcripts, or a list of amino acid FASTAs of the protein products. If a user connects a BLASTP node to this Genscan node, then the edge between these two nodes can be used to get the list of amino acid FASTAs from the Genscan node as input for the BLASTP node.

During the creation of the workflow, the user can modify the parameters of the analysis programs by double-clicking a node. This opens a Node Properties dialogue. An example for BLAST is pictured in Figure 5. The input/output types for each edge must be set during the creation of the workflow. This is done through the Edge Properties dialogue (see Figure 6).

When the user has finished creating the workflow, it can be saved as an XML file representing the DAG. This XML file stores all the parameters for the nodes and edges that have been set by the user during the creation of the DAG. This file can be kept on the local hard drive and retrieved for later modification or distribution, or sent to the server to be executed on the compute cluster. The saved DAG can also be sent to the server using the command-line Java client for high-throughput, or automated processing. When the processing is complete, the results are sent back to the GUI client to be saved as text files.

To ensure that the user's workflow is syntactically correct, the Pegasys client validates the workflow in real time. As the user draw nodes and edges, they are validated for correctness based on their requirements. For example, if a Program Node has a required parameter that is not filled in, the Pegasys client will display that node with a red 'X' beside it. Once this required parameter is filled in, the red 'X' will turn into a green tick mark, indicating the correctness of this node. Invalid edges are displayed in red, while correct ones are displayed in black. Typically, edges will be invalid if the 'output' and 'input' values of the edges are not set or do not match. If the workflow has a red edge or a node marked with a red 'X', the Pegasys client will not allow the user to send the workflow to the server and will output a warning to the 'Console' area.

The GUI component of the Pegasys system is implemented in C++, using QT graphical libraries [32]. The QT libraries offer a "write once compile anywhere" approach. Because the QT components are natively compiled for its target operating system, GUI components written in C++/QT have a more native look and feel and give fast response times to the user. In addition, C++/QT can be compiled on all the major operating systems, giving it nearly the same level of portability as Java and facilitating the distribution of the Pegasys GUI client for most platforms.

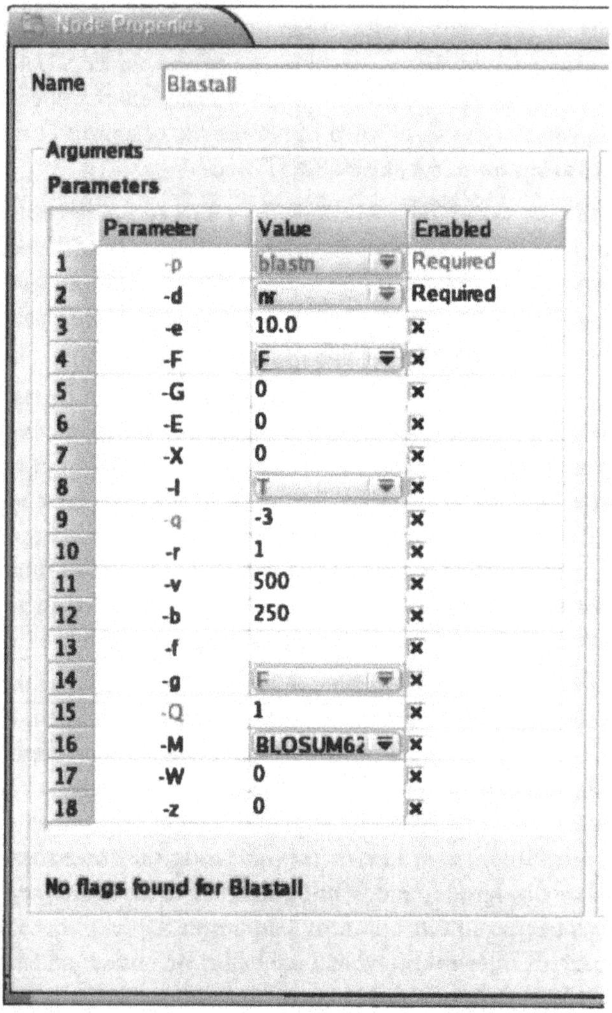

Node Properties

Name: Blastall

Arguments

Parameters

	Parameter	Value	Enabled
1	-p	blastn	Required
2	-d	nr	Required
3	-e	10.0	x
4	-F	F	x
5	-G	0	x
6	-E	0	x
7	-X	0	x
8	-I	T	x
9	-q	-3	x
10	-r	1	x
11	-v	500	x
12	-b	250	x
13	-f		x
14	-g	F	x
15	-Q	1	x
16	-M	BLOSUM62	x
17	-W	0	x
18	-z	0	x

No flags found for Blastall

Figure 5. Screenshot of the Node Properties dialogue window where users can input parameters for the analysis programs. There are three columns—the name of the parameter, its current value and a check box to indicate if this parameter is enabled. Disabled parameters will be excluded from the DAG XML, and consequently from the actual command that is executed on the server. All default values are set in the ProgramList.xml file that the server reads on startup.

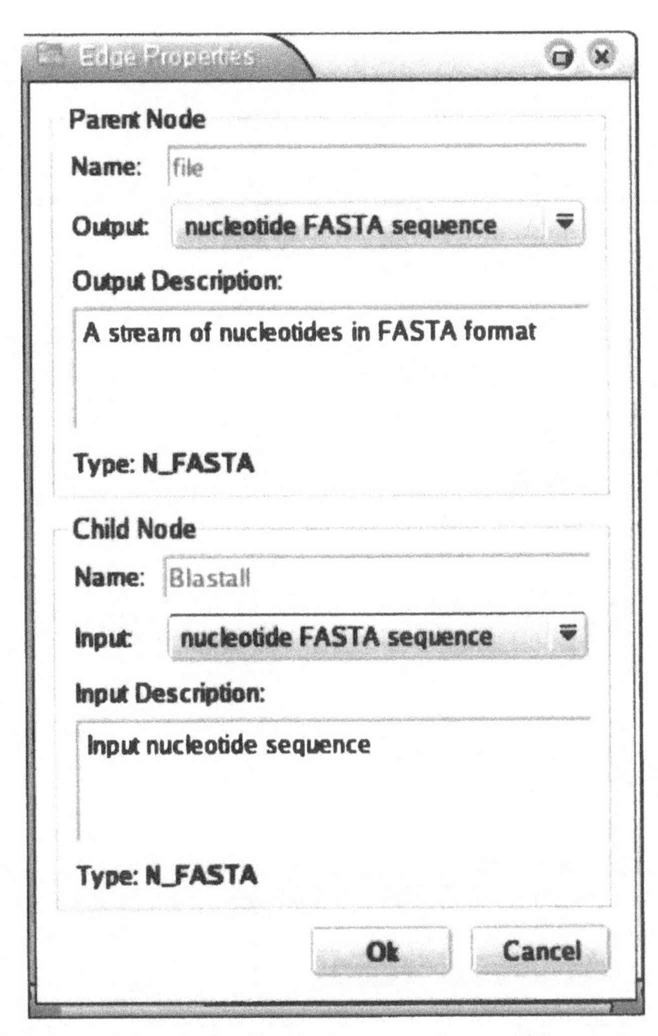

Figure 6. Screenshot of the Edge Properties dialogue window where users set the inputs and outputs of an edge. The input/output values are selected with drop-down select bars so users can only select input/output types that are available to the two nodes. Incompatible input/output types for an edge are not allowed by the GUI and the user is alerted to the error. The input/output lists for each node are set in the ProgramList.xml file that the server reads on startup.

XML Configuration Files

Communication between the client and server is mediated through XML files. There are three key XML files in the Pegasys client. The first XML file, the Pegasys configuration file (PegasysConfig.xml), keeps track of the system settings for default output directories on the server, queuing time for the scheduler, location of Pegasys Java jar files, and database information. This file also contains the path to the second XML file–the program list file which list all of the programs and their associated parameters that are currently available on the Pegasys server (ProgramList.xml). This file needs to be updated whenever a new module is added to the server, or the parameters of an existing module are changed. It is kept on the server and is transmitted to the client every time it starts up to inform the users of the available programs on the server and their associated parameters.

The third XML file is the textual representation of the workflow. This file is generated by saving the workflow using the client. It can be sent to the server where it is parsed and then executed, or it can be re-opened at a later time for further modification. For each node on the canvas, its parameters, flags, and coordinates on the canvas are recorded in the DAG XML file. Edges have their start and end nodes recorded.

Communication via XML is one of the standard ways of disseminating information on the Internet. Both Java for the backend and QT for the client have ready-made parsers for XML. This allowed us to rapidly build the software components that exchange information between the client and the server.

Results and Discussion

To illustrate the flexibility of Pegasys for diverse analyses, we chose three workflows to demonstrate as use cases for the system. The simplest workflow takes an input sequence, runs a single analysis on this sequence and saves the unprocessed results.

Figure 4 shows an example of detecting repeats in a genomic sequence using RepeatMasker. In this example, the unprocessed results are written to a text file. This example is almost as if RepeatMasker were run locally on the command line, except that all information about the parameters used, the input sequence and the results are logged to the Pegasys database.

Figure 7 shows a workflow that has two inputs. The first is a FASTA-formatted nucleotide sequence file. This file is used as input to 'formatdb'—an application that transforms FASTA-formatted databases into a format that can be used by BLAST. The second input is a query sequence that will be used to search the newly formatted database using BLAST. The results of the search are outputted in a GFF-formatted text file.

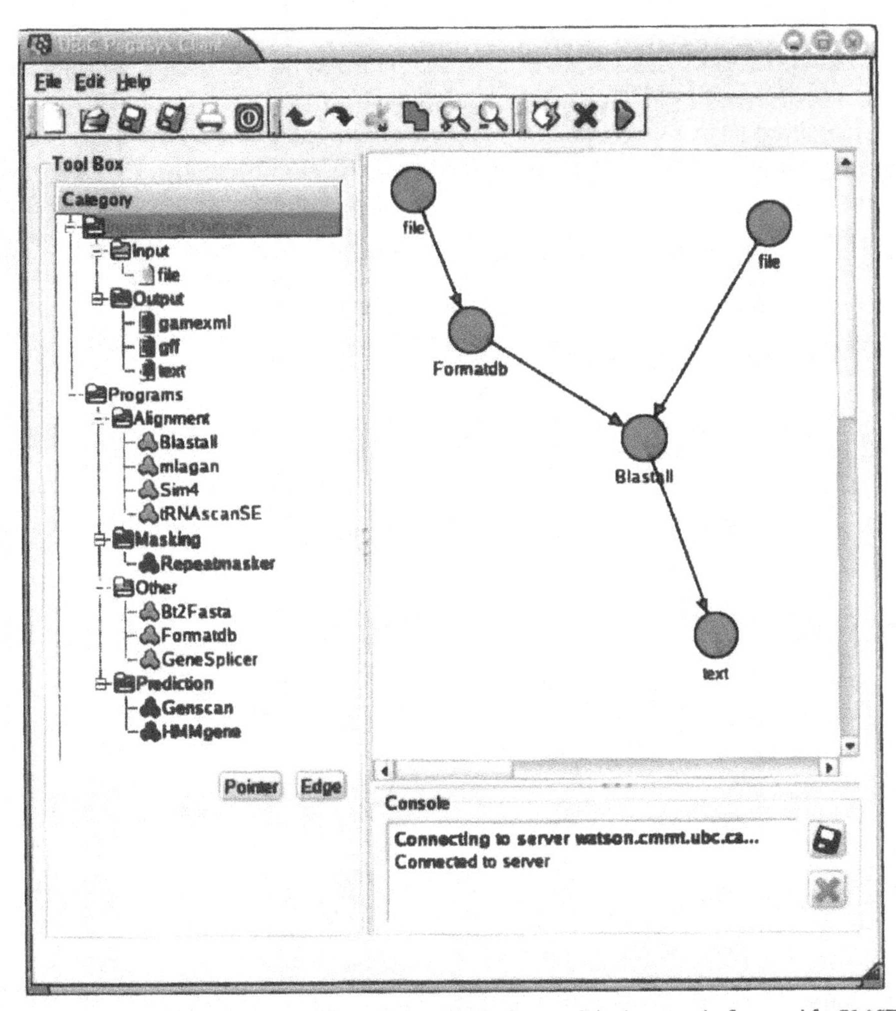

Figure 7. Workflow showing a BLAST pipeline. A FASTA formatted database is to be formatted for BLAST using 'formatdb'. A query sequence is then searched against this new database using BLAST. The results are written to a text file in GFF format.

Figure 8 shows a workflow that would be suitable for annotation of eukaryotic genomic sequence. The output of this workflow would serve as the input for an annotation tool like Apollo. The DAG branches after the input sequence File node into a sub-DAG of analyses that work on the input as is and a sub-DAG that analyzes the input sequence that is masked for repeats with RepeatMasker. The unmasked sequence is analysed for tRNAs using tRNAscan-SE, and for protein coding genes using ab initio gene predictors Genscan and HmmGene. The masked sequence is searched against a database of curated proteins using BLASTX and against a database compiled from ESTs, full-length cDNAs and mRNA sequences (dbTranscript). The results from the latter search are further processed by an application (bt2fasta) that filters all hits based on taxonomy (in this case the user-inputted NCBI taxonid of the source organism of the input sequence) and retrieves their full sequences. This results in an organism-specific database of FASTA formatted sequences consisting of the BLASTN against dbTranscript hits. The unmasked input sequence is then used as input to Sim4, which in turn aligns the input sequence to the entries in the organism specific database. Results for all analyses are then integrated into a GAME XML file for further interpretation using Apollo. The Pegasys XML DAG file that includes the parameters for all programs is available for download at http://bioinformatics.ubc.ca/pegasys/ .

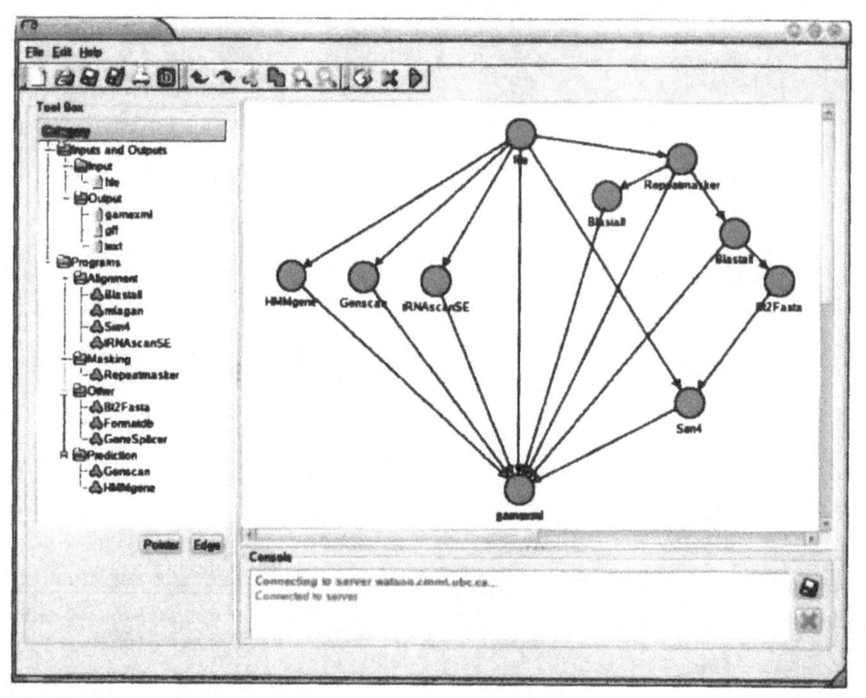

Figure 8. Workflow for genome annotation. This workflow executes ab initio gene prediction, tRNA detection, repeat detection, sequence similarity searching against protein and transcript databases and alignments of transcripts to genomic sequence. Results for all of these analyses are integrated into a single GAME XML output file that can be inputted into Apollo, where a user can create annotations on the original input sequence.

These use cases provide good examples of how Pegasys can be used in sequence-based bioinformatics analyses. The system itself is by no means limited to these examples. In theory any Unix program or script can be incorporated into the system and Pegasys could be used for workflows for systems administration, or other high-level scripting.

Comparison with other Systems

As mentioned above, there are other systems that are similar to Pegasys in philosophy and approach. The DiscoveryNet platform [33] is a system that integrates bioinformatics tools based on Grid computing technologies. This system is a 'middleware' system that can be used to create workflows of annotation tools. Pegasys differs from the DiscoveryNet approach in two major ways. First, Pegasys provides a rigorously defined data model for storing computational features that is mapped by a relational backend database. The use case for DiscoveryNet describes output in the form of text-based flat files. Storing the data in a database allows it to be mined using SQL for selective sub-sets of computational evidence and gives the user more control over what they are interpreting. Second, the Pegasys system is designed to create workflows on the fly using the GUI and XML. The DiscoveryNet genome annotation workflow was programmed and any new workflow would also require programming investment. DiscoveryNet uses the concept of web-services and distributed computing. The architecture of Pegasys is extensible to web service based analyses. We plan on adding the capability of making remote calls to application servers and being able to integrate their analysis results into the Pegasys framework. This would give Pegasys the utmost flexibility and extensibility by combining the power of locally installed applications with remote web services.

The Biopipe framework [31] describes a framework for protocol-based bioinformatics. The protocols are developed with the goal of creating reproducibility of results from computational analyses. This idea complements Pegasys quite well and we envisage using Pegasys to encode protocols by creating workflow standards generated from the Pegasys GUI for specific types of analyses (e.g. genome annotation or mass spectrometry peptide fragment identification) that we can distribute to the Pegasys user community. This will facilitate cross-comparison of results from similar bioinformatics experiments performed on data sources in different research labs, or by colleagues working in the same lab. In addition, Pegasys can be used to compare results of different protocols designed to address similar scientific problems.

Future Directions

The work described in this paper has led us to consider many new challenges for future work on Pegasys. While the specifications, the data model and the software are mature enough to be used in a research setting, there remain many features and enhancements to the system that we are implementing in on-going work. We are adding new modules to Pegasys for distribution to the community. We are implementing Pegasys modules for the Infernal package that is driving the Rfam repository of families of functional RNAs [34]. Our genome annotation work to date has focused largely on eukaryotic systems, and we have therefore devoted most of our development time to applications tuned for eukaryotic animal analysis. We are adding modules for prokaryotic analysis (e.g. Glimmer [35,36]) and plants (Eugene [37]) to complement the current tools in Pegasys.

From a software perspective, we hope to make Pegasys inter-operable and compliant with additional existing Open Source bioinformatics standards and specifications, namely BioSQL and Chado to allow data computed with Pegasys to be used in other systems that employ and interact with these specifications.

Conclusions

We have created a robust, modular, flexible software system for the execution and integration of heterogeneous biological sequence analyses. Pegasys can execute and integrate results from ab initio gene prediction, pair-wise and multiple sequence alignments, RNA gene detection and masking of repetitive sequences to greatly enhance and automate several levels of the biological sequence analysis process. The GUI allows users to create workflows of analyses by dragging and dropping icons on a canvas and joining processes together by connecting them with graphical 'edges'. Each analysis is highly configurable and users are presented with the option to change all parameters that are supported by the underlying program. Data integration is facilitated through the creation of a data model to represent computational evidence which is in turn implemented in a robust backend relational database management system. The database API provides programmatic access to the results through high-level methods that implement SQL queries on the data. The Pegasys system is currently driving numerous diverse sequence analysis projects and can be easily configured for others.

Implemented in Java, the backend of Pegasys is inter-operable with a growing number of bioinformatics tools developed in Java. Pegasys can output text files in standard formats that can then be imported into other tools for subsequent analysis or viewing. We are continually adding to Pegasys through the development of additional modules and methods of data integration. The flexibility, customization, modularity and data integration capabilities of Pegasys make it an attractive system to use in any high throughput sequence analysis endeavour. We are releasing the source code of Pegasys under the GNU General Public License with the hope that the bioinformatics community worldwide will make use of our efforts and in turn contribute improvements in the spirit of Open Source.

Availability and Requirements

Pegasys is available at http://bioinformatics.ubc.ca/pegasys/ and is distributed under the GNU General Public License. Pegasys is designed to run on Unix based systems. Please consult the user manual (available with the distribution) for detailed installation and configuration instructions. The Pegasys server is written in Java and has the following dependencies: Java 1.3.1 or higher, PostgreSQL 7.3.*, JDBC driver for PostgreSQL 7.3.* and BioJava 1.2*. We have tested Pegasys on a distributed memory cluster (recommended) running OpenPBS 2.3.16 to administer the job scheduling. In theory an SMP system running OpenPBS should work, but this has not been tested. The system's analysis programs include the following: NCBI BLAST 2.2.3, WU BLAST 2.0, EMBOSS 2.7.1 (for Smith-Waterman implementation only), tRNAscan-SE 1.23, the LAGAN toolkit 1.2, Sim4, Genscan 1.0, HMMgene 1.1, MaskerAid (2001-11-08) and GeneSplicer. All of the analysis tools are freely available to academics. For details please consult the Pegasys manual available with the distribution. The server has successfully been deployed and tested on a 28 CPU Linux cluster running RedHat 7.3.

The client is written in C++ and requires the QT libraries version 3.11, and gcc version 3.2.2. The client has been tested on Linux Mandrake9.x, Solaris 8, Mac OSX, Windows98/NT/ME/XP.

Authors' Contributions

SS was the lead architect of the system and contributed to the design and implementation and wrote most of this manuscript. DH was the principal developer and contributed to the design and implementation of the server and the GUI. JS contributed to the design of the project and provided requirements to the developers who were designing the system. GQ, GZ, JD, DL and TX all participated in the implementation of various components of the system. BFFO conceived of the project, guided its development, and edited this manuscript.

Acknowledgements

BFFO would like to acknowledge GenomeBC for funding this project. DL is supported by the CIHR/MSFHR Strategic Training Program in Bioinformatics http://bioinformatics.bcgsc.ca. TX is supported by CIHR grant #MOP-53259. We wish to thank Stefanie Butland, Joanne Fox and Yong Huang for critical reviews of this manuscript. We also thank Miroslav Hatas and Graeme Campbell for systems and software installation and maintenance for the Pegasys server.

References

1. Hubbard T, Barker D, Birney E, Cameron G, Chen Y, Clark L, Cox T, Cuff J, Curwen V, Down T, Durbin R, Eyras E, Gilbert J, Hammond M, Huminiecki L, Kasprzyk A, Lehvaslaiho H, Lijnzaad P, Melsopp C, Mongin E, Pettett R, Pocock M, Potter S, Rust A, Schmidt E, Searle S, Slater G, Smith J, Spooner W, Stabenau A, Stalker J, Stupka E, Ureta-Vidal A, Vastrik I, Clamp M: The Ensembl genome database project. Nucleic Acids Res 2002, 30:38–41.
2. Mungall CJ, Misra S, Berman BP, Carlson J, Frise E, Harris N, Marshall B, Shu S, Kaminker JS, Prochnik SE, Smith CD, Smith E, Tupy JL, Wiel C, Rubin GM, Lewis SE: An integrated computational pipeline and database to support whole-genome sequence annotation. Genome Biol 2002., 3(12): RESEARCH0081. Epub 2002 Dec 23.
3. Meyer F, Goesmann A, McHardy A, Bartels D, Bekel T, Clausen J, Kalinowski J, Linke B, Rupp O, Giegerich R, Pühler A: GenDB-an open source genome annotation system for prokaryote genomes. Nucleic Acids Res 2003, 31(8):2187–2195.
4. Yuan Q, Ouyang S, Liu J, Suh B, Cheung F, Sultana R, Lee D, Quackenbush J, Buell C: The TIGR rice genome annotation resource: annotating the rice genome and creating resources for plant biologists. Nucleic Acids Res 2003, 31:229–233.
5. Korf I, Flicek P, Duan D, Brent MR: Integrating genomic homology into gene structure prediction. Bioinformatics 2001, 17:S140–S148.
6. Mathé C, Déhais P, Pavy N, Rombauts S, Van Montagu M, Rouzé P: Gene prediction and gene classes in Arabidopsis thaliana. J Biotechnol 2000, 78(3):293–299.
7. Yeh R, Lim L, Burge C: Computational inference of homologous gene structures in the human genome. Genome Res 2001, 11(5):803–816.
8. Rogic S, Ouellette B, Mackworth A: Improving gene recognition accuracy by combining predictions from two gene-finding programs. Bioinformatics 2002, 18(8):1034–1045.
9. General Feature Format [http://www.sanger.ac.uk/Software/formats/GFF/index.shtml]
10. GAME XML DTD [http://flybase.bio.indiana.edu/annot/gamexml.dtd.txt]
11. Lewis SE, Searle SM, Harris N, Gibson M, Lyer V, Richter J, Wiel C, Bayraktaroglir L, Birney E, Crosby MA, Kaminker JS, Matthews BB, Prochnik SE, Smithy CD, Tupy JL, Rubin GM, Misra S, Mungall CJ, Clamp ME: Apollo:

a sequence annotation editor. Genome Biol 2002., 3(12): RESEARCH0082. Epub 2002 Dec 23. Review.

12. Altschul S, Gish W, Miller W, Myers E, Lipman D: Basic local alignment search tool. J Mol Biol 1990, 215(3):403–410.
13. Altschul S, Madden T, Schäffer A, Zhang J, Zhang Z, Miller W, Lipman D: Gapped BLAST and PSI-BLAST: a new generation of protein database search programs. Nucleic Acids Res 1997, 25(17):3389–3402.
14. Dowell R, Jokerst R, Day A, Eddy S, Stein L: The distributed annotation system. BMC Bioinformatics 2001, 2:7–7.
15. R Development Core Team: R: A language and environment for statistical computing. [http://www.R-project.org] R Foundation for Statistical Computing, Vienna, Austria 2003. [ISBN 3-900051-00-3]
16. Bedell J, Korf I, Gish W: Masker Aid: a performance enhancement to RepeatMasker. Bioinformatics 2000, 16(11):1040–1041.
17. Gish W: WU BLAST 2.0. [http://blast.wustl.edu/blast/README.html]
18. Rice P, Longden I, Bleasby A: EMBOSS: the European Molecular Biology Open Software Suite. Trends Genet 2000, 16(6):276–277.
19. Smith T, Waterman M: Identification of common molecular subsequences. J Mol Biol 1981, 147:195–197.
20. Burge C, Karlin S: Prediction of complete gene structures in human genomic DNA. J Mol Biol 1997, 268:78–94.
21. Krogh A: Two methods for improving performance of an HMM and their application for gene finding. Proc Int Conf Intell Syst Mol Biol 1997, 5:179–186.
22. Brudno M, Do C, Cooper G, Kim M, Davydov E, Green E, Sidow A, Batzoglou S: LAGAN and Multi-LAGAN: efficient tools for large-scale multiple alignment of genomic DNA. Genome Res 2003, 13(4):721–731.
23. Florea L, Hartzell G, Zhang Z, Rubin G, Miller W: A computer program for aligning a cDNA sequence with a genomic DNA sequence. Genome Res 1998, 8(9):967–974.
24. Lowe T, Eddy S: tRNAscan-SE: a program for improved detection of transfer RNA genes in genomic sequence. Nucleic Acids Res 1997, 25(5):955–964.
25. Pertea M, Lin X, Salzberg S: GeneSplicer: a new computational method for splice site prediction. Nucleic Acids Res 2001, 29(5):1185–1190.
26. Stein L, Mungall C, Shu S, Gaudy M, Mangone M, Day A, Nickerson E, Stajich J, Harris T, Arva A, Lewis S: The generic genome browser: a building

block for a model organism system database. Genome Res 2002, 12(10):1599–1610.

27. OpenPBS [http://www.openpbs.org]
28. Booch G: Object-oriented Analysis and Design with Applications The Benjamin/Cummings Publishing Company 1994.
29. Ensj [http://www.ensembl.org/java/]
30. BioJava.org [http://www.biojava.org]
31. Hoon S, Ratnapu K, Chia J, Kumarasamy B, Juguang X, Clamp M, Stabenau A, Potter S, Clarke L, Stupka E: Biopipe: a flexible framework for protocol-based bioinformatics analysis. Genome Res 2003, 13(8):1904–1915.
32. Trolltech–Qt Overview [http://www.trolltech.com/products/qt/index.html]
33. Rowe A, Kalaitzopoulos D, Osmond M, Ghanem M, Guo Y: The discovery net system for high throughput bioinformatics. Bioinformatics 2003, 19(Suppl 1): 225–225.
34. Griffiths-Jones S, Bateman A, Marshall M, Khanna A, Eddy S: Rfam: an RNA family database. Nucleic Acids Res 2003, 31:439–441.
35. Delcher A, Harmon D, Kasif S, White O, Salzberg S: Improved microbial gene identification with GLIMMER. Nucleic Acids Res 1999, 27(23):4636–4641.
36. Salzberg S, Delcher A, Kasif S, White O: Microbial gene identification using interpolated Markov models. Nucleic Acids Res 1998, 26(2):544–548.
37. Schiex T, A M, P R: EUGENE: An Eukaryotic Gene Finder That Combines Several Sources of Evidence. In JOBIM 2000, 111–125.

An Application of a Game Development Framework in Higher Education

Alf Inge Wang and Bian Wu

ABSTRACT

This paper describes how a game development framework was used as a learning aid in a software engineering. Games can be used within higher education in various ways to promote student participation, enable variation in how lectures are taught, and improve student interest. In this paper, we describe a case study at the Norwegian University of Science and Technology (NTNU) where a game development framework was applied to make students learn software architecture by developing a computer game. We provide a model for how game development frameworks can be integrated with a software engineering or computer science course. We describe important requirements to consider when choosing a game development framework for a course and an evaluation of four frameworks based on these requirements. Further, we describe some extensions we made to the existing game development framework to let the students focus more on software architectural issues than the technical implementation issues. Finally, we describe a case study of how a game development framework was integrated in a software architecture course and the experiences from doing so.

Introduction

Games have been used in schools for many years to help children learn skills in math, language, geography, science, and other domains in an interesting and motivating way. Research shows that integrating games within a classroom with children can be beneficial for academic achievement, motivation, and classroom dynamics [1]. There is also evidence that the teaching methods based on educational games are not only attractive to schoolchildren, but also to university students [2]. There have been conducted researches on games concept and game development used in higher education before, for example, [3–5], but we believe there is an untapped potential that needs to be explored. Games can provide teachers in higher education with teaching aids that can promote more active students, provide alternative teaching methods to improve variation, and enable social learning through multiplayer learning games.

Games can be integrated in higher education in three ways. First, games can be used instead of traditional exercises motivating students to put extra effort in doing the exercises and giving the teacher and/or teaching assistants an opportunity to monitor how the students work with the exercises in real time [6, 7]. Second, games can be used within lectures to improve the participation and motivation of students [8, 9]. In this approach, the students and the teacher participate in knowledge-based games. Third, the students are required to develop a game as a part of a course using a game development framework (GDF) to learn skills within computer science or software engineering [10]. This paper focuses on the latter, where game development and a GDF are used in student projects to learn software engineering skills, extending the use of games as a teaching aid in higher education. The motivation of making students develop games to learn software engineering is to bring the students' enthusiasm from playing games to learn courses through game development. In addition, we wanted to investigate if the specific features of a GDF are suitable for teaching software engineering and how game development can be integrated with the education process. More specifically, we wanted to explore how the use of game development and the GDF would affect the learning of software architecture with focus on the technical aspects of the GDF.

This paper focuses on how the technical aspects of a GDF affect the learning of software architecture, the selection of appropriate GDF for a software architecture course, and how a GDF can be applied in a software engineering course. The main contribution of this paper is a presentation of a novel GDF concept that can be used in courses that includes software development, experiences from actual usage of the GDF, and some course design considerations.

The rest of the paper is organized as follows. Section 2 describes and motivates for how a GDF can be used in higher education and what criteria should be considered when choosing one. Section 3 describes a case study of applying a GDF in a software architecture course. Section 4 describes experiences fromusing a GDF in a software course. Section 5 describes similar approaches, and Section 6 concludes the paper.

Game Development Frameworks in Higher Education

This section presents the motivation for applying GDFs in higher education, a model for how GDFs can be integrated with a course, and requirements for how to choose the appropriate GDF for educational purposes.

GDF and Education

The main motivation for introducing GDF in software engineering (SE) or computer science (CS) courses is to motivate students to put more effort into software development project in order to improve software development skills. Game development offers an interesting way of learning and applying the course theory. By introducing a game development project in a course, the students have to establish and describe most of the functional requirements themselves (what the game should be like). This can be a motivating factor especially for group-based projects, as each group will develop a unique application (the game); it will encourage creativity, and it will require different skills from the group members (art, programming, story, audio/music). The result will be that the students will have a stronger feeling of ownership to the project. Furthermore, students also could learn about game development technology. The main disadvantages by introducing a game development project and a GDF into a SE or CS course is that the student might spend too much time on game-specific issues and that the project results might be difficult to compare. It is critical that the students get motivated applying a GDF in a course and that they get increased motivation for learning and applying course theory through a game development project.

Tom Malone has listed three main characteristics that make things fun to learn: they should provide the appropriate level of challenge, they should use fantasy and abstractions to make it more interesting, and they should trigger the player's curiosity [11]. These characteristics can directly be applied when developing a game for learning purposes. However, we can also consider these characteristics when introducing a GDF in a SE or CS course. By allowing the students to develop their own games using a GDF, such projects are likely to trigger students' curiosity as well as provide a challenge for students to design fun games with their knowledge, skills, imagination, and creativity. The level of the challenge can be adjusted according to the project requirements given in courses by the teacher. Thus, the challenge level can not only be adjusted to the right level for most participants, but also tailored for individual differences. As the students will work in groups, group members helping other group members can compensate for the individual differences. An open platform and agile courses requirements should be provided for students to design their own games, combined with their ability, fantasy, and comprehension of lecture content. The main benefit of using a GDF as a teaching aid is that it can be a motivating initiative in courses to learn about various topics such as software requirements, software design, software architecture, programming, 2D and 3D graphic representation, graphic programming, artificial intelligence, physics, animation, user interfaces, and many other areas within computer science and software engineering. It is most useful for learning new skills and methods within a specific domain but also useful for testing and rehearsing theory by applying know skills and knowledge in a project using a GDF.

Circulatory Model of Applying a GDF in a Course

There are several good reasons for introducing a GDF and game development projects in CS and SE courses as described in previous section, but in order to make it a success it is important that the GDF is well integrated with the course. Based on our experiences, we have developed a circular model for how to apply a GDF in a CS or SE course through six steps (see Figure 1). The model is intended for courses where a software development project is a major part of the course.

To choose one appropriate development platform according to the course content, it is important to consider the process of the course related to the development project. This process starts with choosing an appropriate GDF (step A) for the course related to some requirements (described in the next section). Next, the design of exercises and projects (step B) must reflect the limitations and constraints of the chosen GDF. In the initial phase of the student project, it is important that the students get the required technical guidance and appropriate requirements (step C) related to the GDF. It is important that the students get to know the GDF early, for example, by introducing an exercise to implement a simple game in the GDF. It is critical that there is sufficient course staff that knows the GDF well enough to give the required feedback. The next step is for the students to start designing and implementing (stepD) their own game according to the constraints within the course and the GDF. After the students have delivered their final version of their project implementation and documentation, the students should get the chance to evaluate and analyze (step E) their own projects to learn from their successes and mistakes. This information should then be used to provide feedback in order to improve the course (step F). The feedback from the students might indicate that another GDF should be used or that the course constraints on the projects should be altered. The core of this model is that the teacher should encourage the students to explore the course theory through a game development project using a GDF and give the opportunity to improve the game development project through feedback from the students.

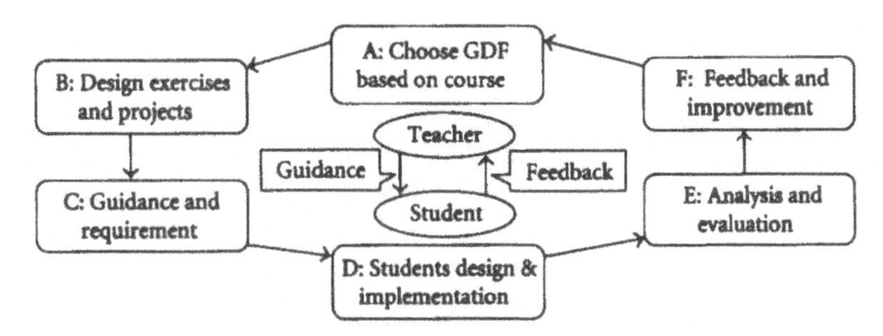

Figure 1. Circulatory model of GDF's application in courses.

Criteria for Choosing the Right GDF

How to choose an appropriate GDF that easily can be integrated with course content should be based on the educational goals of the course, the technical level and skills of students, and the time available for projects and/or exercises. Based on experiences from using GDFs and from student projects in CS and SE courses, we have come up with the following requirements for choosing a GDF for a CS or SE course.

(1) It must be easy to learn and allow rapid development. According to Malone's recommendation of how to make things fun to learn, it is crucial that we provide the appropriate level of challenge. If the GDF is too much of a challenge and requires too much to learn before becoming productive, the whole idea of game development will be wasted, as the student will lose motivation. An important aspect of this is that the GDF offers high-level APIs that makes it possible for the students to develop impressive

results without writing too many lines of code. This is especially critical in the first phase of the project.

(2) It must provide an open development environment to attract students' curiosity. Malone claims that fantasy and curiosity are other important factors that make things fun to learn. By providing a relatively open GDF without too many restrictions on what you can produce, the students get a chance to realize the game of their dreams. This means that the GDF itself should not restrict what kind of game the students can make. This requirement would typically rule out GDFs that are tailored for producing only one game genre such as adventure games, platform games, or board games. In addition, ideally an open development environment should offer public and practical interfaces for developers to extend their own functions. In this respect, open source game development platforms are preferred.

(3) It must support programming languages that are familiar to the students. The students should not be burdened to have to learn a new programming language from scratch in addition to the course content. This would take away the focus of the educational goals of the course. We suggest to choose GDFs that support popular programming languages that the students know like C++, C#, or Java. It is also important that the programming languages supported by the GDF have high-level constructs and libraries that enable the programmers to be more productive as less code is required to produce fully functional systems. From an educational point of view, programming languages like Java and C# are better suited than C and C++, as they have more constraints that force the programmers to write cleaner code, and there is less concern related to issues like pointers and memory leakage. From a game development perspective, programming languages like C and C++ are more attractive as they generally produce faster executables and thus faster games.

(4) It must not conflict with the educational goals of the course. When choosing a GDF it is important that the inherent patterns, procedures, design, and architecture of the GDF are not in conflict with the theory taught in the course. One example of such a conflict could be that the way the GDF enforces event handling in an application is given as an example of bad design in the textbook.

(5) It must have a stable implementation. When a GDF is used in a course, it is essential that the GDF has few bugs so the students do not have to fight a lot of technical issues instead of focusing on the course topics. This requirement indicates that it is important that the GDF is supported by a company or a development community that has enough resources to

eliminate serious technical insufficiencies. It is also important that the development of the GDF is not a dead project, as this will lead to compatibility issues for future releases of operating systems, software components, and hardware drivers.

(6) It must have sufficient documentation. This requirement is important for both the course staff and the students. The documentation should both give a good overview of the GDF as well as document all the features provided. Further, it is important that the GDF provides tutorials and examples to demonstrate 4 International Journal of Computer Games Technology how to use the GDF and its features. The frameworks should provide documentation and tutorials of high quality enabling self-study.

(7) It should be inexpensive (low costs) to use and acquire. Ideally, the GDFs should be free or have very low associated cost to avoid extra costs running the course. This requirement also involves investigating additional costs related to the GDF such as requirements for extra or more powerful hardware and/or requirements for additional software.

The goal of the requirements above is to save the time and effort the students have to spend on coding and understanding the framework, making them concentrate on the course content and software design. Thus, an appropriate GDF could provide the students with exciting experiences and offer a new way of learning through a new domain (games). The requirements above are also important for the course staff, as they will help to find a GDF that would cause less effort spent on technical issues, and incompatibility between GDF and the course contents.

From the requirements above, we acknowledge that there is a conflict between requirements one and two. The level of the freedom the developer is given to make whatever game he likes could be in conflict with providing a development environment that allows rapid development and is easy to learn. A more open GDF usually means that the developer must learn more APIs as well as the APIs themselves are usually of lower level, and thus harder to use. However, it is possible to get a bit of both worlds by offering high-level APIs that are relatively easy to use but still allow the developer to access underlying APIs that give the developer the freedom in what kind of games can be made. This means that the GDF can allow inexperienced developers to just modify simple APIs or example code to make variants of existing games, or to allow more experienced developers to make unique games by using more of the provided underlying APIs. How hard the GDF is to use will then really depend on the ambition of the game developer and not on the GDF itself. This can also be a motivating factor to learn more about the GDF's APIs.

Case Study: Applying a GDF in a Software Architercture Course

This section describes a case study of a software architecture course at the Norwegian University of Science and Technology (NTNU) where a GDF was introduced.

The Software Architecture Course

The software architecture course is a postgraduate course offered to CS and SE students at NTNU. The course is taught every spring, its workload is 25% of one semester, and about 70 postgraduate students attend the course every semester. The students in the course are mostly of Norwegian students (about 80%), but there are about 20% foreign students mostly from EU-countries. The textbook used in this course is the "Software Architecture in Practice, Second Edition", by Bass, Clements et al. [12]. Additional papers are used to cover topics that are not sufficiently covered by the book such as design patterns, software architecture documentation standards, view models, and postmortem analysis [13–16]. The education goal of the course is:

"The students should be able to define and explain central concepts in software architecture literature and be able to use and describe design/architectural patterns, methods to design software architectures, methods/techniques to achieve software qualities, methods to document software architecture, and methods to evaluate software architecture."

The course is taught in four main ways:

(1) ordinary lectures given in English;

(2) invited guest lectures from the software industry;

(3) exercise in design patterns;

(4) a software development project with emphasis on software architecture.

30% of the grade is based on an evaluation of a software architecture project that all students have to do, while 70% is given from the results of a written examination. The goal of the project is for the students to apply the methods and theory in the course to design a software architecture and to implement a system according to the architecture. The project consists of the following phases.

(1) COTS (Commercial Off-The-Shelf) exercise: learn the development platform to be used in the project by developing some simple test applications.

(2) Design pattern: learn how to utilize design pattern by making changes in an existing system designed with and without design patterns.

(3) Requirements and architecture: describe the functional and the quality requirements, and design the software architecture for the application in the project.

(4) Architecture evaluation: use the Architecture Tradeoff Analysis Method (ATAM) [12, 17] to evaluate the software architecture in regards to the quality requirements. Here one student group will evaluate another student group's project.

(5) Implementation: do a detailed design and implement the application based on the created architecture and based on the results from a previous phase.

(6) Project evaluation: evaluate the project after it has been completed using a Post-Mortem Analysis (PMA) method.

In the two first phases of the project, the students work on their own or in pairs. For the phases 4–6, the students work in self-composed groups of four students. The students spend most time on the implementation phase (6 weeks), and they are also encouraged to start the implementation in earlier phases to test their architectural choices (incremental development). In previous years, the goal of the project has been to develop a robot controller for a robot simulator in Java with emphasis on an assigned quality attribute such as availability, performance, modifiability, or testability.

Choosing a GDF for the Software Architecture Course

In Fall 2007, we started to look for appropriate GDFs to be used in the software architecture course in spring 2008.We looked for both GDFs where the programmer had to write the source code as well as visual drag-and-drop programming environments. The selection of candidates was based on GDFs we were familiar with and GDFs that had developer support. Further, we wanted to compare both commercial and open source GDFs. From an initial long list candidate GDFs, we chose to evaluate the following GDFs more in detail.

(i) XNA: XNA is a GDF from Microsoft that enables development of home-brew cross-platform games for Windows and the XBOX 360 using the C# programming language. The initial version of Microsoft XNA Game Studio was released in 2006 [18], and in 2008 Microsoft XNA Game studio 3.0 was released that includes support for making games for XBOX Live. XNA features a set of high-level API enabling the development of advanced games in 2D or 3D with advanced graphical effects with little effort. The XNA platform is free and allows developers to create games for Windows, Xbox 360, and Zune using the same GDF [19]. XNA consists of an integrated development environment (IDE) along with several tools for managing audio and graphics.

(ii) JGame: JGame is a high-level framework for developing 2D games in Java [20]. JGame is an open source project and enables developers to develop games fast using few lines of code as JGame will take care of typical game functionality such as sprite handling, collision detection, and tile handling. JGame games can be run as stand-alone Java games, Java applets games running in a web browser or on mobile devices (Java ME). JGame does not provide a separate IDE but is integrated with Eclipse.

(iii) Flash: Flash is a high-level framework for interactive applications including games developed by Adobe [21]. Most programming in Flash is carried out in Action script (a textual programming language), but the Flash environment also provides a powerful graphical editor for managing graphical objects and animation. Flash applications can run as stand-alone applications or in a web browser. Flash applications can run on many different operating systems like Windows, Mac OS X, and Linux as well as on mobile devices and game consoles (Nintendo Wii and Sony Playstation 3). Programming in Flash is partly visual by manipulating graphical objects, but most code is written textually. Flash supports development of both 2D and 3D applications.

(iv) Scratch: is a visual programming environment developed by MIT Media Lab in collaboration with UCLA that makes it easy to create interactive stories, animations, games, music, and art and to share the creations on the web [22]. Scratch works similar to Alice [23] allowing you to program by placing sprites or objects on a screen and manipulate them by drag-and-drop programming. The main difference between Scratch and Alice is that Scratch is in 2D while Alice is in 3D. Scratch provides its own graphical IDE that includes a set of programming primitives and functionality to import various multimedia objects.

An evaluation of the four GDF candidates is shown in Table 1. From the four candidates, we found Scratch to be the least appropriate candidate. The main disadvantage with Scratch was that it would be very difficult to teach software architecture using this GDF, as the framework did not allow exploring various software architectures. Further, Scratch was also very limited in what kind of games that could be produced, limiting the options for the students. The main advantage using Scratch is that it is very easy to learn and use. JGame suffered also from some of the same limitations as Scratch, as it put some restrictions on what software architecture could be used, and it had little flexibility in producing a variety of types of games. The main advantage using JGame was that it was an open source project with access to the source code and that all the programming was done in Java. All CS and SE students at NTNU learn Java in the two first introductory programming courses. An attractive alternative would be to use Flash as a GDF. Many developers use Flash to create games for kids as well as games for the Web. Flash puts little restrictions on what kind of games you can develop (both 2D and 3D), but there are some restrictions on what kind of software architecture you can use in your applications. The programming language used in Flash, Action Script, is not very different from Java so it should be rather easy for the students to learn. The main disadvantage using Flash in the software architecture course was the license costs. As the computer and information science department does not have a site license for the Flash development kit, it would be too expensive to use. XNA was found an attractive alternative for the students, as it made it possible for them to create their own XBOX 360 games. XNA puts little restrictions on what kinds of software architectures you apply in your software, and it enables the developers to create almost any game. XNA has strong support from its developer (Microsoft) and has a strong community of developers along with a lot of resources (graphics, examples, etc.). The main disadvantages using XNA as a GDF in the course were that the students had to learn C# and that the software could only run on Windows machines. Compared to JGame and other Java-based GDFs, XNA has a richer set of high-level APIs and a more mature architecture.

Based on the evaluation described above, we chose XNA as a GDF for our course. From previous experience we knew that it does not require much effort and time to learn C# for students that already know Java.

XQUEST – An Extension of the Chosen GDF

After we had decided to use XNA as a GDF in the software architecture course, we launched a project to extend XNA to make XNA even easier to use in the student project. This project implemented XQUEST (XNA QUick & Easy Starter Template) [24], which is a small and lightweight 2D game library/game template developed at NTNU that contains convenient game components, helper classes, and other classes that can be used in the XNA game projects (see Figure 2). The goal of XQUEST was to identify and abstract common game programming tasks and create a set of components that could be used by students of the course to make their life easier. We choose to focus only on 2D. There are a few reasons for this. First, the focus of the student projects is software architecture, not making a game with fancy 3D graphics. Second, students unfamiliar with game programming and 3D programming may find it daunting to have to learn the concepts needed for doing full-blown 3D in XNA, such as shader programming and 3D modeling, in addition to software architectures. To keep the projects in 2D may reduce the effect of students focus only on the game development and not on the software architecture issues.

Table 1. Evaluation of four GDF candidates.

Selection requirement	XNA	JGame	Flash	Scratch
(1) Easy to learn	Relatively easy to learn, but requires to learn several core concepts to utilize the offered possibilities.	Easy to learn, but requires to learn a small set of core concepts.	Relatively easy to learn, but requires to learn several core concepts to utilize the offered possibilities.	Very easy and intuitive to learn and supports dynamic changes to the game in run time.
(2) Open develop environment	XNA puts little restrictions on what kind of games can be developed and supports development of both 2D and 3D games. Not open source project.	JGame supports a limited set of games, mainly classical 2D arcade games. Open source project.	Flash puts little restrictions on what kind of games can be developed and supports development of both 2D and 3D. Not open source project.	Scratch limits the options of what kind of games the user can make through the limited options provided in the graphical programming environment. Not open source project.
(3) Familiar programming language	All programming is done in C#.	All programming is done in Java	Some programming can be done using drag-and-drop, but most will be written in Action Scripts.	All programming is done in the visual drag-and-drop programming language Scratch.
(4) Not in conflict with educational goals	XNA puts *little restrictions* on what kinds of software architectures can be used.	JGame puts *some restrictions* on what kinds of software architecture can be used.	Flash puts *some restrictions* on what kinds of software architectures can be used.	Scratch puts *strict restrictions* on what kinds of software architectures can be used.
(5) Stable implementation	XNA has a very stable implementation and is updated regularly.	JGame has a relatively stable implementation and is updated regularly.	Flash has a very stable implementation and is updated regularly.	Scratch has a relatively stable implementation and is updated regularly.
(6) Sufficient documentation	XNA is well documented and offers several tutorials and examples. Many books on XNA are available.	JGame is not well documented, but some examples exist.	Flash is well documented and offers several tutorials and examples. Many books on Flash are available.	Scratch is ok documented and has some examples and tutorials available.
(7) Low costs	XNA is free to use. A $99 for a year of membership is required to develop games for XBOX 360.	JGame is free to use.	The Flash development kit costs $199 per license (university license).	Scratch is free to use.

Teaching Software Architecture using XNA

XNA was introduced in the software architecture course to motivate students to put extra effort in the student project with the goal to learn the course content such as attribute driven design, design and architectural patterns, ATAM, design of software architecture, view points, and implementation of software architecture. This section will go through the different phases of this project and describe how XNA affected these phases.

Introduction of XNA Exercises

In the start of the semester the course staff gave an introduction to course where the software architecture project was presented. Before the students started with their project, they had to do an exercise individually or in pairs where they got to choose their own partner. The goal of the first exercise was to get familiar with the XNA framework and environment, and the students were asked to complete four tasks.

(1) Draw a helicopter sprite on the screen and make it move around on its own.

(2) Move around the helicopter sprite from previous task using the keyboard, change the size of the sprite when a key was pressed, rotate the sprite when another key was pressed, and write the position of the sprite on the screen.

(3) Animate the helicopter sprite using several frames and do sprite collision with other sprites.

(4) Create the classical Pong game in XNA.

Before the students started on their XNA introduction exercise, they got a two-hour technical introduction to XNA. During the semester, two technical assistants were assigned to help students with issues related to XNA. These assistants had scheduled two hours per week to help students with problems, in addition to answering emails about XNA issues.

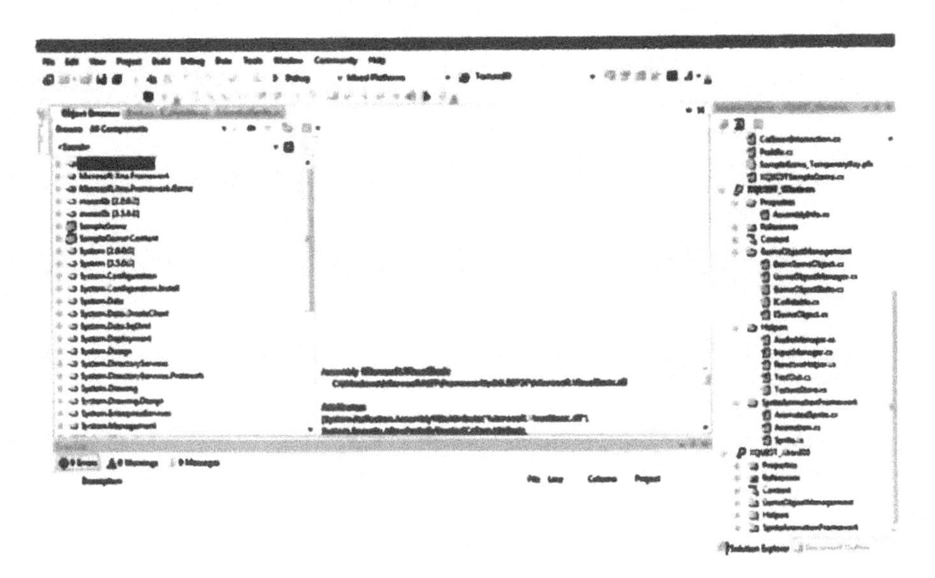

Figure 2. The XQUEST library shown in the XNA development environment.

Requirement and Architecture for the Game Project

After the introduction exercise was delivered, the students formed groups of four students. Students that did not know anyone were assigned to groups. The course staff then issued the project task where the goal was to make a functioning game using XNA based on students' own defined game concept. However, the game had to be designed and implemented according to their specified and designed software architecture. Further, the students had to develop a software architecture that focused on one particular quality attribute assigned by the course staff. We used the following definitions for the quality attributes in the game projects: Modifiability, the game architecture and implementation should be easy to change in order to add or modify functionality; Testability, the game architecture and implementation should be easy to test in order to detect possible faults and failures. These two quality attributes were related to the course content and the textbook. A perfect implementation was not the ultimate quest of this XNA game project, but it was critical that the implementation reflected the architectural description. It was also important that the final delivery was well-structured, easy to read, and made according to the template provided by the course staff.

The first phase of the project was the requirement and architecture phase where the students should deliver requirements and the software architecture of the game along with a skeleton code reflecting the architecture. The requirements document focused on a complete functional requirement description of the game and several quality requirements for the game described as scenario focusing on one particular quality attribute. The architectural description was the most important part of the final delivery of the game project, and the students had to document their architecture according to IEEE 1471-2000 [25]. The architecture documentation could be altered several times before its final delivery. Table 2 lists main attributes required in the architectural description in the game projects.

We also required that the students wrote the code skeleton for the architecture they had designed. This was done to emphasize the importance of starting the implementation early and to ensure that students designed an architecture that was possible to implement.

Evaluation of the Game Project

After the requirements, the architecture and the code skeleton were delivered; the student groups were assigned to evaluate each other's architecture using ATAM. The whole idea was for one project group to evaluate the architecture of the other group's game to give feedback on the architecture related to the quality focus of the software architecture [27]. It included attribute utility tree, analysis of architectural approach, sensitivity points, trade-off points, risks and non-risks, and risk themes.

Table 2. List of architecture description for the game project.

#	Architectural description attributes	Details of the implementation
(1)	Architectural drivers	The main drivers that affect the system mostly, including the attribute on which the students focus.
(2)	Stakeholders and concerns	Stakeholders of the system, and their concerns.
(3)	Selection of architectural viewpoint	A list of the viewpoints used, and their purpose, target audience, and form of description. Places to look in for possible viewpoints including the book [12] and the 4 + 1 article by Kruchten [26].
(4)	Quality tactics	Including all attributes and more details for the focused ones.
(5)	Architectural patterns	The major patterns of your architecture, both architectural and major design ones.
(6)	Views	A separate section for each required views: logic, process, and development views or other views added by students.
(7)	Consistency among views	Discuss the consistency between each described view.
(8)	Architectural rationale	In this section and subsections, add why things are chosen.

Detailed Design and Implementation

The focus of implementation phase was to design, implement, and test the game application. The documentation delivered in this phase focused on the test results from running the game related to the specified requirements and the discussion of the relationship between the implemented game and the architectural documentation [14, 15]. Table 3 lists what should be delivered in the implementation phase. For the test report part in the Table 3, the functional requirements and quality requirements had the attributes like shown in Table 4 and Table 5. The test reports should also include a discussion about the observation of the test unless there was nothing to discuss about the test results.

At the end of this phase, the students had to submit their final delivery of their projects that included all documents, code, and other material from all project phases. The course staff evaluated all the groups' deliveries and gave grades by judging document and implementation quality, document and implementation completeness, architecture design, and readability and structure of code and report.

Table 3. Design & implementation phase description.

#	Implementation deliverables	Details of implementation
(1)	Design and implementation	A more detailed view of the various parts of the architecture description of game design.
(2)	User's manual	To guide the users; the steps to compile and run the game.
(3)	Test report	Contains both functional requirements and quality requirements (quality scenarios).
(4)	Relationship with the architecture	List the inconsistencies between the game architecture and the implementation and the reasons for these inconsistencies.
(5)	Problems, issues, and points learned	Listing problems and issues with the document or with the implementation process.

Table 4. Attributes of functional requirements.

F1: The role in game should be able to jump along happily	
Executor	Super Mario III
Date	23.3.2005
Time used	5 min
Evaluation	Fail: White role cannot jump!

Table 5. Attributes of quality requirements.

A1: The role in game should not get stuck	
Executor	Snurre Sprett
Date	24.3.2005
Stimuli	The role should be able to move around for 10 min
Expected response	Success in 8 of 10 executions
Observed response	Success in 3 of 10 executions
Evaluation	Fail

The Game Project Workshop

In this workshop, selected groups had to give short presentations about the project goal, quality attribute focus, proposed architectural solution with some diagrams or explanations, and an evaluation of how well did the solution worked related to functional requirements and quality focus. Further, the selected groups ran demos of their games, and it was opened for questions from the audience.

The workshop provided an open mind environment to let students give each other feedback, brainstorm about improvements and ideas, and to discuss their ideas to give a better understanding of the course content and game architecture design.

Post-Mortem Analysis

In the final task in the project, every group had to perform a post-mortem analysis of their project. The focus of the PMA was to analyze successes and problems of the project. The PMA was documented in a short report that included a positive (successes) and a negative (problems) KJ diagram (structured brainstorm map), a positive and a negative causal map (a diagram that shows cause-effect relationships), and experiences from using PMA [13]. The PMA made the students reflect on their performance in the project and gave them useful feedback to improve in the future projects and inputs for the course staff to improve the course. The main topics analyzed in the PMA were issues related to group dynamics, time management, technical issues, software architecture issues, project constraints, and personal conflicts.

Experiences of using GDF in Software Architecture

The experiences described in this section are based on the final course evaluation, feedback from the students during the project, and the project reports.

The final course evaluation made all students (mandatory) taking the course answer three questions. The results reported below are a summary of the students' responses related to the project and the GDF.

(1) What have been good about software architecture course?

(a) About the project itself: "Cool project", "Really interesting project", "We had a lot of fun during the project", "It is cool to make a game", "Fun to implement something practical such a game", "Videogame as an exercise is quite interesting", "I really liked the project", and "The game was motivating and fun".

(b) Project and learning: "Good architectural discussion in the project group I was in", "Learned a lot about software architecture during the project", "The project helped to understand better the arguments explained in the lectures, having fun at the meantime", "Fun project where we learned a lot", "I think that the creation of a project from the beginning, with the documentation until the code implementation, was very helpful to better understand in practice the focus of the course", "The game project was tightly connected to the syllabus and lectures and gave valuable experience. The main thing I learned was probably how much simpler everything gets if you have a good architecture as a basis for your system", and "The interplay of game and architectural approaches".

(c) The project being practical work: "I think it was pretty good that you guys made us do a lot of practical work", and "To choose C# as a platformis a good idea as it is used a lot in the software industry; at the same time it is very similar to Java so it is rather easy to learn the language."

(d) Interplay between groups: "It was also good to see the results of the others' projects in the final presentation".

(2) What have been not so good about the course software architecture?

(a) XNA support: "The way the student assistants were organized; during the implementation periods at least they should be available in a computer lab and not just in the classroom", "Maybe the use of XNA Framework XQUEST was very difficult because I never use it. Maybe some extra lecture focus on the use of XQUEST Framework was better", and "We did not have lectures on XNA; could have got some more basic info…Hmm…"

(b) XNA versus software architecture: "Took a lot of time getting to know C#, I liked it, but I did not have the time to study architecture" and "The use of game as a project may have removed some of the focus away from the architecture. XNA and games in general limit the range of useful architectures."

(3) What would you have changed for next year's course?
   (a) Project workload: "Maybe just little more time to develop the game" and "I would change the importance of the project. I think that the workload of the project was very big and it can matter the 50% of the total exam."
   (b) XNA support: "Perhaps have some C# intro?" and "It would be helpful to have some lab hours".
   (c) Project constraints: "Maybe more restrictions on game type, to ensure that the groups choose games suited for architectural experimentation."

The responses from the students were overall very positive. In the previous years, the students in the software architecture course had to design the architecture and implement a robot controller for a robot simulator in Java. The feedback from the XNA project was much more positive than the feedback from the robot controller project. Other positive feedback we got from the students was that they felt they learned a lot from the game project, that they liked the practical approach of the project and having to learn C#, and the interaction between the groups (both ATAM and the project workshop).

The negative feedback from the course evaluation was focusing on the lack of XNA support and technical support during the project and that some student felt that there was too much focus on C#, XNA, and games and too little on software architecture.

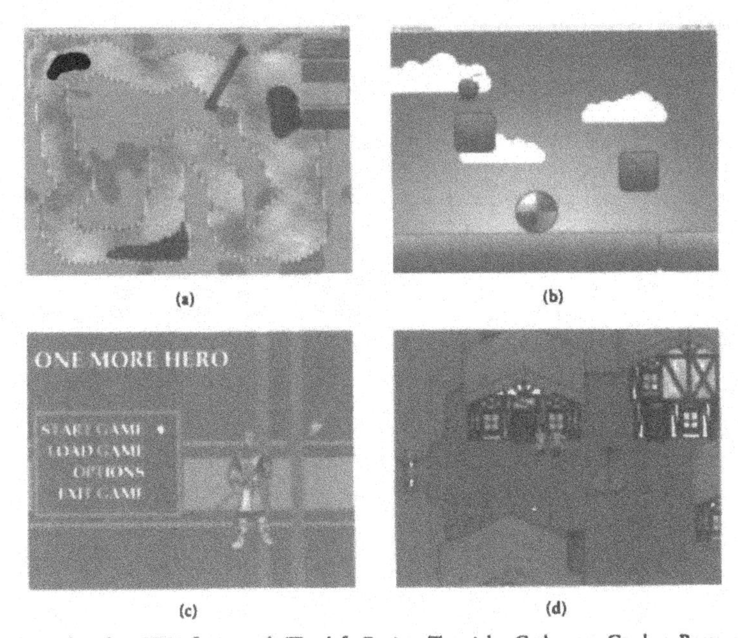

Figure 3. Game based on XNA framework (Top left: Racing; Top right: Codename Gordon; Bottom: RPG).

The suggestions to improve the course were mainly according to the negative feedback, namely, to improve XNA support and to adjust the workload of the project. One student also suggested limiting the types of games to be implemented in project to ensure more focus on software architectural experimentation.

Snapshots from Some Student Projects

Figure 3 shows screenshots from four student game projects. The game at the upper left corner is a racing game, the game at the upper right corner is a platform game, and the two games below are role-playing games (RPGs). Some of the XNA games developed were original and interesting. Most games were entertaining but were lacking contents and more than one level due to time constraints.

Related Work

This paper describes experiences from utilizing the special features of a GDF in a software architecture course. The main benefits from applying a GDF in a CS or SE course is that the students get more motivated during the software development project. As far as we know, there are few papers that describe the usage of a professional GDF concept applied in universities courses that is not directly a target for learning game development, especially no papers about usage of XNA in higher education. However, there are some related approaches in education described in this section.

El-Nasr and Smith describes how the use of modifying or modding existing games can be used to learn computer science, mathematics, physics, and ascetic principles [10]. The paper describes how they used modding of the WarCraft III engine to teach high school students a class on game design and programming. Further, they describe experiences from teaching university students a more advanced class on game design and programming using the Unreal Tournament 2003 engine. Finally, they present observations from student projects that involve modding of game engines. Although the paper claims to teach students other things than pure game design and programming, the GDFs were used in the context of game development courses.

The framework Minueto [28] is implemented in Java, and it is used by students in their second year of undergraduate studies at McGill University in Montreal, Canada. The framework encapsulates graphics, audio, and keyboard/mouse inputs to simplify Java game development. It allows development of 2D games, such as card games and strategy games, but it lacks in support for visual programming and suffers from limited documentation.

The Labyrinth [29] is implemented in Java, and it is a flexible and easy-to-use computer game framework. The framework enables instructors to expose students to very specific aspects of computer science courses. The framework is a finished game in the Pac-Man genre, highly modular, and it lets the students change different aspects of the game. However, it cannot be used to develop different genres of game, and there is little room for changing the software architecture of the framework.

The JIG (Java Instructional Gaming) Project [30] is a collaborative effort between Scott Wallace (Washington State University Vancouver) and Andrew Nierman (University of Puget Sound) in conjunction with a small group of dedicated students. It has three aims: (1) to build a Java Instructional Game Engine suitable for a wide variety of students at all levels in the curriculum; (2) to create a set of educational resources to support the use of the game engine at small, resource-limited, schools; (3) to develop a community of educators that use and help improve these resources. The JIG Project was proposed in 2006, after a survey of existing game engines revealed a very limited supply of existing 2D Java game engines. JIG is still in development.

GarageGames [31] offers two game engines written in C++. The Torque Game Engine targets 3D games, while the Game Builder provides a 2D API and encourages programmers to develop using a proprietary language (C++ can also be used). Both engines are aimed at a wide audience, including students and professionals. The engines are available under separate licenses ($50 per license per year for each engine) that allow full access to the source code. Documentation and tutorials cover topics appropriate for beginners and advanced users.

The University of Michigan's DXFramework [32] game engine is written in C++. The current version is targeted specifically for 2D games, although previous versions have included a 3D API as well. This engine is designed for game programming education and is in its third major iteration. The DXFramework is an open source project. Compared to XNA, DXFramework has no competitive advantage as it has limited support for visual programming, and it is not easier than XNA to learn.

The University of North Texas's SAGE [33] game engine is written in C++, and targets 3D games, not 2D. Like the DXFramework, SAGE is targeted specifically for game programming educational usage. The source code can be downloaded and is currently available without license. Marist College's GEDI [34] game engine provides a second alternative for 2D game design in C++, and is also designed with game programming educational use in mind. Source code can be downloaded and is currently available without license, but GEDI is still in the early phases of development. Only one example game is distributed with the code, and little documentation is available.

For business teaching, Arena3D [35] is a game visualization framework with its animated 3D representations of the work environments; it simulates patients queuing at the front desk and interacts with the staff. IBM has also produced a business game called INNOV8 [36] which is "an interactive, 3-D business simulator designed to teach the fundamentals of business process management and bridge the gap in understanding between business leaders and IT teams in an organization".

Conclusion and Future Work

In this paper we have presented a case study of how a GDF was evaluated, chosen, and integrated with a software architecture course. The main goal of introducing a GDF and a game development project in this course was to motivate students to learn more about software architecture during the game development project. The positive feedback from the students indicates that this was a good choice as the students really enjoyed the project and learned software architecture from carrying out the project.

We will continue to explore the area of using games, games concept, and game development in CS and SE education and evaluate how this affects the students' motivation and performance. The choice of XNA as a GDF proved to be a good choice for our software architecture course. The main disadvantage using XNA is the lack of support for non-Windows operating systems like Linux and Mac OS X. Mono. XNA is a cross platform implementation of the XNA game framework that allows XNA to run on Windows, Mac OS X, and Linux using OpenGL [37]. The project is still in an early phase. An alternative to solve this problem is to let the students choose between different GDFs, for example, XNA and a Java-based GDF. The main challenge for this approach is that the course staff needs to know all the GDFs offered to the students to give proper technical assistance. Based on the feedback from the students; the technical support is very important and must be considered before providing choices of more GDFs.

Acknowledgements

The authors would like to thank Jan-Erik Strøm and Trond Blomholm Kvamme for implementing XQUEST and for their inputs to this paper. They would also like to thank Richard Taylor and Institute for Software Research (ISR) at University of California, Irvine (UCI), for providing a stimulating research environment and for hosting a visiting researcher.

References

1. R. Rosas,M. Nussbaum, P. Cumsille, et al., "Beyond Nintendo: design and assessment of educational video games for first and second grade students," Computers & Education, vol. 40, no. 1, pp. 71–94, 2003.
2. M. Sharples, "The design of personal mobile technologies for lifelong learning," Computers & Education, vol. 34, no. 3-4, pp. 177–193, 2000.
3. A. Baker, E. O. Navarro, and A. van der Hoek, "Problems and programmers: an educational software engineering card game," in Proceedings of the 25th International Conference on Software Engineering (ICSE '03), pp. 614–619, Portland, Ore, USA, May 2003.
4. L. Natvig, S. Line, and A. Djupdal, ""Age of computers"; an innovative combination of history and computer game elements for teaching computer fundamentals," in Proceedings of the 34th Annual Frontiers in Education (FIE '04), vol. 3, pp. 1–6, Savannah, Ga, USA, October 2004.
5. E. O. Navarro and A. van der Hoek, "SimSE: an educational simulation game for teaching the software engineering process," in Proceedings of the 9th Annual SIGCSE Conference on Innovation and Technology in Computer Science Education (ITiCSE '04), p. 233, Leeds, UK, June 2004.
6. G. Sindre, L. Natvig, and M. Jahre, "Experimental validation of the learning effect for a pedagogical game on computer fundamentals," IEEE Transactions on Education, vol. 52, no. 1, pp. 10–18, 2009. 12 International Journal of Computer Games Technology
7. B. A. Foss and T. I. Eikaas, "Game play in engineering education: concept and experimental results," International Journal of Engineering Education, vol. 22, no. 5, pp. 1043–1052, 2006.
8. A. I. Wang, O. K. Mørch-Storstein, and T. Øfsdahl, "Lecture quiz—a mobile game concept for lectures," in Proceedings of the 11th IASTED International Conference on Software Engineering and Application (SEA '07), pp. 305–310, Cambridge, Mass, USA, November 2007.
9. A. I. Wang, T. Øfsdahl, and O. K. Mørch-Storstein, "An evaluation of a mobile game concept for lectures," in Proceedings of the 21st Conference on Software Engineering Education and Training (CSEET '08), pp. 197–204, Charleston, SC, USA, April 2008.
10. M. S. El-Nasr and B. K. Smith, "Learning through game modding," Computers in Entertainment, vol. 4, no. 1, pp. 45–64, 2006.
11. T. W. Malone, "What makes things fun to learn? Heuristics for designing instructional computer games," in Proceedings of the 3rd ACM SIGSMALL

Symposium and the First SIGPC Symposium on Small Systems (SIGSMALL '80), pp. 162–169, ACM Press, Palo Alto, Calif, USA, September 1980.

12. P. Clements, L. Bass, and R. Kazman, Software Architecture in Practice, Addison-Wesley, Reading, Mass, USA, 2nd edition, 2003.
13. A. I. Wang and T. Stalhane, "Using post mortem analysis to evaluate software architecture student projects," in Proceedings of the 18th Conference on Software Engineering Education and Training (CSEET '05), pp. 43–50, Ottawa, Canada, April 2005.
14. J. O. Coplien, "Software design patterns: common questions and answers," in The PatternsHandbook: Techniques, Strategies, and Applications, pp. 311–320, Cambridge University Press, New York, NY, USA, 1998.
15. A. Rollings and D. Morris, Game Architecture and Design: A New Edition, New Riders Games, Indianapolis, Ind, USA, 2003.
16. D. P. Perry and A. L. Wolf, "Foundations for the study of software architecture," ACM Sigsoft Software Engineering Notes, vol. 17, no. 4, pp. 40–52, 1992.
17. R. Kazman, M. Klein, M. Barbacci, T. Longstaff, H. Lipson, and J. Carriere, "The architecture tradeoff analysis method," in Proceedings of the 4th IEEE International Conference on Engineering Complex Computer Systems (ICECCS '98), pp. 68–78, Monterey, Calif, USA, August 1998.
18. Microsoft Corporation, "XNA Developer Center," June 2008, http://msdn.microsoft.com/en-us/xna/aa937794.aspx.
19. B. Nitschke, Professional XNA Game Programming: For Xbox 360 and Windows, John Wiley & Sons, New York, NY, USA, 2007.
20. JGame project, "JGame: a Java game engine for 2D games," November 2008, http://www.13thmonkey.org/~boris/jgame/index.html.
21. Adobe, "Animation software, multimedia software—Adobe Flash CS4 Professional," November 2008, http://www.adobe.com/products/flash.
22. Lifelong Kindergarten Group, MIT Media Lab, "Scratch: Imagine, Program, Share," June 2008, http://scratch.mit.edu.
23. Carnegie Mellon University, "Alice.org," June 2008, http://www.alice.org.
24. T. Blomholm Kvamme and J.-E. Strøm, Evaluation and extension of an XNA game library used in software architecture projects, M.S. thesis, Department of Computer and Information Science, Norwegian University of Science and Technology (NTNU), Trondheim, Norway, June 2008.

25. IEEE Std 1471-2000, "IEEE Recommended Practice for Architectural Description of Software-Intensive Systems," Software Engineering Standards Committee of the IEEE Computer Society, 2000.

26. P. Kruchten, "The 4 + 1 view model of architecture," IEEE Software, vol. 12, no. 6, pp. 42–50, 1995.

27. A. BinSubaih and S. C. Maddock, "Using ATAM to evaluate a game-based architecture," in Proceedings of the 2nd International ECOOP Workshop on Architecture-Centric Evolution (ECOOP '06), Nantes, France, July 2006.

28. A. D. Minueto, An undergraduate teaching development framework, M.S. thesis, School of Computer Science, McGill University, Montreal, Canada, 2005.

29. J. Distasio and T. Way, "Inclusive computer science education using a ready-made computer game framework," in Proceedings of the 12th Annual SIGCSE Conference on Innovation and Technology in Computer Science Education (ITiCSE '07), pp. 116–120, Dundee, Scotland, June 2007.

30. Washington State University Vancouver and University of Puget Sound, "The Java Instructional Gaming Project," June 2000, http://ai.vancouver.wsu.edu/jig.

31. GarageGames, "GarageGames," June 2008, http://www.garagegames.com.

32. C. Johnson and J. Voigt, "DXFramework," June 2008, http://www.dxframework.org.

33. I. Parberry, "SAGE: a simple academic game engine," June 2008, http://larc.csci.unt.edu/sage.

34. R. Coleman, S. Roebke, and L. Grayson, "GEDI: a game engine for teaching videogame design and programming," Journal of Computing Science in Colleges, vol. 21, no. 2, pp. 72–82, 2005.

35. Rockwell Automation Inc, "Arena Simulation Software," June 2008, http://www.arenasimulation.com.

36. IBM, "INNOV8—a BPM Simulator," June 2008, http://www.ibm.com/software/solutions/soa/innov8.html.

37. Monoxna, "Monoxna—Google Code," November 2008, http://code.google.com/p/monoxna.

Copyrights

Index

B

C

D

G

H